P9-CJW-320

THE ASTRONAUT MAKER

THE ASTRONAUT MAKER

HOW ONE MYSTERIOUS ENGINEER RAN HUMAN SPACEFLIGHT FOR A GENERATION

MICHAEL CASSUTT

First edition
Published by Chicago Review Press Incorporated
814 North Franklin Street
Chicago, Illinois 60610
ISBN 978-1-61373-700-2

Library of Congress Cataloging-in-Publication Data
Names: Cassutt, Michael, author.
Title: The astronaut maker : how one mysterious engineer ran human spaceflight for a generation / Michael Cassutt.
Description: First edition. | Chicago, Illinois : Chicago Review Press Incorporated, 2018. | Includes bibliographical references and index.
Identifiers: LCCN 2017060552 (print) | LCCN 2018000759 (ebook) | ISBN 9781613737019 (adobe pdf) | ISBN 9781613737026 (kindle) | ISBN 9781613737033 (epub) | ISBN 9781613737002 (hardback)
Subjects: LCSH: Abbey, George, 1932– | United States. National Aeronautics and Space Administration—Officials and employees—Biography. | Aerospace engineers—United States—Biography. | Government executives—United States—Biography. | United States. National Aeronautics and Space Administration—Management—History. | BISAC: BIOGRAPHY & AUTOBIOGRAPHY /
Science & Technology. | TECHNOLOGY & ENGINEERING / Aeronautics & Astronautics. | SCIENCE / Astrophysics & Space Science.
Classification: LCC TL789.85.A333 (ebook) | LCC TL789.85.A333 C37 2018 (print) | DDC 629.4092 [B] —dc23
LC record available at https://lccn.loc.gov/2017060552

Typesetting: Nord Compo

Printed in the United States of America
5 4 3 2

For Bert Vis

Dave Shayler

Tony Quine

Francis French

Colin Burgess

In memory of

Rex Hall

The real book about the manned space program would be a book about George Abbey.

—Richard H. Truly,
two-time shuttle flier, former NASA administrator (1989–92)

Contents

Part III: "The Quintessential Staffer"

Part IV: "Ten Consecutive Miracles Followed by an Act of God"

Part V: "Go at Throttle Up"

Part VI: "A National Quasi-Emergency"

Part VII: "Together to the Stars"

Abbreviations

AFIT	Air Force Institute of Technology
AFWL	Air Force Weapons Laboratory, Kirtland Air Force Base, New Mexico
ALT	Approach and Landing Test, first phase of space shuttle flight tests, 1977
ARPS	Air Force Research Pilot School, interim name for the Air Force Test Pilot School
AS	Apollo-Saturn mission designation
ASCAN	Astronaut candidate
ASTP	Apollo-Soyuz Test Project, 1972–1975
ATLAS	Atmospheric Laboratory for Applications and Science, shuttle Spacelab
CCB	Configuration control board
CDR	Commander of a shuttle crew
CFES	Continuous flow electrophoresis system, McDonnell Douglas payload carried on the shuttle
CM	Command module, Apollo spacecraft
CSM	Command and service module, Apollo spacecraft
DC-X	Delta Clipper, McDonnell Douglas experimental vehicle
DOD	Department of Defense
ESA	European Space Agency
ET	External tank, part of the shuttle stack at launch
EVA	Extravehicular activity, also known as "space walk"
FCOD	Flight crew operations directorate, NASA Johnson Space Center
FLAT	First Lady Astronaut Trainee
FOG	Friends of George [Abbey]
GASSER	George Abbey Saturday Review
GCTC	Gagarin Cosmonaut Training Center, Russia

GE	General Electric Corporation
GEO	Geosynchronous orbit, approximately 22,500 miles of altitude
GT	Gemini-Titan, mission designation
HQ	NASA headquarters, Washington, DC
ICBM	Intercontinental ballistic missile
JSC	NASA Johnson Space Center, Houston, Texas
KSC	NASA Kennedy Space Center, Cape Canaveral, Florida
ISF	Industrial Space Facility
ISS	International Space Station program
IUS	Inertial upper stage, propulsion unit for satellites
JPL	NASA Jet Propulsion Laboratory, Pasadena, California
LEM/LM	Lunar excursion module, later lunar module, Apollo spacecraft
LLTV/LLRV	Lunar landing training vehicle, lunar landing research vehicle
MA	Mercury-Atlas, mission designation
McDAC	McDonnell Douglas Aerospace Corporation
MECO	Main engine cutoff, a key moment in any shuttle launch
MMU	Manned Maneuvering Unit, a powered backpack for space walks
MOL	Manned Orbiting Laboratory, USAF-NRO program 1963–69
MR	Mercury-Redstone, mission designation
MS	Mission specialist, nonpilot shuttle astronaut
MSC	NASA Manned Spacecraft Center, Houston, Texas (after 1973, Johnson Space Center)
MSE	Manned Spaceflight Engineer, air force shuttle payload specialist cadre
NACA	National Advisory Committee on Aeronautics, 1915–58 (NASA predecessor)
NAR	North American Rockwell Corporation
NASA	National Aeronautics and Space Administration
NBL	Neutral Buoyancy Laboratory, NASA Johnson Space Center
NRC	National Research Council
NRO	National Reconnaissance Office

NROTC	Naval Reserve Officer Training Program
OFT	Orbital flight test, early designation for first shuttle orbital test flights
OV	Orbital vehicle, NASA designation for individual shuttle orbiters
PAM	Payload assist module, shuttle upper stage
PEAP	Personal Egress Air Pack, worn by shuttle crews from 1981 to 1986
PLT	Pilot on a shuttle crew
PS	Payload specialist, nonastronaut shuttle crew member
RKA	Russian Space Agency
RMS	Remote manipulator system, the shuttle's robot arm
ROTC	Reserve Officer Training Program
RSLS	Redundant set launch sequence, NASA term for on-pad abort
SAFSP/SP	Secretary of the Air Force Special Projects / Special Projects, the air force element within the NRO
SAIL	Shuttle Avionics Integration Laboratory
S&ID	Space and Information Systems Division, North American Rockwell
SCA	Shuttle Carrier Aircraft
SDI	Strategic Defense Initiative
SDOM	Station Development and Operations Meeting
SEAL	Sea, Air, and Land US Navy elite special operator
SED	Systems and Engineering Division, Apollo program
SRB	Solid rocket booster, element of the space shuttle
SSF	Space Station Freedom program
SSME	Space shuttle main engine
STA	Shuttle Training Aircraft
STEM	Science, technology, engineering, and mathematics education
STS	Space Transportation System
TANG	Texas Air National Guard
TDRS	Tracking and Data Relay Satellite
TFNG	Thirty-Five New Guys (in one formulation), nickname for the 1978 astronaut candidate group

TPAD	Trunnion pin acquisition device, satellite capture unit carried on mission 41-C
TRW	Thompson Ramo Wooldridge, aerospace company
TsUP	Tsentr Upravleniya Polyotami (Russian for "flight control center")
VAFB	Vandenberg Air Force Base, California
VASIMR	Variable Specific Impulse Magnetoplasma Rocket
WET-F	Weightless Environment Training Facility, NASA Johnson Space Center
XS-11	"Excess Eleven," ironic self-selected nickname for the 1967 astronaut group

Prologue

ON NOVEMBER 11, 1982, the crew of STS-5, the first operational flight of the Space Transportation System (or space shuttle), emerged from the crew quarters at the Kennedy Space Center headed for launchpad 39A. Trailing the four astronauts in their blue flight gear was a dark-haired middle-aged man in a jacket and tie. One of the crew held up a joking sign pointing to this man: NASA OFFICIAL GEORGE ABBEY.

George William Samuel Abbey bore the title director of flight operations, NASA Johnson Space Center (JSC). He was in charge of mission control, astronaut selection and training, and the all-important flight assignments. But while Abbey was well known within the JSC family, he was a mystery—rarely interviewed, shy of the limelight. Even at NASA, few knew that he had been an air force pilot, specifically a helicopter instructor, before turning to engineering, or that he had worked on the air force's Dyna-Soar space plane.

Or that he was with the Apollo 1 astronauts the night before the fatal fire that killed them in January 1967, then took part in the grim task of disassembling the charred vehicle in search of a cause and went on to play a key role in managing NASA's recovery from that tragic incident, resulting in the Apollo 11 landing little more than two years later. Or that he was in mission control the night of the Apollo 13 accident and helped organize the famous recovery effort. Or that he led NASA's recruitment of women and minorities as space shuttle astronauts.

On that November 1982 day of the STS-5 launch, Abbey was only halfway through his NASA career. Over the next nineteen years, he would suffer through the agony of the *Challenger* disaster, ultimately losing his powerful position in Houston and being exiled to NASA HQ—where, in a miracle of bureaucratic maneuvering and sheer will, he not only survived but prospered, helping to reshape NASA while reinventing America's human space program.

He would return to Houston in 1994 to run the Johnson Space Center, guiding the space shuttle program and leading the development of the International Space Station, all the while mentoring a new generation of NASA and aerospace leaders. (In 2010 the head of NASA, the director of the Johnson Space Center, and the director of Russia's Gagarin Cosmonaut Training Center were all astronauts or cosmonauts he had assigned to spaceflights.)

The week after the International Space Station was declared complete, Dan Goldin—the same NASA administrator whom Abbey had served so completely that the pair had become known within the agency by a single moniker, "Chief Dan George"—fired Abbey.

No one claims that Abbey is the most important figure in the history of America's human spaceflight—Abbey himself vigorously disputes such a characterization, insisting that he was always a small player on a large team. His idol George Low was far more central to the success of Apollo. Vital decisions about the space shuttle and International Space Station were made by others. Pioneering astronauts risked their lives—and died.

What makes Abbey unique is that during thirty-seven years at NASA he worked for or with all of them. He was an engineer, an administrator, a department head, a staffer, and finally a center director and program manager whose control over the Johnson Space Center was so complete that it became known as his "fiefdom."

He was also brilliant and hardworking, sometimes sentimental but also cold eyed, blessed with a fantastic memory and a willingness to learn lessons, then apply them. A family man and father, he sometimes treated his subordinates "like children"—guiding their careers whether or not they knew it or wanted it. Astronaut Thomas Kenneth Mattingly said, "I've never known anyone who had a better read on people than George Abbey."

He did this all in the service of a single but universal goal: allowing humans to live and work in space.

Much like F. Scott Fitzgerald's movie mogul Monroe Starr from *The Last Tycoon*, one of those rare persons who "had the entire equation of motion pictures in their heads," George Abbey held the entire equation of human spaceflight in his head.

At every step of his professional journey, Abbey created a mystique that divided his colleagues, subordinates, and superiors. One group labeled him "the dark lord" and accused him of being "a godfather type who runs a patronage system," comparing his management style to J. Edgar Hoover's. A NASA flight surgeon called Abbey a "dictator." Another astronaut dubbed him "a rapacious power monger." Yet another dismissed him with a single epithet: "snake."

But there were many others who said, for example, "George Abbey saved the space program four times." Or "George Abbey was the single human being most responsible for the ISS being in space and successful." Another veteran astronaut called Abbey "the best program manager NASA ever had." One coworker said Abbey was not only a "real advocate for the human spaceflight business" but also one of the few "who also understood it." Spaceflight, he concluded, "was just in George's blood."

It could be argued that, after the *Columbia* disaster of 2003, there would have been no human spaceflight at all if not for decisions made by George Abbey a decade earlier.

Which is it? Was George Abbey some kind of space-age Thomas Cromwell, skillfully using his intelligence and experience to convince an entire nation to embrace his vision for human spaceflight? Or was he a shadowy puppet master, a cold manipulator like some space-age Iago, bending not just astronauts but all of NASA to his iron will for over thirty years? And both camps, pro and con, wonder openly, *How did he do it? Where did the power come from? His considerable intelligence? His fantastic memory? His experience? His methods? Did he have a philosophy? If so, what was it?*

The story begins on a deserted roadside.

Part I

"Don't Send Me There"

1 | October Sky

OCTOBER CAN BE LONELY on the plains of Montana. Temperatures might reach sixty degrees Fahrenheit during the day, but they fall fast at sunset, to freezing and below. The wind can be biting. And US Route 12 on the long eastbound stretch between Missoula and Helena is not the place you want to pull to the side of the road. Even during daylight, there's not much to see, except scrub, clusters of pines, low rounded hills. Unless you're looking at the sky.

It is Sunday, October 6, 1957, early evening. A 1955 Oldsmobile sits by the eastbound side of the highway. There are pockets of early snow on the darkened dirt to either side, but there's no traffic beyond the odd slow-moving semi or pickup truck. A young man leans against the driver's side, staring to the west. He is short and slim with dark hair and is wearing blue slacks and a Windbreaker. The clothing is not suited to the weather. His name is George Abbey; he is a first lieutenant in the US Air Force, a helicopter instructor at Randolph Air Force Base near San Antonio. He is driving back to Texas from his hometown of Seattle, where he was visiting family. It's a trip he's made from one starting point or another many times.

Just a few years earlier Abbey was a midshipman at the US Naval Academy in Annapolis, coming off the summer cruise with thirty days' leave and three whole dollars a month spending money. In those days, his travel options were military transport (great if it was going where you wanted when you needed it) or family funds (not available). So Abbey and his roommate, Gordon Iver "Gordy" Dahl, another Seattle boy, found a third method: Knowing that car dealers in Detroit were eager to sell used vehicles in Seattle, the pair hitchhiked to Michigan, took a car, and headed out on US Route 12 to Chicago, then across Wisconsin, Minnesota, the Dakotas, Montana, and Idaho into Washington State. Gas was covered; the trip was free, costing them only their time.

Now, three and a half years after graduating from Annapolis, Abbey is a married man with a car of his own. Between his basic officer's salary plus flight pay, he is pulling down $400 a month. Life is better.

But Seattle remains a frequent destination. Abbey's parents still live there, and so do two of his three brothers and a sister. Abbey has grown to love the long drive from Texas north to Wyoming, then west across Montana, and back. A fan of Western movies, especially John Ford's, Abbey likes the bleak scenery and the famous Big Sky. Especially on this particular evening. Tonight, Montana's Big Sky will allow Lieutenant Abbey a once-in-a-lifetime opportunity—to see the flight of an astronomical object by the name of Sputnik, the Earth's first artificial orbiting satellite.

Abbey has always had an interest in space travel, the happy result of a youth spent reading Buck Rogers comics and watching Flash Gordon movie serials. He also knows that the United States is preparing to launch a small scientific satellite named Vanguard. But the news that broke the day before, as he was driving across western Washington State, has shocked him. The Soviet Union, the Red Communist enemy halfway around the world, fired a giant rocket from some mystery location in central Asia, putting Sputnik into orbit. As the miles passed, Abbey heard more details: Sputnik was a silvery sphere twenty-three inches in diameter and weighing 184 pounds, small enough to fit inside the trunk of his Olds—though the Soviet satellite is a giant compared to the planned three-pound Vanguard. Sputnik is circling the Earth every ninety-two minutes at an altitude ranging from a low point of 139 miles to a high of 900, flying as far north as Alaska and Greenland and as far south as Antarctica. It emits a beep that has been tracked all over the world.

Abbey knows what this accomplishment implies: the Soviets do indeed possess the giant rockets they bragged about earlier in the year, missiles capable of flinging H-bombs over the North Pole to blow up New York and Washington, DC. As a cold warrior, he's troubled. But, as a pilot and Buck Rogers fan, he's also fascinated. Hearing on the radio earlier this day that Sputnik might be visible to Montana residents this very night, Abbey has pulled off the road to see.

He glances at his watch—

And *there's Sputnik*, moving swiftly from northwest to southeast, a small, bright dot. Abbey has seen shooting stars before, but this is different. Shooting stars vanish in a second or less . . . This light in the October sky remains steady. *A satellite!* He is stunned by the sight of it—and surprised by the depth of his reaction, feeling a bit like Saint Paul on the road to Damascus. He has grown up with rockets, seen newsreel footage of German V-2s rising to incredible altitudes over New Mexico. He knows about the V-2's descendant, the Redstone, and its rival cousin, the Atlas. But it isn't until this moment that he connects them all—satellites, rockets, high-speed aircraft. The chance to see Earth from orbit, to walk on the Moon, to visit Mars. The Space Age is no longer that thing-to-come—it's *here*. If the navy and aviation are in George Abbey's blood, his bones are space and rocketry.

He gets back in his car. Texas is still many hours away.

2 | The Abbeys

ABBEY'S PARENTS, Sam and Brenta, met on a London double-decker bus during World War I. Sam, recovering from wounds suffered while serving with the Canadian Army at the Battle of St. Eloi Craters in April 1916, was a passenger; Brenta was the ticket taker.

Of Scottish descent, Sam had been born in 1890 in London but spent his early years with his father, who was working in Bloemfontein, South Africa. Sam recalled watching one of the battles of the Second Boer War from a hillside there. His father fell ill and returned to London in 1902, where Sam completed secondary school. He then became a professional soccer player in Amsterdam. Eventually, he moved to the United States to work on an uncle's farm near Boston. He followed the wheat harvest to western Canada, becoming, at one time or another, a member of the Royal Canadian Mounted Police, a boxer, and a fireman.

In August 1914 Sam was working for the Canadian Pacific Railway in Kamloops, British Columbia. A train loaded with fellow workers from down track came through Kamloops. When asked where they were going, they said the war had started and they were bound for Vancouver to go fight the kaiser. Sam jumped on and, arriving in Vancouver, enlisted in the Canadian Army and was subsequently shipped to France.

Brenta, born Bridget Gibby in Laugharne, Carmarthenshire, Wales, was four years younger. Her mother had died when she was ten, and Brenta had been left to take care of a younger sister, Edith, for several years. She also worked as a servant for a family in Carmarthen until moving to London in 1912, where she went to work for the bus company.

The two married in England and soon had a son, James Robert David, born October 8, 1917, in Cardigan, Wales, while Brenta was visiting an older sister. (Sam had been given only one name; to make up for that lack, all of Brenta and Sam's children would carry three given names.)

After Sam served a year in the army of occupation in Europe, the couple moved to Canada in 1919, where Sam enrolled at the University of British Columbia in Vancouver, working his way through college as a farmer and salmon fisherman. In 1922, at the urging of friends in Washington State, the Abbeys moved to Seattle, where Sam joined the Continental Can Company, eventually becoming the company's transportation manager for the Pacific Northwest and Alaska.

The family expanded: A second son, John Lloyd Richard "Jack," followed on March 28, 1922. Then came Vincent Hugh Donald on July 10, 1923, and a fourth son on August 21, 1932, George William Samuel, born in the house at 505 North Seventy-First Street. A fifth child, Phyllis Gwendolyn Brenta, was born there on March 28, 1934.

The Abbeys had the typical first-generation immigrant virtues: work hard, study hard, become part of the community, vote for Democrats. Brenta made dinner for the family every Sunday (a practice George himself would continue well into his eighties). Her dishes tended to be traditional, like Yorkshire pudding. There were jokes, a lot of teasing and pranking. Also music: Brenta made sure there was a piano in the house. Sam had a fine singing voice and appeared in local operas. He also had an entrepreneurial side, investing in several cottages in the town of White Rock, on Semiahmoo Bay, a few miles north of the US-Canadian border. When school was out for the summer, George and Phyllis and Brenta could be found in White Rock, where the children spent long days on the beach, swimming, rafting, crabbing, and fishing—or visiting a nearby dairy farm in Milner that belonged to one of Sam's friends.

Unusual for Americans in those days, Sam was not a member of any church. In fact, he was openly critical of organized religions. Family members ascribed this to his experiences in the trenches, where he had been surrounded by the sudden and daily death of friends and companions. Agnostic or not, Sam Abbey had all kinds of rules for proper behavior, most of them easily summed up by three words: do what's right.

Brenta was Presbyterian and raised the children in that faith, though young George made his own religious conversion. In order to spend time with high school friends of Scandinavian descent, he began to attend their Lutheran church one block from the Abbey house—and would remain a regular Lutheran churchgoer for the rest of his life. One lure was the church's slate of activities—George joined the Boy Scout troop there, in order to go camping in the mountains. Another attraction was a youth choir "with several pretty blonde singers."

Sam lived a healthy life: no smoking, no drinking, no cursing. He was squarely built and remained fit into his eighties. He kept playing football—meaning soccer; he contemptuously referred to the American game as "carry ball." Immigrants from England, Wales, and Italy—football nations—kept their sporting traditions alive by forming teams centered around Pacific Northwest mining communities like Black Diamond and Ravensdale. George remembered watching his father playing for the Maple Leafs at Woodland Park during the 1930s, spending many a Sunday afternoon on damp fields with the cold seeping through the thin soles of his shoes. Sam and his Maple Leafs were repeat winners of the Washington State Championship.

When Sam got too old to play right wing or even goalie, he would referee matches, especially those involving sailors from the British battleship HMS *Warspite*, which had been severely damaged in a battle off Crete in May 1941 and sent through the Suez Canal, then across the Indian and Pacific Oceans for repairs at the Puget Sound Naval Shipyard.

It was while watching Sam and the *Warspite* sailors on a cold Sunday, December 7, 1941, that nine-year-old George learned of the Japanese attack on Pearl Harbor.

George was also an athlete, playing American football in spite of his slight stature, though he preferred basketball, hockey, and—when he could find Seattle boys who wanted to play the game—soccer. He was pretty much a straight arrow like his father—not that he had much opportunity for mischief. Large families have their own means of discipline. In the Abbey family, in addition to father Sam, two of George's older brothers, Vince and Jack, often provided day-to-day authority. After Sam joined the US Army for World War II service, and with his brothers also away serving in the military, George inherited the responsibility of helping Brenta and Phyllis.

During the 1940s, the skies of Seattle were filled with Boeing B-17 and B-29 bombers—as many as 350 rolled out of the plant each month during peak production—inspiring a whole generation of fliers and engineers, including George Abbey.

But Seattle was also a navy town, home to the Pacific Fleet. George would remember Sam taking the family to Elliott Bay to see the big battleships anchored there. George was fond of *Clear for Action!*, a 1940 novel by Stephen Meader dealing with the adventures of a young midshipman aboard the USS *Constitution* in the War of 1812. He also read H. Irving Hancock's two series about a cadet at West Point and a midshipman at Annapolis, and was a fan of C. S. Forester's Hornblower stories in the *Saturday Evening Post*.

When Sam volunteered for duty with the US Army in World War II, he shaved ten years off his actual age in order to enlist. (Years later, he would volunteer for Korea and Vietnam. "If there was a war," his granddaughter Suzanne said, "he wanted to fight in it.") After being commissioned a captain in the Transportation Corps, Sam arrived in Antwerp, Belgium, in October 1944, just as the Germans began shelling the port city with V-1 and V-2 rockets—the V-2s were in fact the brainchild of George's future NASA associate Wernher von Braun.

London was a famous target for the V-2, but Antwerp was hit by more missiles. Belgian fatalities numbered over thirty-seven hundred, with six thousand wounded, during the V-weapon blitz, which lasted into March 1945. There were over six hundred American casualties, too, and Sam Abbey came close to joining that list. His office was in the city's King's Cross, which took at least two direct hits and suffered several near misses. While Sam was away searching for replacement windows, another V-2 struck. He returned to his office to find it destroyed.

His last assignment in Europe was overseeing freed Soviet prisoners of war who had been used as slave laborers by the Nazis. With the European war ending in May, Sam returned to the United States in late June with orders to the Pacific to fight the Japanese. The announcement of the Japanese surrender on August 14, 1945, found him on leave in Seattle before his departure for his Pacific assignment.

The entire family served in the military or in support of it. George's older brothers Jim and Jack had been in military service even before the start of

World War II—Jim as a navy warrant officer, Jack as a lieutenant junior grade in the US Naval Reserve assigned to the merchant marine, seeing combat during most of the island landings in the Pacific, and at Kiska and Attu in the Aleutians. Vince had enlisted in the army when the United States entered the war in December 1941. And Brenta worked on a Boeing assembly line during the course of the war.

When George graduated from Lincoln High in 1950, it was only logical that he would want to follow the family tradition—in his case, to become a naval aviator. (The navy had "pilots," but they guided surface ships, not aircraft.) This in spite of the fact that the closest he had come to flying was seeing a Ford Trimotor up close at Boeing Field when he was six. George was enrolled at the University of Washington as a US Naval Reserve Officers' Training Corps (NROTC) student when Vince decided to "help" him out.

The Abbeys lived within walking distance of UW—Sam had moved them there from North Seventy-First while Jack and Vince were students at the university. George hoped to live at home and have an easy commute. But Vince, now at law school in Washington, DC, felt that George ought to be away from home—it would be good for him, and he would be less of a burden to Sam and Brenta. So Vince looked into appointments to the service academies and discovered that West Point's quota was already filled—but not the Naval Academy at Annapolis.

Vince sent George's grades and NROTC information to the appropriate political powers, and suddenly George found himself scheduled to take the Annapolis exams—for which he was completely unprepared. (Refusing his brother was not an option.) He reported to the Federal Building in downtown Seattle on the appointed date for a miserable experience, failing to complete two of the five sections and guessing at the answers to far too many questions.

But when he received his grades shortly thereafter, he found he had passed—barely. A quick redo of his NROTC physical followed at Fort Lawton, and he was off to Annapolis to be sworn in on June 17, 1950—a little more than one week before the outbreak of the Korean War—for four years of sometimes unhappy schooling and indoctrination.

Years later, when running the astronaut office and later the Johnson Space Center, George Abbey would be criticized for rearranging people's lives and careers; he had been the victim of the same sort of paternalistic manipulation himself.

3 | Annapolis

PARTLY BECAUSE OF THE circumstances of his enrollment, partly to Annapolis's awful climate, but mostly due to hazing at the hands of upperclassmen and general Naval Academy discipline, Abbey quickly concluded that he had made a huge mistake, envying friends who had gone to UW or other civilian universities. "I don't know anyone who enjoyed his plebe year," he would say later.

He was assigned to a room in Bancroft Hall with two roommates—Gordy Dahl and another prep school athlete, Dave McGinnis. Dahl and McGinnis immediately landed on the plebe football team—and McGinnis later started for the academy baseball team. The trio would become lifelong friends despite some early misunderstandings. "You faced an exam every day in every class," Abbey recalled. "Dahl and McGinnis told me that there was no way you could study for them, since the subject would always be a surprise. You either knew the answer or you didn't. I followed that advice for a month, and it was disastrous. So I began studying and did better."

By his second and third years, Abbey grew to appreciate Annapolis. He made other friends, though largely within his company. Through John Sollars, another Seattle boy in the class of '53, he got to know Frank Liethen (also '53), a Wisconsin native.

In addition to playing soccer, intramural football, basketball, and softball, Abbey excelled academically and graduated 128th out of 855, with high marks (top five) in history, his favorite class. (His least favorite was thermal engineering, the useful if dull study of the boilers that powered naval vessels.)

Nevertheless, Annapolis changed him much as it changed the noted science fiction writer and prophet of space travel Robert A. Heinlein—an Annapolis grad who described the experience like this:

> Take a young boy, before he has been out in the business world. . . . Tell him year after year that his most valuable possession and practically his only one, is his personal honor. . . . Tell him that he will never be rich but that he stands a chance of having his name inscribed in Memorial Hall. Entrust him with secrets. . . . Feed him on tales of heroism.

These methods produced officers and gentlemen who adhered to a stringent code of behavior—which is what allowed them to command expensive, armed vessels sailing alone in some distant ocean, frequently in bad weather. The honor code at Annapolis was strict when it came to lying, cheating, and stealing and required midshipmen to report such violations should they learn of a fellow's transgression. Unlike at West Point Military Academy, however, the Annapolis code did not require a midshipman to report violations of the other academy rules and regulations that were usually punishable by demerits.

Largely abiding by the regs, Abbey nevertheless accumulated demerits, though he couldn't rival the exploits of some of his associates, several of whom routinely went "over the wall" in search of less formal interactions with women, or just to drink and raise hell. George Knutkowski frequently vanished over the wall, among other risky activities. Stocky and uninhibited, Knutkowski played guard on the academy's junior varsity, which frequently worked out against its varsity counterparts. On one of the JV's more successful plays, Knutkowski leveled his opposite number. While performing a chest-thumping victory dance over the opponent's prostrate form, he exuberantly proclaimed, "I hope you die!" Unfortunately for Knutkowski, Navy's coach Eddie Erdelatz was standing close enough to observe the display. Which contributed significantly to the extension of Knutkowski's long and illustrious career . . . on the junior varsity.

As for other misbehavior, Abbey knew several midshipmen—including Gordy Dahl—who were secretly married, against academy rules. During their senior year, midshipmen were required to complete an application for security clearances. One of the questions was "Are you married?" Other married midshipmen Abbey knew answered no. But Dahl felt he had to answer honestly, and he wrote yes. Obligingly, he even filled in the blank that asked for his wife's name. The form was processed through the academy and in Washington, DC, before someone discovered the yes and the wife's name on the form. Dahl had to resign just before graduation. (He joined the US Air Force ROTC at the University of Washington and received a commission upon graduation, becoming a pilot.)

By senior year, Abbey had his own girlfriend—"drag" in the academy's brutal slang—Joyce Rokel Widerman, the only daughter of Melvin and Gertrude Widerman. Melvin was general manager of Rosenthal's, a well-known Baltimore department store. Four years younger than George, Joyce was a student at the exclusive Bryn Mawr School and, like many of her classmates, frequently drove down to Annapolis for social events. George had spotted her at a dance. Taking advantage of his status as an upperclassman, he'd asked one of his plebes to cut in on Joyce's partner—then cut in himself on the plebe.

He and Joyce began to see each other and during that year grew serious, eventually planning to marry once George had completed flight training and Joyce graduated from high school.

The 1954 *Lucky Bag*, the academy's yearbook, featured brief bios of each new officer. It described Abbey, "Lists among his likes popular music, traveling, reading, aviation and sports. He has many hobbies, mostly along athletic lines such as skiing and golf. His military career got off to an early start as a Boy Scout. Even at such an early age, predictions were made as to how far he would go." It noted one other trait that Abbey would become famous for: "He's well known for his unusual knack of falling asleep whenever the idea strikes him." It finished, "One of his foremost goals is to obtain a million dollars and live in style. His good nature and his quiet, easy manner as well as his 'never say die' spirit stand him in good stead. He should prove a credit to his chosen profession"—which would not, however, be naval officer.

He still believed in his teenage goal of flying. During his plebe year he had made his first flight ever, a brief hop over Chesapeake Bay in the backseat of an ancient N3N biplane. The weather was suitable; he suffered no airsickness. He enjoyed the experience and was eager to repeat it.

But Abbey had served on three summer cruises while at the academy, first to Europe and Cuba (where he had his first beer) on the battleship USS *Missouri*, then to Newfoundland on the carrier USS *Midway*, and finally to Rio de Janeiro and Cartagena, Colombia, on the destroyer USS *Brownson*. While the cruises provided a welcome break from the academic year, he found the daily routine and standing watches—"four hours on, eight hours off"—to be tedious and wearisome, particularly the midwatches. The navy required academy graduates to serve eighteen months or two years on a surface ship before they could even apply to attend flight school.

Service in the US Marine Corps was equally undesirable if you wanted to fly. The US Marine Corps required young graduates to serve a year and a half or longer in infantry units before they could apply for flight school.

Fortunately, there was third possibility.

The US Air Force, which lacked a service academy, had the right to select up to 25 percent of each year's Annapolis or West Point class. (Perhaps motivated by the surface duty requirement, 30 percent of Abbey's class opted for the air force, Abbey and McGinnis among them.) This relatively new branch of the armed forces not only promised immediate pilot training and the lure of supersonic fighter jets but also offered faster promotions.

The faster promotions, Abbey wrote later, "would turn out to be a myth." But that minor betrayal wouldn't be evident for a few years.

4 | Air Force

TO REWARD HIMSELF FOR completing four years at Annapolis, 2nd Lt. George W. S. Abbey bought his first car, a 1954 Mercury, and, after managing to get to Detroit, drove it home, from its factory to Seattle. There he traded it for his father's 1951 Mercury.

Then he was off to Hondo Air Base, Texas, a World War II–era facility that had closed in 1945, then reopened as a civilian flight school soon thereafter. In 1951, with the Korean War increasing the need for pilots, the base had been reactivated by the air force—even though the base operator was a civilian contractor, Texas Aviation Industries.

The facilities were aged and crude, hot and dusty. Abbey was quartered in an old barracks that, like most structures in those days, lacked air-conditioning. He disliked it but believed he could tolerate it for a few months—which is all the time he expected to spend in the state of Texas.

As a member, at first, of Class 55-Sierra, Abbey was assigned to the Applejack flight section and to a civilian instructor named Alfred Sluyter. On his first day, he found himself in the cockpit of a Piper PA-18 Super Cub at Hondo's "auxiliary" runway, a dirt airstrip in the town of Castroville.

Under Sluyter's patient guidance, Abbey quickly soloed and moved on to the North American T-6 Texan, the revered tail-dragging trainer with its six-hundred-horsepower engine. Turning final in the T-6, the pilot opened the canopy. Flying off dirt strips and this being Texas, the wind blew, kicking up dust and making takeoffs and landings interesting. "We wore goggles," Abbey later recalled. "When we finally stopped, we'd take them off and see that our faces were dirty everywhere. We looked like raccoons."

Days at Hondo were regimented, with flying in the morning, academics in the afternoon. There was also physical conditioning. Free time was limited, not that there was much for the students to do: beers at the officers' club and forays to San Antonio. Abbey stayed in touch with Joyce Widerman, writing frequently and occasionally splurging on long-distance phone calls.

He graduated with Class 55-Victor on April 2, 1955. However, he was not yet qualified to wear wings.

On April 21, 1955, Abbey arrived at Webb Air Force Base at Big Spring, Texas, between Midland and Abilene, to commence basic single-engine training on the North American T-28A and then the Lockheed T-33.

The day of his arrival he learned that his instructor would be 1st Lt. William G. Kirk, Annapolis class of 1953. He and Kirk had not gotten along, and Abbey realized that Kirk had probably requested him as a student, "relishing the chance to put me through my paces."

Later that day, however, Kirk took off with a different student on a training flight. Their T-28A collided with another aircraft flown by an instructor and student; both crashed, and all four pilots were killed.

Conditions at Webb weren't much better than at Hondo. Abbey was assigned to another ancient barracks, where he and his roommate were forced to stuff towels into cracks around the windows in order to keep out the blowing dust, which didn't really work. "Some mornings I'd wake up and find that our blankets were both covered with a fine sand."

Abbey enjoyed the T-28, and fell in love with the jet-powered T-33, logging dozens of happy hours flying at speeds up to 450 miles an hour, much faster than the prop-powered trainer.

He also made a lifelong friend in Lt. Bob Alpert, an ROTC graduate from Colby College in Maine, who would leave the air force within a decade to become a success in rail, real estate, and steel. For years, Alpert would tell anyone who asked that "George Abbey was the best pilot I've ever seen."

As the basic phase came to a close, pilots in Abbey's group prepared to consider their next assignments. In the late summer of 1955, the air force needed B-47 pilots for the rapidly expanding Strategic Air Command (SAC) bomber fleet. Abbey was ranked high enough to be allowed to make a choice—and would have loved to fly the F-86D. Like many, possibly most, of the new pilots at Webb, he didn't want to fly B-47s. That assignment meant a long-term commitment to the SAC, and worse yet, during that first year every pilot would be trained to fly as a navigator.

He had already decided that he wanted to be a test pilot and felt he would be better qualified with an advanced engineering degree as well as varied flying experience. An alternative arose, however: Abbey could become a helicopter pilot. It would be just as far removed from single-seat fighters as the SAC job, but it also promised to expose him to an entirely new type of flying machine. Surely that would be useful in his flight test career. The commitment was only three years, as opposed to five. So he volunteered for what the military called rotary wing flight and received his wings on September 15, 1955.

Leaving his friends, "an unhappy bunch" now in the pipeline for the B-47, Abbey packed up and drove the Mercury out of Webb headed for Baltimore, where he and Joyce were married three days later, on Sunday, September 18. They spent their honeymoon in Quebec, then drove south to Abbey's new post, Edward Gary Air Force Base in San Marcos, Texas, the Air Training Command's 358th Pilot Training Wing (Liaison-Helicopter).

In 1955 Edward Gary Air Force Base was the US military's primary helicopter training facility, with twenty-two squadrons and close to five thousand personnel. Abbey and his new bride rented a room in a home that belonged to a professor at Southwest Texas State Teachers College. Their first days as a couple were happy ones, though life in rural Texas was a big change for Joyce.

Her husband enjoyed the challenges of transitioning from fixed wing to rotary wing, first qualifying on the H-13 Sioux, known to a later generation of Americans as the helicopter in the television series *M*A*S*H*. The H-13 was built by Bell beginning in the 1940s. It was a two-bladed vehicle with one engine and a bubble-like cockpit. Underpowered, it was difficult to fly on hot

Texas days with high temperatures. It was especially hard to keep it in the air when there was no wind. After thirty-five hours in the H-13, Abbey went on to fly the Sikorsky H-19 Chickasaw, a multipurpose utility helicopter that was largely used for rescue work.

Once he had completed H-19 training, Abbey's performance was sufficiently stellar that he was offered the chance to remain at Gary as an instructor. He was happy to accept, assuming that he and Joyce would be spared a change of station. (He was wrong about that.) The assignment had other attractions. Gary trained pilots from not only the air force but also the army and NATO partners, so Abbey was able to meet students from diverse backgrounds and traditions. At Gary, he was also able to fly aircraft other than helicopters. In the 1950s, the air force had no restrictions on the number and type of aircraft a pilot could remain current on, so Abbey returned to flying the T-28 while branching out to the Cessna L-19 (which belonged to the army liaison at Gary) and the twin-engine Beechcraft C-45. He also gained experience in a third helicopter, the Piasecki H-21 Workhorse/Shawnee, originally developed as an Arctic rescue helicopter.

There was also a North American B-25 bomber at Gary, and after one of the World War II vets at the base asked him to serve as copilot, Abbey got checked out in the venerable twin-engine bird. His experiences in multiengine flying were educational, since both the C-45 and B-25 were bigger than the trainers he had flown.

As a helicopter flight instructor, Abbey's first goal was teaching his students to hover; new pilots tended to overcontrol and "get wild." A key element in successful hovering, of course, was "avoiding the ground." Once a student proved he could hover, then it was on to approaches, landings, and the vital autorotation—landing a helo without power. Eventually Abbey had to be brave enough, and confident enough in each student, to let him solo. Essentially, he just hoped he had taught the student "enough to keep him from killing himself." One lesson he learned, and would apply later at NASA: Once your student knew the basics, let him fly! As much as possible, hands off!

Abbey's time at Gary turned out to be brief. In early 1956 the air force announced that it would be closing the base in June and moving helicopter training to Randolph Air Force Base at San Antonio. This was but a minor disruption: San Antonio was only fifty miles from San Marcos.

Construction on the huge facility at Randolph had begun in the 1920s, when the US Army Air Corps decided it required an airfield designed specifically for flight training—a "West Point of the Air"—and incorporated a then-radical new layout. Buildings were generally constructed in the Spanish Colonial Revival style, of concrete bricks covered with stucco and roofs of Mission red clay tile. Even the hangars were art deco, with roofs painted in checkerboard. The base's most prominent structure was its administration building, known as the Taj Mahal.

In 1956 Randolph was home to the air force's B-29 bomber training units; other aircraft based there included the T-33, F-84E, and F-86, as well as helicopters.

Abbey's senior officers at Randolph were World War II vets who had mostly flown fighters in that conflict. Some had flown helicopters in Korea. One of them, Bob Ferris, became a good friend. He and Abbey had the fun of flying an H-19 from Texas back to the Sikorsky factory in Connecticut, a journey of eighteen hundred miles or so. The H-19 was good for three hundred miles between refuelings. So every time they needed gas—or even to grab a couple of Cokes—Abbey and Ferris would set down in some field near a country store, surprising the locals. Another sport for instructors and senior students was dropping low to chase jackrabbits.

Abbey also flew search-and-rescue missions in the San Antonio area during massive floods that took place in the spring of 1957.

His tour as an instructor would end in the fall of 1958, and throughout 1957 Abbey considered his next potential station: topping the list was reassignment to an operational helicopter unit at Spangdahlem, Germany. Another possibility was F-86D interceptor pilot training, then a tour with the Continental Air Defense Command. There was, however, a third option: graduate school in engineering. And not long after his Sputnik sighting, Abbey requested an educational assignment to Purdue or possibly the University of Michigan. He was told that all the fall 1958 slots in aeronautical engineering had been filled, that the only openings at Purdue and Michigan were in electrical engineering. Rather than wait another year, he accepted the placement in a master's program in electrical engineering. But instead of being sent to a civilian university, he was assigned to the Air Force Institute of Technology (AFIT) at Wright Field in Dayton, Ohio.

He worked through the final months of his helicopter tour, watching as the Soviets launched the first living being into orbit, a dog named Laika, on Sputnik 2. He watched as America's first attempt to launch the Vanguard satellite ended in public failure on December 6. But a team under Wernher von Braun at the army's Redstone Arsenal in Huntsville, Alabama, successfully placed America's Explorer 1 satellite in orbit on January 31, 1958.

And the United States began to expand its space program, notably creating a new civilian space agency, the National Aeronautics and Space Administration (NASA), using the facilities and personnel of the National Advisory Committee for Aeronautics (NACA). Since World War II, much of the NACA's work had been in rocketry and high-speed flight. The NACA's head, Hugh Dryden, would be NASA's first deputy administrator. T. Keith Glennan, president of Case Western Reserve University, was nominated as the incoming administrator.

NASA opened for business on October 1, 1958.

The AFIT's winter quarter began in early January 1958, and Abbey immediately regretted trading helicopter flying hours for hours of classwork in electronic circuits, radar, math, and communications, and for late evenings of intense study. He grew familiar with the Aeronautical Research Laboratory and for the first time used a computer, to complete a thesis titled "Near Infrared Transmission by Cadmium Sulfide," a topic suggested by one of his professors.

Abbey's electrical engineering class at AFIT numbered sixteen, one of them his old Annapolis buddy Frank Liethen, who had also gone into the air force after graduating a year ahead of Abbey. At six feet three and 220 pounds, rarely without a cigar, Liethen was an outstanding pilot and—except for a preference for polka music as opposed to Celtic—his interests matched up nicely with Abbey's. They would undertake a good number of T-33 "proficiency flights" to other air force bases throughout the country. On occasion they visited Gordy Dahl, who was stationed at Truax Air Force Base in Madison,

Wisconsin—Liethen's home state. In addition to his T-33 hops, Abbey had occasional access to an orphan H-13 that nobody else was qualified to fly.

Abbey had his first encounters with potential astronauts at Wright, meeting Gus Grissom (assigned to the same base as a test pilot), Alan Shepard, and John Glenn when they were among thirty candidates undergoing psychological tests in March 1959. All were senior to Abbey in age and experience, but meeting them was a reminder that humans would indeed be flying in space within a year or two. And Abbey could be one of them.

As he had with all his past schooling, Abbey did well at AFIT, scoring near the top of his class. By summer 1959 he was pondering his next post within the air force's massive Air Research and Development Command, then exploring not only advanced aircraft like the F-108 fighter and B-70 bomber but also guided missiles.

Liethen, who had come to AFIT from single-seat fighters, would go to the F-108 program office. He subsequently applied and was accepted to the US Air Force Test Pilot School. With his helicopter background, Abbey knew he didn't qualify for test pilot training. A visit to the personnel office alerted him to an immediate opening in the Dyna-Soar program office at Wright Field. Dyna-Soar—its name a contraction of the term *dynamic soaring*—was a radical new step in high-altitude flight in which a dart-shaped piloted vehicle would be launched atop a Titan rocket to complete a single loop around the Earth, flying through reentry to a controlled landing at an air base. Abbey was drawn to the idea because it combined a spacecraft and an aircraft—launching atop a rocket but landing on a runway. (NASA's Mercury project used bell-shaped blunt-body craft that would launch atop a rocket but reenter the atmosphere like a warhead, to splash down in the ocean. Some test pilots sneered at the idea, calling it "spam in a can.") Dyna-Soar was officially approved in October 1959, just as Abbey signed on. Boeing was to build the winged vehicle, and the Martin Company in Baltimore was to provide a launcher based on its Titan missile.

When Abbey arrived at building 14, "just down the hill" from AFIT, there were only ten people in the program under Col. Walter M. "Mickey" Moore, a forty-three-year-old West Point graduate who had flown combat missions in Europe during World War II, then served as a test pilot, a wing commander, and more recently, a student at the War College. He was energetic, blunt, well liked, and determined to get Dyna-Soar flying.

Moore also understood the value of mentorship and career building. He made Abbey the junior officer on the program, his "man Friday." Soon Abbey was handling a variety of other tasks, from defining and negotiating Boeing's official statement of work, which dealt with the tedious details of government-furnished equipment and systems, to communicating with the Martin Company on the Titan launch vehicle—and with NASA.

5 | Dyna-Soar

THE LANGLEY RESEARCH CENTER was a collection of ancient brick buildings, hangars, and wind tunnels on the Virginia coast near Hampton. Founded in 1917, it was the oldest of the former NACA labs that were absorbed into the new agency, NASA. Within the aeronautical community, Langley was famed for the dependable precision of its discoveries and reports (the so-called technical notes) and for the eccentricities of some of its staffers.

In April 1960 it was home to the Space Task Group, a cadre of seventy engineers, secretaries, and "computers" (women who operated bulky calculating machines*) running Project Mercury, America's first human spaceflight program. The Space Task Group was led by Robert Gilruth, a balding forty-seven-year-old engineer. Born in Michigan, Gilruth received bachelor's and master's degrees in aeronautical engineering from the University of Minnesota in 1935 and 1936. He joined NACA in January 1937, where he pioneered standards for flying qualities in aircraft as well as data recording, and later worked in hypersonic research including satellites. The seven Mercury astronauts, as members of NASA's Space Task Group under Gilruth, were based at Langley.

Langley also housed a team that worked on the air force's Dyna-Soar. By spring 1960, Dyna-Soar had reached a critical phase in its design, moving from theory to hardware. All interested parties, from the air force and NASA to major contractors like Boeing and Martin, needed to agree on the path forward, to certify such basics as the shape of the vehicle, its control systems,

---

* Later to be celebrated in the book and feature film *Hidden Figures* (2016).

its aerodynamics, its thermal protection system, and its weight. With that end in mind, the air force and NASA scheduled the Joint Conference on Lifting Manned Hypervelocity and Reentry Vehicles. Flying into nearby Langley Air Force Base that first week of April 1960 was Capt. George W. S. Abbey.

On Abbey's first day he arrived at the center's gymnasium for a rehearsal of the conference, which was to be held the following week. He expected a dry run-through of all the papers, possibly with a helpful comment or request for clarification. What he found was veteran NASA engineer John Stack sitting at a desk on a stage at one end of the gym, facing an audience of unruly NASA engineers who blithely wandered in and out as they wished—no military discipline here.

The first engineer approached a podium to talk through his slides on the Dyna-Soar thermal protection system, designed to protect the spacecraft from reentry temperatures of thousands of degrees. The audience leaped on that first slide, and every other slide, questioning each conclusion, each equation, each underlying assumption, even the typeface. There was shouting. There were red faces. There were expressions of violent disbelief. Finally, mercifully, Stack hammered his gavel. The presenter was told the changes that must be made; then he slunk off the stage.

Abbey had barely recovered from this display when a second presenter threw a slide up on the projector, and the free-for-all resumed.

This went on all day.

With the assistance of his NASA contact at Wright Field, Peter Korycinski, Abbey began to identify some of the participants. Bob Gilruth glided in and out of the conference, much like the bemused father of an unruly family. Another man in his forties, George Low, the aerodynamics specialist who was head of manned spaceflight at NASA's HQ in Washington, watched the bloodbath with serene detachment. One of the most persistently vocal critics was a tiny man who could barely sit still. Korycinski told Abbey, "That's Max Faget," who happened to be the genius behind the basic Mercury spacecraft.

Abbey realized that engineering wasn't purely numbers and science, that there was always room for interpretation and even opinion. But this wire brushing? It would *never* have happened in the air force. Briefers in the Dyna-Soar program office were treated with respect and deference! Colonel Moore was occasionally blunt and a practitioner of chain-of-command management, but he was willing to listen and, if motivated to comment on what he heard, to do so quietly. Bill Lamar might have been one of the world's experts on aerodynamics, but he would *never* have challenged Moore in public the way Max Faget did his NASA boss.

The session ended. Abbey emerged into the Langley afternoon to find that kegs of beer had been set up outside the gym. And there was Max Faget, joshing with one of his many victims as if nothing unpleasant had happened! The drinking and kidding continued well past dinnertime, and Abbey had to drag himself back to bachelor officer quarters.

The next day, the battle resumed. By this time, Abbey was able to enjoy the spectacle. What surprised him most, however, was the conference itself, when the NASA presentations—the product of those painful batterings—sailed through with no criticism or challenges. It was the presentations from non-NASA scientists and contractors that meandered or failed or otherwise led to hushed rebuttals in the hallways, that would have to be discarded or rethought.

Abbey learned a vital lesson from this experience—engineering worked best when subjected to immediate face-to-face peer review. Also, you could have serious disagreements about data and still be friendly.

And beer had its uses, too.

6 | Boeing

SHORTLY AFTER EIGHT IN the morning on Friday, May 5, 1961, Captain Abbey stopped by the Wright Field cafeteria on his way to work. As he was getting a cup of coffee, he noticed that the radio was playing NBC from Florida, with coverage of the launch of navy commander Alan Shepard on the first manned Mercury mission. Abbey had expected the launch; he knew that NASA had had to scrub the first attempt on May 2. He hadn't anticipated being able to hear the live coverage.

The Soviet Union had made the first manned spaceflight, sending twenty-seven-year-old pilot Yury Gagarin on a single orbit of the Earth in a spacecraft called Vostok (meaning "east") on April 12. The event had surprised Abbey, but since he, like most space professionals, had expected a Soviet flight at some point, it wasn't the shock Sputnik was.

Abbey sat down alone at a table, listening as NASA worked through several holds in the countdown. Finally, at 8:34 AM (CST), Mercury's Redstone rocket ignited, and Shepard was on his way, rocketing to an altitude of 116 miles, heading 303 miles downrange. Only when Abbey heard word of Shepard's safe splashdown did he rise to go back to work.

Two months later, on July 19, 1961, former Wright test pilot Gus Grissom repeated Shepard's flight, arcing to an altitude of 116 miles and flying 300 miles downrange from Cape Canaveral. The early ejection of the Mercury capsule hatch caused the vehicle to flood; Grissom was forced to swim for his life.

In mid-August, the Soviets launched a second cosmonaut, Gherman Titov, on a Vostok mission that lasted an entire day.

In January 1961, newly elected president John F. Kennedy took office, bringing new leadership not only to NASA (where wily political operator James Webb replaced the more academic Keith Glennan) but also to the Department of Defense.

The new secretary, Robert McNamara, former president of the Ford Motor Company, examined America's massive and expensive military complex of bombers, submarines, missiles, and ground forces for potential cost reductions. He questioned the need for a new supersonic fighter, the F-108; a supersonic bomber, the B-70; and a nuclear missile launched from a bomber, the GAM-87 Skybolt—and soon canceled all three. Dyna-Soar was in McNamara's sights, too, but there the news was better. The new secretary expanded Dyna-Soar's mission from hypersonic research to actual orbital flight, a goal that seemed more achievable in the spring and summer of 1961 than it had earlier. Abbey welcomed the decision to make Dyna-Soar an orbital program, though it now required a more powerful booster than the original Titan I, creating a host of technical problems.

On May 25, 1961, President Kennedy had proposed, and Congress had supported, the goal of landing a man on the Moon—and returning him safely to Earth—by the end of the decade. This $20 billion effort, Project Apollo, would be managed by NASA—the organization Abbey had been working with since 1959. By September 1961 he would have been assigned to Wright Field for three years, including his time at AFIT. This Apollo project was something entirely new and exciting, far surpassing even Dyna-Soar's goal of winged orbital flight. Abbey looked at Apollo with a bit of envy but saw no way to get involved.

Air Research and Development Command had decided to assign development engineers to a contractor's facility, Boeing in the case of Dyna-Soar. Mickey Moore and the program leaders felt that Abbey might be a good choice.

When the possibility was first raised, Abbey—having grown canny about these matters—protested, "Oh, don't send me there."

His ploy worked. Assigned to the new Development Engineering Office at Boeing, Abbey made the long drive to Seattle the first week in September. Joyce, son George Jr., and a baby girl, also named Joyce Brenta Kathryn (Joyce BK), flew out a few weeks later. Returning to Abbey's beloved Seattle, they first moved in with Sam and Brenta, which allowed grandparents and grandchildren to get to know each other. George Sr.'s older brother, Vince, and younger sister, Phyllis, and their families were in the area too.

Abbey eventually found a house in Seattle's Magnolia district, not far from Sam and Brenta's. It rested on the edge of a hill so steep that riding up and down the incline frightened George Jr. Abbey immensely enjoyed being back in his hometown. He was surrounded by family and old friends, had important work, and could take advantage of the city's outdoor activities. "I could leave Boeing at 5:30 in the afternoon and be in the mountains within an hour. They had night skiing, so I could ski until 9:30."

Moving to Seattle at the same time was Lt. Col. Jim Wiley, forty-three, who held a physics degree from the University of Pittsburgh and had served in World War II as a fighter pilot with the Tuskegee Airmen.

The Abbeys were raised to be free of racial prejudice, an ideal that had been severely tested in the fall of 1944. Vince was then training to be a Japanese linguist in army intelligence. He telephoned home in mid-November to tell Brenta that two of his friends would be passing through Seattle at Thanksgiving. Would she invite them for dinner? Of course, she replied.

Twelve-year-old George and ten-year-old Phyllis were sent to meet the two soldiers at the streetcar downtown. They discovered that the two were Japanese Americans, known in those days as nisei. The kids brought them to the house and watched as Brenta welcomed them as if they were members of the family, a lesson George would remember all his life.

Sixteen years later, Abbey offered to help Lieutenant Colonel Wiley find a place to live. For several evenings, the pair drove to different parts of Seattle looking at new homes and developments. Abbey had telephoned ahead and received eager encouragement from landlords happy to host an air force officer and pilot. But when Abbey and Wiley showed up, things changed. Whenever Wiley expressed interest in a property, there was some new problem, or even a suggestion that the colonel might consider locating in a different part of the city—an obvious hint that he should look in neighborhoods more hospitable to people of color.

Abbey had witnessed other officers in the Dyna-Soar program treating Wiley poorly, often ignoring his suggestions. This latest snub from hometown folks left Abbey embarrassed and angry.

On February 20, 1962, astronaut John Glenn, a Marine Corps lieutenant colonel, made the first American manned orbital flight, circling the Earth three times in dramatic fashion, splashing down safely in spite of a fear that his vehicle's heat shield had come loose.

Navy lieutenant commander Scott Carpenter followed in May 1962, largely repeating Glenn's three-orbit mission with his own dramatics: running short of fuel to control his spacecraft's attitude and splashing down 250 miles from recovery forces. Navy commander Wally Schirra launched on a six-orbit Mercury mission on October 3, 1962, and completed it without heart-stopping moments. The Soviets conducted two more Vostok missions too.

At the same time, NASA was hiring a new group of astronauts for Gemini, a spacecraft capable of carrying two astronauts, and Apollo. Abbey had seen the announcement back in May and would have loved to apply, but applicants had to be graduates of military test pilot schools or employed as civilian test pilots. He was neither.

Like the Mercury program, Dyna-Soar had astronauts, though they were not given that title but instead called "pilot-consultants." Based at Edwards Air Force Base, they were required to support other Edwards development, flying pace and chase, for example. One of the men would visit Wright, and later Boeing, for a week at a time.

The senior pilot was Maj. Jim Wood, a finalist for Mercury who had been rejected. Nevertheless, he was highly regarded as a pilot and engineer. There were three other air force test pilots: Pete Knight, Hank Gordon, and Russ Rogers. Abbey would grow to know all of them well, finding Rogers especially capable.

NASA had three pilots in the Dyna-Soar program, the most senior being Milton Thompson, a navy World War II veteran from Minnesota. The others were Bill Dana and Neil Armstrong. Dana left the program in early 1962 to concentrate on other NASA programs at Edwards. Then, in September, Armstrong was selected for the 1962 NASA astronaut group. Unhappy with the number of civilian pilots in their military program, Dyna-Soar officials replaced Dana and Armstrong with Capt. Albert Crews.

That same month Dyna-Soar—now designated X-20A—had its public unveiling at a meeting of the Air Force Association in Las Vegas. Its future seemed bright. But Dyna-Soar faced an enemy, and it was neither the Soviets nor some insurmountable technical challenge; it was the Aerospace Corporation, a nonprofit research and engineering company that had been created to provide technical management to all air force missile and space systems. It was able to pay higher salaries than the air force offered its engineers and to keep people assigned to programs for years at a time, providing continuity.

While in Los Angeles for yet another Dyna-Soar review early in 1963, Abbey learned firsthand that Aerospace wanted to scrap the space plane in favor of Blue Gemini, a series of military manned orbital missions that would use NASA's Gemini spacecraft. Blue Gemini would lead to a military manned space station to be launched by the Titan IIIC. This idea fit perfectly with McNamara's philosophy, expressed by the defense secretary in a briefing in January 1963. If the United States was researching controlled reentry from orbit, NASA should be the responsible agency. And migrating the civilian-developed Gemini to operational military uses would save vast sums of development money.

Abbey didn't accept McNamara's reasoning; NASA had been involved with Dyna-Soar from the beginning. And if the program's scope had evolved beyond pure hypersonic research, it was because McNamara wanted it that way.

Dyna-Soar lived on, managed from Wright Field, but George Abbey had learned a new and valuable lesson: there wasn't just one air force.

In May 1963 air force major Gordon Cooper flew Mercury-Atlas 8, the fourth and final orbital mission in the program, circling the Earth eighteen times in thirty-three hours. It struck Abbey as an impressive achievement—after all,

Mercury had originally been designed to put an astronaut in space for three orbits, little more. It was a testament to the space agency that engineers had found a way to improve and enhance the vehicle.

Three weeks later, Cooper's mission looked less notable. On June 14, the Soviets launched cosmonaut Valery Bykovsky aboard Vostok 5, and he remained in space for five days. Worse yet for American space prestige, on June 16 Vostok 6 was launched carrying a woman cosmonaut named Valentina Tereshkova. The twenty-six-year-old wasn't even a pilot; she was a parachutist recruited for space training. She logged as much time in space during her seventy-hour flight as all the Mercury astronauts combined.

Tereshkova's flight also triggered a great many newspaper editorials and letters to NASA, asking when the agency would put an American woman in space—or why there were no woman astronauts. Abbey knew the answer, of course: NASA desired experienced test pilots for its program, and under the restrictive practices of the day, there weren't many women who had acquired such a background.

In spite of battles between Wright and Los Angeles, by November 1963 the first Dyna-Soar flight vehicle was taking shape on the factory floor at Boeing. The first piloted orbital flight would be in July 1966, one of eight that would continue into 1968.

On Friday, November 22, 1963, the Dyna-Soar office at Boeing hosted a group of students from the Aerospace Research Pilot School Class 63B. Among them was Abbey's friend Frank Liethen, and also Jim Irwin, another pilot he'd met and flown with at Wright Field.

Three members of the class were absent. A month earlier David Scott and Ted Freeman had been selected as members of the third group of NASA astronauts, and they were planning to relocate to Houston. (Liethen had applied but had been rejected as too tall. Abbey had considered applying but wasn't eager to leave Dyna-Soar at this stage and felt his high-performance jet time was insufficient.) A third member also failed to make the trip: Capt. Ed Dwight, an African American pilot whose enrollment in the US Air Force Aerospace Research Pilot School (ARPS) had been ordered by the Kennedy administration over the vehement objections of ARPS commandant Chuck Yeager, who didn't believe that Dwight (whose flying background was in bombers) qualified.

Suddenly, a secretary received a phone call that President Kennedy had been shot in Dallas. Everyone turned on radios and televisions, and soon Abbey and the others heard the word: Kennedy was dead. The news turned what would have been a reunion of old friends into a wake as the pilots discussed the impact of Kennedy's death on the world, the nation—and the space program.

On Tuesday, December 10, Abbey arrived at the Dyna-Soar office at Boeing to learn, via a phone call from Col. Moore's office, that McNamara had canceled the program, making the announcement in a Pentagon news conference. There would be no reprieve: the air force had had four super program offices when McNamara became secretary of defense. Dyna-Soar was the final one to be canceled. The pilots simply returned to their duties at Edwards. The Boeing engineers would be assigned elsewhere within the company.

After closing out several Dyna-Soar contracts and research efforts, Abbey quickly found new work at Boeing. He supported the supersonic transport (SST) program being built by the company and, thanks to his contacts at NASA Langley, also joined the team building the Lunar Orbiter, a series of unmanned probes scheduled to be launched to the Moon in 1965 to take high-resolution photographs of possible Apollo landing sites.

He was also able to fly acceptance tests on new KC-135 tankers and C-135 transports as they rolled off the Boeing production lines, then to deliver the vehicles to their new units. When not doing acceptance or delivery flights, he flew with the Air Defense Command's 325th Fighter Wing at nearby McChord Air Force Base in Tacoma. He hoped that the increased flight time and the variety of flying experience would qualify him for acceptance at ARPS. He wasn't building a winged spacecraft any longer, but he might have the opportunity to fly in NASA's Apollo spacecraft.

Elsewhere, Maj. Gen. Samuel Phillips, program manager for the Minuteman intercontinental ballistic missile (ICBM), became Apollo program director at NASA headquarters. Abbey had met Phillips during one of his trips to Boeing. Abbey already knew the Phillips legend—how one Friday he had briefed the new president Kennedy on the progress of the Minuteman missile and on Monday was promoted to brigadier general. Phillips felt that NASA needed engineers with experience in program management. With his experience at Dyna-Soar's program office and prior experience with NASA, Abbey became a candidate. He received orders to visit Houston to interview for a position in the Apollo program office's system engineering division. He flew there in October 1964 to meet with his potential new boss, Owen Maynard, and other engineers assigned to the program, including some he'd met on Dyna-Soar.

During his week in Houston, Abbey developed immediate concerns about the state of the Apollo program and the number of yet unresolved issues and, more surprisingly, the absence of any plan to correct them. And, coming from Seattle, he was also taken aback by the weather, unable to believe that humans would choose to live in such heat and humidity. During his previous assignments in Texas he had experienced the heat, but not Houston's humidity.

On Friday, his last day, Maynard asked him if he wanted to move to Houston and join the team.

Abbey told him, "I don't want to work here."

"Oh come on. At least think about it over the weekend."

"I won't change my mind. I don't want to come here."

"Call us Monday."

"Why? I don't want to come here."

Abbey flew home, greatly relieved, and put NASA and Houston out of his mind. He didn't telephone Maynard on Monday. But as he was preparing to leave Boeing at about 5:30 that evening, the phone rang. It was Maynard. He said, "You haven't called." Abbey said, "I haven't changed my mind."

"That's too bad," Maynard said, "because we told the air force last Friday that we wanted you, and they said fine, and they will be sending you orders reassigning you to the Manned Spacecraft Center in Houston."

Abbey knew that he was inevitably going to NASA, but if he had to join the space agency, he preferred to go to Langley in Virginia. He asked his contacts in the Lunar Orbiter office at Langley if they would write to General Phillips and explain the necessity of Captain Abbey's presence there.

Phillips wrote back: "I'm very pleased to hear all that Captain Abbey is doing to support the Lunar Orbiter Program. He has thirty days to complete all those activities and to report to Houston."

Part II

"Before This Decade Is Out"

7 | Destination: Moon

IN THE FIRST WEEK of November 1964 George Abbey drove to Houston with two friends from the Seattle area who were seeking jobs at NASA. He moved into bachelor officer quarters at Ellington Air Force Base, four miles up the road from the Manned Spacecraft Center (MSC). His family—Joyce, six-year-old George Jr., four-year-old Joyce BK, and ten-month-old Suzanne—would remain in Seattle until he got settled.

On Monday, November 9, Abbey reported to Owen Maynard on the sixth floor of building 2, the tallest building at the MSC. The buildings and sidewalks were new, but the site looked raw: there was no real yard, just sprigs of St. Augustine grass in what was largely mud.

Like Abbey, the forty-year-old Maynard was a pilot, having flown Mosquito fighters for the Royal Canadian Air Force in World War II. He had then joined aircraft manufacturer Avro Canada while working his way through the University of Toronto, becoming one of the Canadian engineers recruited by NASA in 1959 after the cancellation of the CF-105 Arrow fighter jet. During that recruitment interview, Maynard had volunteered as a pilot in the American manned space program—and was, of course, instantly told no, making him "the first Canadian to volunteer to become an astronaut, and the first to be rejected."

In the fall of 1964 he was head of the Systems Engineering Division (SED) within the Apollo Spacecraft Program Office, just one office in the giant effort that was Apollo.

Flights to the Moon had been the staple of science fiction novels for over a century, and for two generations of motion pictures, pulp magazines, and comic books. George Abbey had devoured such stories. After World War II, the concept began to receive serious consideration from rocket pioneers like Wernher von Braun and—after Sputnik—from the US Army, which authorized Braun to develop a series of heavy-lifting Saturn boosters.

In 1959, with the Mercury "manned satellite" officially in progress, leaders of the new space agency agreed on a series of nine future steps that began with Mercury's Earth orbital flight and culminated in the Mars-Venus landings. Lunar landing was step 7 and was being pushed hard by George Low of NASA Lewis and Max Faget of NASA Langley.

By May 1961, Apollo's conceptual vehicle had begun to take shape. Its command module would be a blunt-body cone equipped with a thermal protection system that would shield it from the heat of reentering the Earth's atmosphere at twenty-four thousand miles an hour. It would carry a crew of three (the navy-minded men of the Space Task Group assumed round-the-clock watches of eight hours each), so it would need an environmental system that could support them in safety, if not comfort, for two weeks. The crew would need to be able to operate or "fly" the vehicle, so controls were needed in a workable cockpit layout. They would also need to be able to navigate, necessitating a guidance system. And a communication system. And cameras and scientific instruments.

The command module was only one element. The vital cylindrical unit below it, the twenty-five-foot-long service module (SM), had its own requirements and challenges, many of them complicated by the delay in selecting the landing mode. (As originally conceived, the SM would carry a cluster of solid rocket motors to return Apollo from lunar orbit.)

Apollo's potential booster rockets, the Saturn family, were in development too.

With President Kennedy's address to Congress on May 25, 1961, calling for a manned landing on the Moon and safe return to Earth "before this decade is out," the modest Apollo effort was suddenly raised to a major national program with a projected $25 billion price tag (over nine years). North American

Aviation had won the contract to build Apollo. It took another year for NASA to decide on the mode of landing—lunar orbit rendezvous. The Grumman Corporation was then awarded a contract to build a lunar lander.

Within NASA, the effort had been distilled to three words: "man, Moon, decade."

Apollo's budget exploded from $1 million in 1961 to $162 million a year later. When George Abbey arrived in Houston in November 1964, Apollo was a $2 billion a year program employing 350,000 people at an estimated twenty thousand different companies working on spacecraft, boosters, facilities, and other technology. (NASA itself was consuming 2.5 percent of America's gross national product.)

In addition to the command and service module (CSM), lunar module, and Saturn launchers, Apollo was supported by a number of unmanned programs too. Ranger, out of the Jet Propulsion Laboratory (JPL), was designed to beam back photos of the lunar surface as it dove to an eventual crash. (Seven Rangers had been launched by the time Abbey reached Houston, with only one—Ranger 7 in July 1964—returning useful imagery. Two more were to be launched early in 1965.)

Imagery of the lunar surface from orbit—a more challenging mission—was the goal of Langley's Lunar Orbiter program, one Abbey obviously knew well. The Lunar Orbiter was scheduled to make its first launch in the summer of 1966.

The most ambitious unmanned program was Surveyor, a JPL-led effort to land a vehicle on the lunar surface, where close-up photos could be taken—and a remote-controlled claw could dig into the lunar regolith to provide information on its consistency. (A small but powerful group of scientists had warned for years that the lunar surface was soft dust that would swallow up an Apollo.)

When he first walked into building 2, George Abbey was familiar with the broad strokes of the Apollo effort—the yearlong battle over lunar orbit rendezvous, for example. He was given a stack of reports and studies and advised to start reading. He also met his new coworkers. There were three dozen people in Maynard's office, all white males in their twenties, thirties, and forties. The only women were secretaries.

Abbey officially worked directly for Owen Maynard and his deputy, Bob Williams, who had come to NASA from Thompson Ramo Wooldridge (TRW), where he had worked with Joseph Shea in support of air force unmanned missile and satellite programs. Williams would become a good friend. Harry Byington, who specialized in propulsion, and Charles Haines, who worked on displays, controls, and the human interface with spacecraft systems, were Abbey's officemates. He would quickly grow to like and respect Maynard, though he judged him to be "not the most organized person in the world." At first he was surprised that Maynard expected him to be an expert on configuration management and the means and methods of coordinating and scheduling the development of cutting-edge technology such as ballistic missiles. It turned out that Chet McCullough, another engineer in Maynard's office, had worked with Abbey on the Dyna-Soar program and knew of Abbey's prior experience in that subject. McCullough had been urging Maynard to get Abbey assigned to Houston.

It would take time, but eventually Abbey would judge Maynard's team to be "very technically competent" and more accomplished than those he had known in the air force. In those first weeks, however, he found it difficult to judge the program's status. For example, in September, NASA had performed a review of mock-ups of the Block II CSM with North American, yet two months later, in mid-November, it was still uncertain what had been agreed upon in that review, and if North American was proceeding on an agreed-upon configuration and schedule of work.

During his earlier visit, Abbey had been surprised at the lack of oversight of the spacecraft's configuration and of a specification. And he was concerned about the existence of a Block I *and* a Block II Apollo command module. (The Block I spacecraft would only be flown in Earth orbit and would not have the same configuration as the Block II, which would be flown to the Moon.)

He brought his concerns to Maynard, who had the ideal solution. After all, this was why he had insisted on having Abbey on his team. "Why don't you take this over?" As in, go to North American and Grumman and sort this out.

"But I just got here."

"That's why you'd be perfect."

8 North American Aviation

"READY OR NOT, WE'RE coming out."

It was the first week in December 1964, and George Abbey and Owen Maynard were on a conference call with James R. Johnson of North American Aviation in California. After exchanging pleasantries with Johnson, Maynard introduced Abbey and said he would be visiting Downey the following week for Johnson's update on the Block II CSM. Johnson told Maynard that would be fine, and Abbey made plans to fly to California the following Monday.

Founded in 1928, North American was famed for aircraft and missiles, from the T-6 Texan trainer, B-25 bomber, and P-51 Mustang of World War II to the more recent F-86 and F-100 jet fighters and the X-15 hypersonic research aircraft.

By December 1964, North American had twenty-five thousand employees at work in Southern California: at a facility at Los Angeles airport known as the "Brickyard," at a large plant in Downey, and also at sites in Seal Beach, Palmdale, and Canoga Park.

Abbey flew to Los Angeles alone, rented a car, and drove to Downey. His first stop was customer relations in the headquarters building of the Space and Information Systems Division, where he was badged and, after a phone call, escorted to Johnson's office. He had just grabbed a welcome cup of coffee and sat down there with Johnson when two men, Alan Kehlet and Gerald Fagan, burst in. They were both unhappy at seeing this unknown person from NASA. "Who are you and what are you here for?"

Abbey explained that he was in Downey to meet with Johnson to discuss the Block II CSM.

Kehlet and Fagan insisted emphatically, "He doesn't have a damn thing to do with it. We're in charge of Block II."

Abbey turned to Johnson. "Is that right?"

"Yes."

Abbey couldn't believe it. Owen Maynard hadn't known who at North American was in charge of the Block II! Johnson, it turned out, was a structural engineer on the company's engineering staff and had little to do with Apollo.

The company was used to working for the air force, "who pretty much left North American alone," Kehlet recalled.

> North American would tell them when they needed money—not quite this way, but almost—and would tell them when the product was ready for acceptance by the air force. And they expected NASA, the same thing.
>
> Well, it was a big surprise to find out NASA was going to send troops in, and you were going to have technical reviews every two weeks and design reviews and contract reviews and everything else.

Fortunately, Kehlet was a former NASA engineer who had worked under Max Faget. He appreciated NASA's expertise and its experience on Mercury and Gemini. With apologies to Johnson, Abbey went off with Kehlet and Fagan to get his answers about the Block II. He asked the pair if they had agreed upon a configuration and a schedule at the September mock-up review. He was told no in both cases and that Abbey was the first NASA visitor Kehlet and Fagan had had since that review. "They were quite angry about it." Both knew how important it was to agree on the configuration and schedule and the support activities. Over the next week, Abbey worked with Kehlet and Fagan to lay out a series of progressive reviews to support the manufacture and production of the Block II command module.

Abbey's first encounter with North American was one of thousands that had taken place between NASA engineers and contractors since 1961. He himself would downplay his role, claiming that in Apollo he was merely one member of a vast team. But years later, shuttle astronauts such as Rick Searfoss would hear stories about Abbey in his early days on Apollo: "There was a meeting where Abbey, who was quite junior, grew frustrated that no one was making a decision. So he did—and it made his reputation."

Returning to Houston after a week in California and briefing Maynard on his work with Kehlet and Fagan, Abbey received a new assignment: complete the specification for the command and service modules. Abbey surrendered to the inevitable and asked who he should see at North American to conduct the negotiation. "Norm Ryker," Maynard said, naming the head of systems engineering at the company. Abbey telephoned Ryker's office in California. There was no answer. He tried again. Still no answer. This went on for three days.

Maynard finally got through to Ryker after Abbey complained. Hearing that Abbey would be the NASA point of contact, Ryker enthusiastically endorsed the plan.

One month into his tour in Houston, Abbey found himself (under Maynard's deputy Bob Williams) handling the CSM specification and managing the program reviews. There was another major contractor for Apollo spacecraft, too, and soon Maynard told Abbey that he would be doing the same work on the lunar module built by Grumman.

Based on Long Island, Grumman was a venerable airplane manufacturer famed for its rugged World War II carrier aircraft, the F4F Wildcat and F6F Hellcat, and for its TBF Avenger dive bomber. It had moved into jet aircraft with the F9F Panther.

With NASA, however, Grumman's history was spotty. The company had competed for every major contract the civilian agency had offered since 1958, including Mercury and Apollo. But until the lunar excursion module (LEM), Grumman had won only a single program: the unmanned Orbiting Astronomical Observatory. Thanks to the foresight of Joseph Gavin, its chief missile and space engineer, however, Grumman had developed a lunar module on its own, winning the contract in 1962.

Now, three years in, the strains were showing. Like every element of Apollo, the LEM had proved to be far more technically challenging than anyone had expected. Grumman had originally begun work on a LEM with a cylindrical descent stage supported by five fixed legs, and a cylindrical ascent stage complete with a helicopter-style cockpit for two astronauts. It had evolved into a boxy, nonaerodynamic vehicle with an octagonal descent stage and four legs.

The LEM was considerably overweight. Grumman's lead for the program, Thomas Kelly, instituted a rigorous weight-loss program that lightened the vehicle by over two thousand pounds, putting it under the design weight for the first time. The trade-off was to leave the LEM fragile—an ironic state for Grumman, which had been famous for its rugged, robust naval aircraft. (The fear was that a dropped screwdriver in the crew cabin would punch a hole through the floor and the descent stage below.)

In spite of the technical challenges, as a company, Grumman was popular with the NASA team, which admired the work ethic of the Long Island crew. By 1965, according to Abbey and other NASA folks, the hours and productivity declined as one traveled west across the United States. Grumman in New York was the best, with McDonnell in St. Louis, builders of Mercury and Apollo, seen as adequate. The California team at North American was judged to be arrogant, complacent, and too eager to leave work early on Fridays for weekend recreation.

Still, there were problems everywhere. "With Grumman, it was their way or no way," Abbey said. "They were a close-knit family operating almost by word of mouth," which made it difficult to control drawings and to make changes—which were estimated to be a thousand per month across both Grumman and North American.

Nevertheless, Abbey worked well from the very beginning with the LEM team under Tom Kelly and with Kelly's point of contact, Bill Craft.

9 Houston and the Manned Spacecraft Center

ABBEY FLEW BACK TO Seattle in late January, loaded up his family, and took to the road, arriving back in Houston on the thirtieth.

A city girl at heart, Joyce was not pleased by the new environment. George himself found the new MSC area a bit grim at first: "There was nothing here, very few homes in Clear Lake. A restaurant called the Flintlock, which is where everybody went. The Nassau Bay Hotel and the Kings Inn were about the only places to stay. There were no grocery stores, no shops, not even gas stations. No easy way to get bread or milk. You had to drive up the freeway to Houston. The first store of any kind was a 7-Eleven, and even that took some time to arrive. It was like living out in the country." It was worse than the air base communities in Texas.

Motivated to get settled, Abbey visited a model home that had just been built across the street from the MSC. "Because it had the right number of bedrooms, I bought it."

The community the Abbeys now called home began to grow, with housing developments, shops and stores, gas stations, and new schools sprouting like dandelions around Clear Lake.

The center itself was new and unfinished, with some engineers on site and others in office buildings nearby. All were concentrating on two major programs being developed in parallel: Apollo and Gemini, scheduled for its first flight in March 1965.

The Gemini program office was located in Building 2, one floor down from Apollo systems engineering, but Abbey had no direct dealings with the team there. He would follow all the Gemini missions on radio and television. He would later judge this lack of communication and interaction to be a problem. There were lessons to be learned from Gemini, both from the NASA side and the contractor, and these were just not being passed along, forcing the Apollo team to reinvent the wheel.

Soon after arriving in Houston, Abbey offered his services to the Texas Air National Guard (TANG)—"defending the coast of Texas." He was a pilot, after all, and loved to fly. He also needed to remain current, since logging four hours in the cockpit every month qualified him for bonus pay, badly needed for an air force captain supporting a growing family in a civilian environment.

He was first assigned to Ellington's base operations, flying the C-47 cargo plane and the Cessna 310, just to satisfy his monthly requirements. But Abbey was eager to continue to fly jet aircraft, so he approached Col. George McCrory, commander of the guard's 111th Fighter-Interceptor Squadron. Abbey knew that McCrory needed an instructor pilot and liked the idea of a "free" pilot on his team. Abbey wouldn't count against McCrory's numbers, and he would be paid by the air force.

Abbey flew back to McChord Air Force Base in Tacoma to get certified as a T-33 instructor, a process that took exactly one flight. (He had already flown T-33s there and was certified by one of the pilots he knew.) He presented McCrory with his certificate; the colonel said, "We're doing exercises Friday night. Be there."

"Do I need to do a local area checkout?"

"You're a pilot, aren't you? Show up for the briefing."

By summer 1965 Abbey was flying TANG T-33s on weekends and making friends with other guard pilots like C. B. Doss (a teacher), Jerry Killian (a full-time guardsman), Tom Griffin (a stockbroker), and Fred Davis (a pilot for Delta Air Lines). He also got to know two NASA staff pilots who happened to be neighbors—Stu Present and Arda "A.J." Roy. Present and Roy worked for Joe Algranti, a longtime NASA test pilot who headed the expanding aircraft operations unit. Abbey kept in shape by playing basketball and softball with his Ellington friends as well as colleagues from the MSC.

Joyce and the children were adjusting to a community that not only grew larger and more populous every week—it also had buses taking tourists around to astronauts' homes.

Abbey generally arrived at his office at 7:30 AM—the house was just south of the Manned Spacecraft Center, across the main road, NASA One, so he had a short drive—carrying his briefcase containing whatever reports he was reading. He didn't carry a slide rule, though he had one in his desk. His life was paperwork and meetings on that paperwork. A specification for the command or service module, or a summary of a review or equipment test required discussion with NASA engineers and contractors, either to settle on a course of action or to plan a next step. One rule of every meeting was that it was late. The workdays didn't end at 5:00 PM; they usually ran until 7:00 or 8:00 PM or later.

Many weeks included travel, to North American in Los Angeles or to Grumman on Long Island. It was a brutal schedule, tolerable only because most of the staffers were young and unmarried.

Joyce and the children accepted it for now, but it was a big change from life in Seattle.

Development and construction of Apollo systems and hardware began to surge on a staggering variety of fronts: high-altitude launches at White Sands, to test abort systems; modifications to launchpads at the Cape; work in California on a lunar landing research vehicle, which would allow astronauts to perform Earth-based rehearsals for lunar landings; configuration of communications systems at Motorola in Chicago; tests of astronaut portable life-support systems at ILC in Delaware; docking simulations at Langley; and on and on. Tracking the development, and approving changes, was a huge issue—and it fell to thirty-nine-year-old Joe Shea, head of the Apollo spacecraft program office.

The Bronx-born Shea, son of a New York subway mechanic, had graduated from high school at age sixteen, at the height of World War II, and enlisted in the navy, where it was discovered that he had an aptitude for math. Rather than assign him to a ship or submarine, the navy sent him to the University of Michigan for a degree in engineering. He did his postgraduate studies at Dartmouth and MIT and with a doctorate in engineering went to work for Bell Labs in 1950, where he acquired a reputation for technical brilliance and managerial competence. Later posts at General Motors, working on the guidance system for the Titan II missile, and at TRW only confirmed this.

Shea was a popular leader, willing, when programs hit snags, to move into a factory to share round-the-clock labors with a technical team. He was famed for his ability to make puns and for an eccentricity of dress: at important meetings, he wore red socks.

It was a huge job wrangling two prime contractors and dozens of significant subcontractors, all of them struggling to produce state-of-the-art propulsion, electronic, environmental support, thermal protection, and structural hardware as well as innovative computer systems on a brutal schedule with never enough money. Shea chaired a configuration control board that met infrequently; in reality Shea ran the Apollo spacecraft development on his own, collecting reports every Thursday, annotating them on Friday, and returning them to Tom Markley, his chief of program control, on Saturday, for distribution.

This one-man band allowed Shea to communicate with the hundred members of his staff and thousands of others across the nation. It avoided the chaos of conflicting change orders originating with supervisors at North American or Grumman, or with astronauts. If Shea hadn't signed off, an action wasn't taken.

But there were significant gaps in Shea's reporting chain. Originally even Max Faget was out of the loop. Abbey knew that initial encounters with Faget could be bumpy, but Shea had needlessly antagonized the temperamental engineer when, just after arriving in Houston in 1963, he had remarked that he wasn't impressed with Faget's operation. Faget, whose team had just pulled off the successful Mercury program, had exploded at Shea: "What have *you* done?" This early conflict degenerated into an all-out battle of wills between two brilliant men, each convinced of his own genius.

With time, Shea came to appreciate Faget's particular skills and style, but he still couldn't hide his disdain for the rest of the Langley-MSC team. Compared to people Shea had worked with in electronics, guidance, and missiles, the Langley team was quaint, eccentric, old-fashioned—Shea called them "the flyboys."

Most telling, however, Shea refused to include the chief "flyboy," MSC director Gilruth, in his reporting, sending most information directly to General Phillips and George Mueller (the head of NASA spaceflight) at HQ.

Abbey knew nothing of this during his first months in Shea's department, largely dealing with Markley, Williams, and Maynard. In midyear, however, Abbey was given the job of evaluating North American's proposal describing all the engineering work to be accomplished on its CSM contract. Abbey's report went into great detail and apparently showed such a solid grasp of the overall program that Shea began to treat him as his Block II specialist. He also involved him in Block I decisions.

In person, Abbey found Shea to be a demanding manager, willing to argue with a specialist on the individual's own specialty on the most trivial technical matter. He once questioned some poor engineer about whether the screwdriver used for a particular procedure was flat head or Phillips—not because Shea really cared but because he wanted the engineer to admit that *he didn't know*. It was all right if you didn't know something; nobody knew as much as Shea. But woe to those who didn't *admit* what they don't know. Shea was merciless to the unprepared. The only individuals who escaped his wrath were the astronauts, his contemporaries in age and drive if not in background.

Maynard's opinion of Shea, delivered years later: "He was a genius, but sometimes you'd rather have three men who were merely smart."

Apollo was in deep trouble by September 1965. A month earlier, Shea, Phillips, and Mueller had laid out a series of seven milestones each vehicle (CSM or LEM) would have to complete in order to be launched. These started with a preliminary design review (PDR) followed by a critical design review (CDR) and leading through hardware inspections and vehicle certification up to the actual flight readiness review a month before launch. In some cases, the first two steps had been completed. But not the others.

Problem number one was the giant Saturn V booster and especially its North American S-II second stage. But progress on the command and service modules wasn't much better. Owen Maynard listed the many problems for Shea, beginning with spacecraft weight that exceeded the original limits, substandard performance by all engines, the failed integration of scientific experiments into the program, and the unresolved issue of controlling spacecraft temperature in flight to and from the Moon. There were ongoing concerns about the deep space environment and its effect on an Apollo crew—for some months there was fear that the command module would require thousands of pounds of additional shielding against radiation. And more: Would the vehicle survive a water landing? The space suit and portable life-support backpack were incomplete. The Block II hatch needed to be redesigned to support extravehicular activity (EVA) from the command module. Costs were not being controlled or predicted. Every element, in fact, was five months behind schedule, threatening planned launches—unmanned Block I vehicles 008 and 011 in early 1966, and 012 and 014 carrying astronauts later that year.

Grumman's lunar module program wasn't doing much better. In addition to myriad weight, environmental, and flight control problems, its planned descent engine had exploded during a test in Tennessee on September 1. The bad news about both prime contractors reached HQ, where Sam Phillips concluded that North American needed a visit from a Tiger Team.

Phillips had used Tiger Teams with great success in the Minuteman program. NASA was familiar with the concept too. In fact, the definition of a Tiger Team originated in a 1964 paper coauthored by NASA's Walt Williams and North American's Scott Crossfield: "a team of undomesticated and uninhibited technical specialists, selected for their experience, energy and imagination, assigned to track down relentlessly every possible source of failure in a spacecraft subsystem."

In mid-October, Phillips and Shea created two Tiger Teams, one from the Marshall Space Flight Center to examine the S-II and the other from Houston to go after the CSM. There were five subteams evaluating program planning and control, contracting, engineering, manufacturing, and quality assurance. Abbey was one of five on the engineering team.

10 | Astronaut Abbey

ON THE AFTERNOON OF Friday, October 29, 1965, a little more than eight years since that fateful Sputnik sighting on a Montana highway, George Abbey took his first step toward flying in space himself.

He flew a T-33 from Ellington Air Force Base, south of Houston, to his old station of Randolph Air Force Base in San Antonio. The weather was clear; his spirits were high. He was personally delivering his application to join the NASA astronaut team to the Air Force Personnel Center.

A month and a half earlier, on September 6, 1965, just days after the successful end of Gordo Cooper and Pete Conrad's eight-day Gemini V mission, NASA had announced that it would select a new group of pilot astronauts in the spring of 1966. The air force piggybacked on the announcement, saying it would also select astronauts for its new program, the Manned Orbiting Laboratory (MOL), which was scheduled to start flying in 1968.

Military candidates—like Maj. George Abbey (he had been promoted in summer 1965)—could apply for NASA or MOL, or both. The requirements?

- Be born after December 1, 1929, for NASA applications, after December 1, 1931, for the MOL assignment.
- Be under six feet tall.
- Hold a bachelor of science degree in engineering or natural, physical, or biological sciences, or be a graduate of a military academy.

- Have logged a minimum of one thousand hours of flying time, or have attained test pilot status through the military or aircraft industry.
- Be able to pass a Class I physical.

Abbey met all the requirements. He was under the max age for NASA by three years and under six feet tall. He held not only a degree from Annapolis but also a master's in electrical engineering plus experience on the Dyna-Soar and now Apollo. He had three times the required flying hours, thanks to his time at Boeing and additional hours logged in Houston. He was physically fit and qualified to fly a high-performance jet aircraft. Since he lived in Houston, NASA wouldn't even have to transfer him.

He also had the endorsement of his supervisor, Owen Maynard. (Maynard would admit, years later, that he knew from the beginning that Abbey was more interested in being an astronaut than in being a systems engineer.) He also had an endorsement from the senior air force officer detailed to the center, Lt. Col. Dick Henry—a flying buddy too.

Abbey felt that his helicopter experience would be an advantage. Believing that mastery of vertical landings would help in lunar landings, NASA had already begun to send astronauts to the naval air station in Pensacola to qualify as helo pilots. Abbey not only was qualified but also could instruct.

Astronauts were the stars of the MSC community, of course. The most famous, John Glenn, was no longer in Houston, having gone back to Ohio in 1964 to make an aborted run for a US Senate seat before joining the Royal Crown Cola Company as an executive. Glenn's Mercury comrade Scott Carpenter was back in the navy, detailed to the SeaLab underwater research program. But the other Mercury astronauts were present, notably Deke Slayton, the MSC official in charge of astronaut selections, flight assignments, and aircraft. Judged to be a superior pilot, the straightforward and well-liked Slayton operated in two modes. In supervising the other astronauts he was notoriously hands-off, telling them they were "big boys" and depending on them to do their jobs and not make trouble. With entities outside the astronaut office—NASA doctors, the press, and the scientific community—he was fiercely protective if not downright uncooperative.

Slayton was aided and abetted by the chief of the astronaut office, Mercury veteran Alan Shepard. Brilliant and arrogant, Shepard had made America's first manned spaceflight in 1961 and had been selected to command the first

Gemini mission but was diagnosed with Ménière's disease, an inner ear condition that grounded him—allowing Slayton to assign him day-to-day supervision of the astronauts.

Where Slayton had two modes, Shepard had two moods: warm and glad-handing, or icy and intimidating. (His secretary eventually began to display a status photo for astronauts approaching Shepard's office: a smiling photo meant "Go on in" while the frowning version suggested a wave-off.) This pair acted as squadron leaders for two dozen astronauts, all of them experienced pilots, most of them experienced *test* pilots, with their own internal hierarchy that manifested itself in competition for coveted flight assignments. Abbey knew Gus Grissom from Wright-Patterson and Neil Armstrong from Dyna-Soar but didn't consider them close friends. Since he had not attended either the air force or navy test pilot schools, he was not a member of the fraternity of test pilots.

In the fall of 1965 astronauts began to be more involved in Apollo. Unflown astronauts like Donn Eisele and Roger Chaffee took part in equipment tests. And with the splashdown of Gemini V in August, veteran Jim McDivitt began showing up in meetings in Houston, at the North American plant in Downey, and at Grumman's facility in Bethpage too. He was Slayton's choice for first lunar module commander.

Abbey began to form opinions about the astronauts. Some of these men—and they were all men, all white—were workaholics, while some were less willing to put in long hours. Their engineering skills ranged from first-rate to don't ask. Some were skirt-chasers—"wild" in Abbey's term. (Those who were relatively straight arrows he called "serious.") They were rarely home. Astronaut Walt Cunningham would calculate that one year he had spent 270 days on the road. The one thing all had in common was confidence in their ability to fly any aircraft in any situation better than anyone else.

Take Dick Gordon, a thirty-six-year-old navy test pilot noted for his high spirits, swagger, and confidence. Gordon, like Abbey, was from Seattle and offered Abbey a ride to their mutual hometown. The two of them took off from Ellington in a T-33 jet, flying to Denver, where they landed to refuel. "There's a hell of a thunderstorm in the area," Gordon said later. "Weather's bad all the way to Seattle. So I said to George, 'Do you want to press on?'

"'If you want to,' he says.

"So we take off, and we fly to Seattle, but let me tell you: I've never bounced around the sky that badly. We were both pale and shaking by the time we landed. And George never flew with me again."

Abbey believed that he could be selected and begin training as an astronaut in July 1966—too late to hope for a flight on Gemini or the early Apollo missions, but not for later landings on the Moon or visits to space laboratories. He had seen engineers and astronauts testing prototype space suits for walking on the lunar surface—it didn't take much imagination to picture himself wearing one, happily bouncing across the gunmetal soil.

Which is why he jumped into that T-33 to fly to Randolph to personally deliver his application, making sure that it was stamped RECEIVED ahead of the deadline. Mission accomplished, he flew back to Houston to await the next step in the process, and then soon on to Downey for what became a monthlong assignment.

11 | The Phillips Report

NASA'S TIGER TEAM DESCENDED on North American on November 22, 1965. Most of them remained there until December 6. There were two targets: the troubled S-II program and the command and service modules, all under the Space and Information Systems Division (S&ID).

It was tedious, painstaking work reviewing North American's status reports, comparing them to orders from Shea and the many subsystems people, and correlating these orders with the actual work performed in the shop. It also required Abbey to sit for hours with Kehlet, Fagan, and other members of the North American team, which was most illuminating. Kehlet was fully aware of his company's shortcomings—its arrogance, its discomfort with the amount of NASA supervision, its matrixed S&ID management structure (which hindered the flow of information and decision making), and its personnel. Many of the lapses identified in the Tiger Team survey came directly from Kehlet and his colleagues.

The S-II team did not have the same experience as Abbey's CSM group did, relying more on the paper trail—and earning scorn from North American leaders on that program, who claimed that their hardware was working just fine, given the challenges.

Abbey was tasked with writing the report on the spacecraft survey for the eyes of Phillips and Mueller, among others—but not NASA administrator James Webb, a decision that would have famous repercussions a year later. (The report was classed confidential and meant for limited distribution.) The final result wasn't lengthy—the overview and conclusions required fewer than twenty pages of text, with support from half a dozen appendixes that included the inevitable charts and graphics.

The introduction stated that NASA had reviewed North American's S&ID for program planning and control (including logistics), contracting, pricing, subcontracting, purchasing, engineering, manufacturing reliability, and quality assurance on the S-II and Apollo command and service modules. Milestones had slipped. Costs had tripled. Technical performance was inadequate. The report was laced with such statements as: "We believe that S&ID is overmanned and that the S-II and CSM Programs can be done, and done better, with fewer people." And on and on in blunt detail.

Phillips sent ten copies of the Tiger Team notes to North American's president, Lee Atwood, on December 19, saying, "I am definitely not satisfied with the progress and outlook of either program and am convinced that the right actions now can result in substantial improvement of position in both programs in the relatively near future."

George Mueller added his own message to Atwood: "I submit that the record of these two programs makes it clear that a good job has not been done. Based on what I see going on currently, I have absolutely no confidence that future commitments will be met." He wanted immediate improvements. He wanted a response from Atwood and North American by the end of January, and he warned that Phillips's team would return to Downey in March.

Abbey followed the Gemini program from Downey and from Washington, greatly admiring the accomplishments of the astronauts and their mission control teams. His feelings about his place within NASA were otherwise conflicted, since learning the first week of December—via mail—that he had been rejected as an astronaut candidate. He would not be invited for the medical examinations at Brooks Air Force Base, which meant he would not be sent to Houston for interviews with the selection board. He was finished. That was a disappointment, but the reason was infuriating.

Frank Liethen, who would have been a good candidate except for his height, had a test pilot friend on the board. Liethen told Abbey that the air force selection board decided late in the game that it would only forward applicants to NASA who were graduates of the Aerospace Research Pilot School, currently under the command of Chuck Yeager.

Many in the air force believed Yeager's reputation to be seriously inflated. Everyone acknowledged Yeager's stick handling and dogfighting skills, but his arrogance, his resentment of pilots with college degrees (Yeager had only a high school diploma), and his penchant for self-promotion were equally well known. Mickey Moore's wife, Mary Ann, told Abbey that "Yeager got credit for things Mickey did." Yeager had also earned the ire of other pilots at ARPS for his destruction of one of the school's three rocket-assisted NF-104 jets on December 10, 1963—the same day Abbey learned of Dyna-Soar's cancellation. Abbey doubted that Yeager had been the first person to fly faster than sound.*

But Abbey's opinion didn't matter. Yeager's involvement in the selection board was bad news. For two years NASA astronaut Bill Anders had been gleefully telling the tale of his encounter with the famed test pilot. In 1963 Anders had applied for ARPS, and his paperwork was still under review when NASA selected him as an astronaut. When Yeager called Anders to tell him, "Too bad, you didn't make it," Anders was able to shoot back, "That's all right, I'm going to NASA." Yeager had blown up. "You're going to be an astronaut without going through the school?"

Abbey had heard later that Yeager was also angry that Buzz Aldrin, another pilot who had never attended ARPS, made it through the air force board to NASA.

The board's decision to forward only ARPS gradates' applications was especially difficult for Abbey to accept because this new restriction only applied to *air force* pilots who wanted to become *NASA* astronauts. The board happily accepted candidates for MOL who weren't test pilots or ARPS grads—they simply designated them "contingency selects" and planned to send them to ARPS to qualify. And the navy, which used its own review board, did not insist on the test pilot requirement.

---

* Abbey believed that George "Wheaties" Welch, a North American Aviation test pilot, had broken the sound barrier in an XP-86 Sabre on October 1, 1947, two weeks before Yeager's epochal X-1 flight—and that Welch had repeated the feat the very day Yeager did.

Angry or not, Abbey had no means of redress. He could resign from the air force and resubmit an application as a civilian—but a move like that would leave him unemployed. (His job under Shea and Maynard was an air force billet.) It was moot, anyway: the deadline had passed. All he could do was hope for another chance a year or two down the line. Meanwhile he had his work on Apollo. If he couldn't fly to the Moon, he could at least help NASA astronauts get there safely—and back.

A sharp letter from Sam Phillips to Lee Atwood was one thing. NASA's development team also needed to make its displeasure known directly to its counterparts at North American. On January 11, 1966, Joe Shea and his boss, George Mueller, met at Downey with Harrison Storms (head of the Space and Information Systems Division), Dale Myers (head of the CSM program), and two dozen others. Because he had done so much work on the final Phillips Report, Abbey was designated as the presenter on the Apollo CSM critique. The North American managers naturally resented the review and its stinging rebukes. They were particularly incensed about the S-II statements, since they felt that the NASA Marshall reviewers spent too much time on schedules and budgets and not enough looking at the actual hardware. The CSM review was less contentious but equally embarrassing.

Storms turned to his corporate engineer, George Jeffs, who had done his own study of the company's Apollo and S-II efforts. "We found a number of the same kind of problems, and probably a few others here and there that didn't get into the Phillips Report."

The thirty-nine-year-old Jeffs had worked on the Hound Dog missile program. He had also spent a challenging two years on the Rogallo wing, an inflatable paraglider intended to return a Gemini spacecraft to a land landing. Though the concept had proved useful, it had been abandoned in 1965. Jeffs had worked closely with NASA on the Rogallo project, however, becoming good friends with agency staffers, so Storms made him chief engineer for Apollo.

Abbey was pleased by the change and expected improvement from North American, but not immediately: Jeffs had months of catching up ahead of him.

Back in Houston, with Apollo launches looming, Maynard formed a group of staff engineers and set about working with them and with the flight operations and flight crew organizations to plan these upcoming missions. Bob Williams took on more day-to-day management of systems engineering.

Maynard would later speculate on why Shea gave him a new role in mission planning and dealing with the operations teams. "He was an unmanned kind of guy. He wasn't a pilot, he wasn't an aeronautical engineer. . . . He figured that I had the entrée into that whole system." Shea found him to be the "perfect interface between him and this hostile group of manned people."

And Abbey himself found that he was being asked to travel with Shea—yet another "interface."

In addition to Abbey, other MSC engineers like Kenny Kleinknecht and Aaron Cohen began to take more visible roles in Apollo too. Now Abbey found himself associating with senior NASA leaders, notably George Mueller.

Born in St. Louis, Missouri, in 1918, Mueller (pronounced "Miller") had grown up reading science fiction and building model ships and airplanes and radios. He had earned a degree in electrical engineering from the Missouri School of Mines and Metallurgy; then, finding it impossible to get a job during the Depression, he entered graduate school at Purdue, where he wound up working on an early television system. Graduating with his master's, he joined Bell Labs, where his peacetime television work led him to wartime radar. He became head of the company's research and development team, and also program manager for the Pioneer spacecraft. By the time he joined NASA, he had built a fearsome reputation for intellectual arrogance—a trait he shared with Shea and several other NASA figures—tempered by personal affability. He never yelled or lost his temper. One NASA engineer described working with him as "a piece of cake."

Now, in January 1966, Mueller was still the single most powerful and persistent advocate for keeping Apollo on schedule. Abbey respected Mueller's intelligence but didn't warm to him. For one thing, it was obvious that Mueller was unhappy with Bob Gilruth and with Chris Kraft, the head of flight operations at MSC, and would have fired both if allowed. (Fortunately for Gilruth and Kraft, they had support from Texas congressman Olin "Tiger" Teague.)

Sam Phillips, brought into NASA by Mueller, was quite different. Abbey felt that Phillips was an outstanding leader who understood how to manage major programs, having led two successful air force developments, the B-52 Stratofortress bomber, and the Minuteman intercontinental ballistic missile.

As he worked with Shea, and later George Low, Abbey would come to know Phillips quite well. He would often pick up the general from his Houston hotel and give him rides to the center for meetings. These were largely fruitful encounters, except when Abbey made a hurried dash across a crowded highway and earned a gentle rebuke.

12 | Flight

BY THE END OF 1965 North American had built twenty command module mock-ups, twenty-two boilerplates, and eight completed spacecraft, all used for various tests. The ninth, known as 009, arrived at the Cape on October 25 to begin the tedious but necessary integration onto its service module and Saturn IB booster.

Spacecraft 009 was not equipped with crew couches or an instrument panel, and had a special control system and an emergency detection system. The command module rode atop the first Apollo service module, which varied from the basic design by using batteries for power rather than fuel cells, and not carrying an S-band communication system.

Launched on February 26, the Apollo-Saturn 201 mission tested the command module's basic flight worthiness atop the Saturn IB—launch dynamics, separation, capabilities of both the service module's propulsion system and its reaction control motors, and the integrity of the command module's heat shield—all in a suborbital flight.

Abbey wasn't present. He had no role in mission support and considered the mission a test of the booster, not the spacecraft.

It wasn't perfect—no mission ever was—but it was a start.

With Apollo-Saturn 204, the first manned mission, now tentatively scheduled for as early as November 1966—eight months hence—it was time for Deke Slayton and NASA to assign the flight crews.

Mercury and Gemini veteran Gus Grissom would command the two-week shakedown mission. His senior pilot would be Gemini space walker Ed White. The pilot would be a rookie astronaut from the 1963 group, Roger Chaffee.

Backups would be Gemini IV commander Jim McDivitt, with Gemini VII veteran Dave Scott as senior pilot and 1963 group rookie Rusty Schweickart as pilot.

The crews were introduced to the public at a press conference in Houston on Monday, March 21; three days later, they made their first trip to Downey.

It was one thing having astronauts involved in various systems; having actual flight crews that included four famous space fliers was a different matter. Unflown astronauts were treated much like ordinary NASA engineers, housed in the Tahitian Village Motel and left to arrange their own social lives, but Grissom, White, Chaffee, and their backups were blessed with access to the fabulous North American executive dining room, and were entertained at Harrison Storms's equally fabulous home in exclusive Palos Verdes. (They also had first call on the services of Bud Mahurin, the World War II ace fighter pilot and Korean War prisoner of war who was North American's director of customer relations—aka chief party planner.)

Social activities aside, Grissom and McDivitt were both capable engineers and began clamoring for changes in the spacecraft, something they had done frequently with McDonnell on Gemini.

But sleek, arrogant North American was nothing like the mom-and-pop shop of McDonnell, where "old man" James McDonnell was eager to indulge the astronauts. At North American, engineers wouldn't automatically do what the astronauts wanted. And Harrison Storms, Dale Myers, and other senior execs were too far removed from the shop floor to get involved.

Resistance also came from inside NASA, where Shea had been struggling for years to throttle unauthorized fixes.

Shea spent time with the astronauts—even playing handball with Grissom—and in meetings would nod his head and listen. But he was under tremendous pressure from Mueller and HQ to keep to the schedule, and slowing things down to redo an instrument to satisfy Gus Grissom was not on his list of priorities.

Although both Grissom and White had overlapped with him at Wright Field, Abbey did not know either man well. Nor did he grow close to them during 1966. Although a hard worker, Grissom wasn't especially outgoing. (His Gemini copilot, John Young, thought that Grissom had a terrific dry sense of humor, but the astronaut rarely showed it.)

White was aloof in a different way. His Gemini experience had raised his profile outside NASA, and it was obvious that he was positioning himself to be the next John Glenn—the most famous astronaut in America.

It was a different matter with Roger Chaffee. The young navy lieutenant, a Purdue grad like his boss Grissom, began to ape Grissom's manner of speaking, using the Indiana native's phrases—and engaging in wilder personal activities.

For example, Chaffee conspired with Abbey to "welcome" Joe Kotanchik—newly appointed as Shea's deputy for the Apollo command and service module—to Downey by inviting the staid, stolid fifty-eight-year-old engineer to join them one evening at a bar featuring attractive go-go dancers. Kotanchik bore up well under the hazing, but the outing was not repeated.

On April 4, 1966, NASA announced the names of nineteen new pilot astronauts, members of the group that Abbey had applied for. Included were pilots Abbey knew, like Jim Irwin. Many of them were his exact contemporaries: Marine Corps pilot Jerry Carr, air force test pilots Joe Engle and Al Worden, and navy aviator Paul Weitz were all born in 1932.

Abbey would get to know many of them over the next year, as the new astronauts moved into Apollo support roles—Ed Givens, Jack Swigert, and Ron Evans in particular. Ken Mattingly, Vance Brand, and Joe Engle would all become close associates too.

If NASA's operational tempo—the increasing rate at which missions were being launched—was any indicator, the new astronauts would have a lot of opportunity to fly in space. Gemini IX-A flew in June, then Gemini X in July.

Just before Gemini X, on July 5, NASA launched AS-203, the second mission in the Apollo schedule, a test of the Saturn IB booster that did not carry a command and service module. The first stage of the Saturn IB placed the S-IVB stage in its proper orbit, where its engines were fired, stopped, and then refired. The mission was deemed a success, even though a tank on the S-IVB ruptured four hours after launch.

On August 10, NASA's unmanned Lunar Orbiter, the program Abbey had worked on at Boeing in 1964, launched its first mission. LO-1 took up a parking orbit around Earth, just as Apollo would, then used an Agena upper stage to fire itself toward the Moon. Four days later, it used its own smaller rockets to insert itself into orbit around the Moon—a first for the United States or the Soviets.

The spacecraft began taking pictures of potential Apollo landing sites on August 18, continuing through August 29, and acquired 42 high-resolution and 187 medium-resolution images that were downlinked to Earth through September 14.

They not only proved that the Apollo sites were smooth enough for landings but also provided the first data on a new phenomenon, so-called mass concentrations within the Moon itself that caused perturbations in spacecraft orbits. That was vital information for Apollo.

Another notable event occurred that summer. George Mueller decided that "excursion" in "lunar excursion module" made the program sound too "frivolous." In June 1966, NASA officially changed LEM to LM, plain old "lunar module." Nevertheless, for the life of the program, engineers, flight controllers, and astronauts would pronounce it "lem."

The splashdown of Gemini X triggered an expansion of the astronaut team at Downey.

In the spring Mercury and Gemini veteran Walter Schirra was informally named to command the second Block I Apollo mission. His crew included air force test pilot Donn Eisele, and civilian pilot-scientist Walt Cunningham. This trio began to make trips to Downey a couple of months after the Grissom and McDivitt crews.

Schirra, Eisele, and Cunningham were joined by Schirra's backup, Gemini VII veteran Frank Borman. In late July Borman's two other crew members arrived, senior pilot Tom Stafford, coming off Gemini IX-A, and pilot Mike Collins, after Gemini X.

In his acclaimed memoir *Carrying the Fire*, Collins described Gemini as "simplicity itself" compared to Apollo. "There were over three hundred of one type of switch alone, not to mention the scores of pipes, valves, levers, brackets, knobs, dials, handles, etc." And these were just the controls. Lockers and other compartments filled with food, household supplies, and scientific experiments turned the command module interior into a "dynamic mess, never static, as workmen came and went with alarming regularity, rerouting wires, replacing black boxes, rearranging equipment.

"It was a bit disconcerting," he continued, "because no one I talked to seemed to know exactly what was, or was not, in the spacecraft at any one time." In contrast to the McDonnell teams who had honed their skills on Mercury and Gemini, the North American team was "Amateur Hour," and even the NASA engineers working the program were "supercilious," denigrating any possible lessons Gemini might hold.

Collins was talking about spacecraft 014. Troubled as 014 might have been, it was in better shape than Grissom's 012 vehicle—so much so that the Downey team placed an image of a leaping frog on 014, suggesting that it would be ready to fly before 012.

By late summer 1966 Abbey concluded that the Phillips Report and personnel changes had not resulted in a great deal of improvement at North American. Jeffs was making a difference, though he was hampered by corporate inertia. He was most effective on the Block II spacecraft, where changes could be made. The Block I models were too far along, and there was too much pressure to get them shipped.

The whole NASA organization was feeling pressured because it was clear that the agency's budget had peaked.

For fiscal year 1966, the agency was authorized to spend $5.993 billion, almost 4.5% of the federal budget. The workforce, civil servants and contractors, had reached 411,000 in 1965 and was already declining, to less than 400,000.

The reason? The growing conflict in Vietnam. At the end of 1964, America had 23,310 military personnel in the country; by December 1965, that number had shot up to 184,314, as the country responded to increasing strikes by North Vietnamese and Vietcong forces against the South.

The American presence was still growing during 1966, and by the end of the year would reach 385,000.

At the same time, President Lyndon Johnson's Great Society programs—federal spending to address education, medical care, and poverty—were also increasing.

NASA could no longer count on increasing support from the White House or Congress—or, with the war dominating the news, the public. The "decade" element of "man, Moon, decade" was threatened.

North American's Space Division officially delivered Apollo spacecraft 012 to NASA at a ceremony in front of Downey's Building 290 on Friday, August 19, 1966—the same day LO-1 was taking its first pictures.

By this time, North American had handed over several earlier spacecraft, but 012 would be the first intended to carry astronauts, and the routine event evolved into a giant presentation attended by NASA's Joe Shea, Robert Gilruth, and Max Faget—and dozens of other MSC staffers, including Chris Kraft and Abbey. The astronauts—Grissom, White, and Chaffee—were also present.

North American was represented by its senior leaders: Storms, Myers, Feltz, and Jeffs.

The acceptance review began at 8:00 AM, with representatives from both teams examining 012 in its assembly bay, then adjourning to the company conference room. Here, in his brisk, thorough fashion, his briefing book in front of him, with Abbey nearby, Shea took charge, running down his list of fifty items yet to be addressed, better known as "squawks."

Some involved the troubled environmental control system within the command module and potential chafing of wiring bundles on equipment bay doors, and a good percentage of the others concerned such mundane items as placement and number of Velcro fasteners inside the cockpit. (Grissom, White, and Chaffee had, as astronauts would, personalized their workspace, and had grown especially fond of adding Velcro strips to hold harnesses and documents in place.)

No hardware was changed that day, of course. The squawks were each tagged with an action to be taken, either by North American before shipping, or NASA and North American at the Cape.

The whole process took six hours. When it was over, Gus Grissom stepped up with a pair of manila envelopes in his hands. He opened one and presented a special crew portrait to Harrison Storms, and a second one to Joe Shea.

The picture showed all three astronauts in NASA flight suits, heads bowed and hands clasped in prayer. They had inscribed Shea's photo: "It isn't that we don't trust you, Joe, but this time we've decided to go over your head."

There was general laughter, and the CARR ended.

The same day that 012 arrived at the Cape—August 25—AS-202 launched on a second, though longer and more ambitious, suborbital test of the service propulsion system. All firings went well, and the command module's heat shield survived in good shape.

Abbey missed this one, too—as with AS-201, the mission wasn't using a Block II spacecraft, and even the Block I used this time lacked real systems, like crew couches or cockpit displays.

The Apollo command and service modules were judged ready to carry astronauts.

After completing his tour as an instructor at ARPS in December 1965, Abbey's friend Frank Liethen had become executive officer of the Thunderbirds, the air force aerobatic demonstration team.

On October 12, Liethen took off from Nellis Air Force Base in Las Vegas, headed for Indian Springs Auxiliary Airfield, riding backseat in an F-100 Super Sabre piloted by Capt. Robert H. Morgan. This was intended as an orientation flight for Liethen, who was about to assume command of the Thunderbirds—and take part in demonstrations.

Morgan's plane was performing maneuvers with another F-100 piloted by Capt. Robert Beckel. One was the "half Cuban eight," in which the two jets flew directly at each other, crossed, nosed up, then repeated the speedy approach but now spiraling as they crossed and nosed up.

According to witnesses, the two aircraft scraped each other at the top of their loop, damaging both. Beckel was able to nurse his Sabre to an emergency landing at Nellis. Morgan ejected from the other Sabre, but too low for his parachute to open.

Liethen didn't get out and was killed when the Sabre slammed into the desert.

Death by aircraft was inescapable in the air force and NASA. Abbey had lost colleagues as far back as his time at Webb in 1955. On February 28, 1966, Elliot See and Charles Bassett, the prime crew for Gemini IX, had been killed when their T-38 crashed during a poor weather landing attempt in St. Louis.

Abbey had known See and Bassett, but not well. Neither man was a close friend the way Liethen had been. He heard the news while assigned to the Pentagon, part of a two-week refresher tour designed to remind personnel and other officers that he was still an air force officer, and to catch up on aircraft and space developments.

Gemini XII, the last of the series, lifted off on November 11, 1966, with commander Jim Lovell and pilot Buzz Aldrin, who accomplished a rendezvous and docking with an Agena. Aldrin conducted three EVAs that proved that it was indeed possible for astronauts to work in pressure suits.

Lovell and Aldrin splashed down safely on November 15.

And now the entire Manned Spacecraft Center could concentrate on Apollo.

One lingering issue for Mueller, Phillips, and Shea was the question of flying the Block I command and service modules. There had originally been four such missions, back when the NASA program expected to leap from Mercury to Apollo.

But Gemini had come along to bridge the gap in spectacular fashion. Those four Block I missions had been reduced to two in late 1963.

And now the question was, Why even fly that many? Block II vehicles would be equipped with docking systems; they would fly the lunar landing missions. Block I was already outmoded and incomplete.

Just leaving the second Block I—AS-205, to be launched in May 1967—in the schedule was an invitation to further delays. It was only going to repeat the AS-204 mission, with its crew spending another two weeks in space. The major difference was that Schirra, Eisele, and Cunningham would spend more time on scientific experiments—many of them kicked off the AS-204 flight plan by Gus Grissom, who never met a science experiment he liked.

Wally Schirra, the 205 commander, was no scientist, either, and had been quietly, then not so quietly, lobbying for a more challenging assignment for his crew—the first flight of a Block II with a lunar module, for example.

The day Gemini XII splashed down, word spread through the Manned Spacecraft Center that AS-205 was being canceled. Having received no advanced word from Deke Slayton, Schirra was furious—doubly so when Slayton told him about his new assignment: Eisele, Cunningham, and he were now going to be backups to Grissom, White, and Chaffee for AS-204.

Schirra exploded. He believed that as a Mercury astronaut he didn't back up anyone, not even Grissom, another Mercury astronaut. And for several days, he didn't—with rookies Eisele and Cunningham left wondering about their astronaut careers. It took a personal appeal by Slayton and Grissom to convince Schirra to accept the backup assignment, which he finally did.

Personnel matters were only one challenge facing AS-204 as it assumed the coveted "number one on the runway" position. The command module simulator at the Cape only barely resembled the 012 flight vehicle in its configuration and operations, and was so out of whack that Gus Grissom brought a lemon from his Houston backyard and hung it on the structure.

Biomedical harnesses to be worn by the crew during flight failed so often that Slayton had them declared "unfit for flight." The NASA and North American teams were still working through an endless list of squawks as the launch date of February 14 grew closer.

Abbey had a more pointed question: Why fly the Block I at all?

No one was asking his opinion. As 1966 turned into 1967, all he could do was follow orders from Shea and Maynard.

13 | Countdown

"I DON'T THINK WE should do this test."

Wally Schirra tossed a thirty-page document onto the table.

It was Thursday, January 26, 1967, just after six in the evening. Schirra was one of five men seated in a conference room in the astronaut crew quarters on the fifth floor of the Operations and Checkout Building at the NASA Kennedy Space Center. With him were two other astronauts from his Apollo 1 backup crew, Donn Eisele and Walt Cunningham; Apollo program director Joe Shea; and George Abbey. Knowing they had an early morning, the prime crew of Gus Grissom, Ed White, and Roger Chaffee had already gone to bed.

Abbey and Shea had arrived earlier in the week for meetings with a NASA team lead by director of launch operations Rocco Petrone and North American's CSM test operations group under Jim Pearce, a veteran test pilot. Shea had concerns about the readiness of the spacecraft and launch teams for the launch of the AS-204 mission in three weeks, and those to follow.

Three Apollo-Saturns had been launched in the past year, but none of them carried astronauts. Three weeks from now, on February 21, the crew of Gus Grissom, Ed White, and Roger Chaffee were scheduled to launch aboard Apollo spacecraft 012, a Block I model, spending as many as fourteen days in space.

For Abbey, it was a mission without purpose.

The Block I *looked* like Block II, and shared basic environmental and control systems, but little else. It had no docking system, no rendezvous radar, and its basic access hatch was a crude three-piece design that took far too long to open. Block I was where North American's manufacturing team learned to build to NASA specs by trial and error—too much error, in Abbey's view.

A second Block I mission, scheduled for May 1967, had already been canceled. Nevertheless, George Mueller, Joe Shea, and even Sam Phillips all seemed convinced that this Block I ought to be flown.

If successful, the flight by Grissom's crew would be followed midyear by the first unmanned flight of Grumman's lunar module. Later in 1967 NASA hoped to launch a second LM and the first Block II with a crew led by astronaut Jim McDivitt, for a rendezvous and docking—the first real Apollo test flight.

Even before arriving at the Cape, Abbey doubted that the 204 launch would take place on schedule. Nothing he had seen or heard since had changed his mind.

The crew wasn't ready, either. They had never had a chance to train in a simulator that actually matched their spacecraft.

Among the astronauts, commander Grissom had been openly critical of spacecraft 012's state, complaining to anyone he could reach, from Dale Myers to Joe Shea.

Changes had been made, though not quickly and not completely, and now another aspect of Grissom's persona emerged, a fatalistic determination to do the job no matter the circumstances. At an Apollo press conference, when asked how he would define a successful mission, Grissom simply said, "If all three of us get back."

Grissom wasn't going to ask for a delay, or even for the postponement of a test. Nor were his crewmates, Ed White and Roger Chaffee.

In addition to the prime crew, Apollo 1 had a backup crew of three, led by Schirra. The former Annapolis graduate (class of 1946, eight years ahead of George Abbey), naval aviator, and test pilot was considered the most polished of the Mercury Seven, and also the least willing to waste time on pointless procedures.

And now, in front of Shea and Abbey, he was urging cancellation of tomorrow's test.

Schirra and his crew had just emerged from a six-hour stint inside 012 that afternoon, patiently running through a prelaunch simulation called a "plugs-in" test, which meant that the heavy, three-part hatch was propped open, allowing external power cables to be connected.

Tomorrow's test, with Grissom's crew, was going to be "plugs-out," meaning that spacecraft 012 would be sealed and operating on its own power.

Each test of a manned spacecraft required a specification—what was being tested and why, what support was required, how long the event would run, who would be involved. (Friday's plugs-out test would require the participation of *one thousand* engineers at the Cape, in Houston, and in California.)

Those specs filled that thirty-page document that called for the spacecraft to be pressurized to sixteen pounds per square inch—higher than sea level in order to keep the spacecraft from deforming. (Apollo in flight would use an atmosphere of pure oxygen at five pounds per square inch. Sealed on the launchpad, however, it was subject to damage at five pounds. Hence the need to overpressurize the cabin.)

The test was considered hazardous, since the crew would be sitting in 012 atop a Saturn IB booster. What no one apparently considered—Abbey included—was that in a pure oxygen atmosphere at sixteen pounds per square inch, the tiniest spark could erupt into a conflagration.

The preliminary version of this test had been scripted in July 1966. It had undergone four major revisions, with continual minor changes over the past months. A fifth major revision arrived one hour before Shea and the astronauts gathered, and yet more changes were promised in the morning.

This was why, according to Abbey, Schirra objected: "With all these redlines, we're not ready to proceed." He and his crew were having trouble just *understanding* the changes.

After about a half hour of discussion Shea told Schirra, "I'll talk to Rocco about postponing."

Abbey was left with Schirra and his crew to continue reviewing the procedures.

An hour later, Shea was back with a decision. Petrone agreed that the test procedures were not in good shape but felt that the ground and flight support teams needed the experience. If things didn't work as planned, Petrone would schedule a repeat test. Hearing that, Shea agreed to let the Friday test go ahead.

Schirra reluctantly resigned himself to the decision. He, Eisele, and Cunningham would continue to work on the updates so they could brief Grissom's crew in the morning. As Shea was leaving, Schirra invited him to join the six astronauts for breakfast.

The men split up at that point, and Abbey and Shea returned to their hotel. Shea, of course, was toting a thick binder filled with weekly Apollo status reports: these would occupy him for hours.

The next morning, Friday, January 27, Abbey went off to more meetings with Petrone's team while Shea met with the two crews and Slayton. In midmorning, Shea telephoned Abbey to say he would not be returning to Houston with him that day. Grissom had invited him to join his crew inside spacecraft 012 for the test. A communications headset would be rigged for Shea so he could follow the action and judge its success firsthand. But it soon became apparent that a fourth headset jack couldn't be installed in the spacecraft in time. Shea gave up the idea.

Grissom still wanted Shea inside 012, but Shea offered a deal instead: he would return to the Cape on Monday and go through the next test in the simulator with the crew.

The astronauts left to get suited, a process that would take an hour.

Shea called Abbey around 11:00 AM and said he wasn't going to sit in on the test after all. "Let's get back to Houston."

They caught the 2:30 PM flight out of Melbourne, arriving at Houston's William P. Hobby Airport two hours later, 3:30 Houston time. Florida had been sunny and warm, but this Friday afternoon in Houston was cold and dismal. Shea wanted to stop at his MSC office to finish annotating his Apollo binder. Abbey went home.

One of the kids had the television playing just before dinner when the first incomplete news bulletin aired. Noting the cessation of music and laughter from the television, Abbey stepped out of the kitchen and heard the announcer saying that there had been an accident at Cape Canaveral and that "one of the Apollo astronauts" had been killed.

Within moments the report was corrected: *three* astronauts had died at the Cape.

Grissom, White, and Chaffee, Abbey's friend, all gone in an instant. As he struggled to accept the news, Abbey thought again of Frank Liethen, killed just three months earlier. What had happened? And what it would mean for Apollo?

Given the sketchy information, Abbey knew that he would glean little from a phone call—assuming he could even get through to the Cape or mission control. Forgetting about dinner, he ran to his car to drive to the center.

Shea and Markley were both in mission control, where most of the Apollo team had gathered. (The center had been staffed for the plugs-out test, and in contact with the crew.) Chris Kraft, the head of flight operations, was there too. Everyone looked pale and shaken. Three astronauts had died in the past couple of years, but in aircraft accidents—this was the day NASA had been dreading, the loss of a crew in a spacecraft. Only they were still on the ground! It didn't seem real.

The terrible news arrived in fragments, but eventually the narrative emerged. Grissom, White, and Chaffee had entered spacecraft 012 after 1:00 PM. There had been one early distraction: the crew smelled an unusual odor like sour milk, so a "sniffer" team was dispatched to level A-8, to the White Room, a portable structure that enclosed the spacecraft on its booster, twenty stories off the ground. After sampling the atmosphere inside the cabin, the team found that the odor seemed to have dissipated, so the countdown resumed with the hatches finally closed at 2:40 PM.

Then there were further delays, especially with the communication system, which kept flooding with static. The communications problem forced another hold in the count at 5:40 PM.

The count resumed, but then a microphone got stuck in the On position, forcing Chaffee to devote several more minutes to flipping switches in the cockpit to resolve the problem. Grissom lost his temper. "How are we going to get to the Moon if we can't talk between two or three buildings?"

Finally, at 6:30 PM, the test countdown reached T minus eleven minutes, the point at which 012 would be on its own internal power. Once that milestone had been reached, the astronauts would perform an emergency egress—and the test would be over.

There were twenty-seven technicians at level A-8 by this time, waiting for the end of the test to help the crew with egress. The sun was setting; it would be dark when the astronauts emerged to make their way from the White Room across the swing arm to the gantry.

At 6:31:04, according to a medical harness he was wearing, White's heart rate suddenly spiked—as if the astronaut had seen or heard something surprising. Suddenly, one of the astronauts, probably Chaffee, called out something like "Hey!" followed by "We've got a fire in the cockpit."

The astronauts frantically tried to open the hatch, a time-consuming and complicated business under ideal circumstances. The hatch consisted of three parts—an outer covering (which was off), a middle plate, and an inner element that required White to insert a ratchet and unscrew six different bolts.

There was another transmission: "We have a bad fire!" followed by what sounded like yelling. "We're burning up! Get us out!"

Then another longer yell or scream.

Temperatures and pressure rose inside the cabin—even "fire-resistant" material readily ignited. Then the cabin ruptured, spewing molten metal into the White Room. The fire began to produce carbon monoxide, creating smoke and depositing soot and ash, driving off the pad crew—which had been trying to reach 012 and open the hatch from the outside.

This phase lasted only five seconds, since there was insufficient oxygen for continued combustion. It was, however, likely the most lethal.

A mere eighteen seconds had passed from first alert to final agony.

The shocking news had reached Washington, DC, just as NASA's highest-ranking officials were at a cocktail party celebrating the signing of an international treaty that forbade nations from placing weapons in space. Webb, Mueller, von Braun, and Phillips were present, along with senior contractors like Lee Atwood of North American and members of the Johnson administration and the diplomatic community. Astronauts Carpenter, Cooper, Armstrong, Lovell, and Gordon were also present. Upon hearing word of the tragedy at the Cape, the astronauts hurried off to their hotel to work the phones, and several officials boarded corporate planes to head for the Cape.

In Houston, after conferring with Markley, Shea flew there too. By the time his plane arrived, flight surgeons and the pad crew had removed the astronaut's bodies from 012, a grim task complicated by the fact that the rubber fabric of their suits had melted in the high heat, welding the bodies in place. (Chaffee was found in his couch, the right-hand one of three. White was lying on the floor of the spacecraft below the hatch. Grissom was under White's center couch with his feet still on his couch.) The extraction took an hour and a half.

Autopsies eventually showed that while the crew had suffered burns where their suits had failed, none were lethal. They died from asphyxiation and smoke inhalation.

Abbey remained at the center late into the night, listening to the progress report on the removal of the bodies. Once that was complete, he went home, since Shea and Markley had not given him an immediate assignment. All he could do was keep up with the news, and with NASA's plans to investigate the accident.

In the close-knit neighborhood around the Abbeys, astronauts and other NASA officials, many of them familiar with this sad duty from their days in flight testing, were paying calls on the widows—Betty Grissom, Martha Chaffee, and Patricia White.

Local memorials for the 204 astronauts were held at their churches over the weekend. Abbey was an usher for one at Webster Presbyterian on Sunday. The public funerals for Grissom and Chaffee were held at Arlington on Tuesday, January 31. White's family insisted that he be buried at his beloved West Point, and his ceremony was held there the same day.

By that time, Abbey knew that an investigative board had been formed under Floyd Thompson, director of NASA Langley. On Monday morning, January 30, he was called into Markley's office. He was not surprised when Markley told him, "I've just spoken with Joe [Shea] and you need to go down to the Cape to help him out."

14 | Ashes

BY THE FIRST WEEK in March 1967, Abbey had been at the Cape for five straight weeks. Working with astronaut Frank Borman as well as engineers Scott Simpkinson and Sam Beddingfield and a handful of others, he had been putting in long days, working until collapse, grabbing sleep, then back to the job. He had bunked in the crew quarters in the Operations and Checkout Building, not at one of the Cape motels. There had been no trips back to Houston.

The reason? The lengthy, challenging, necessary, and painstaking disassembly of charred spacecraft 012. It was neither a simple nor a safe process. When the work began, 012 was still mounted atop the Saturn 1B booster, still under its boost protective cover, with a live escape rocket on its nose. That escape rocket had to be removed. There was also the matter of controlling the scene of the accident—the charred 012 cockpit. How could one conduct even preliminary examinations without damaging the evidence?

Simpkinson, a veteran of launch operations, proposed the creation of a false floor that could be inserted into spacecraft 012 (once the astronauts' couches were removed), allowing technicians to work there without touching the floor or panels.

Abbey found it to be a brilliant concept, and even more brilliant in construction and installation, since the fake floor had to be supported by the struts that held the escape rocket on the nose of 012, then folded for insertion into the cockpit.

Conception, design, construction, and installation were complete in four days. Within a week, enough critical components had been removed from 012 that the spacecraft itself could be lifted off the booster and moved to the Pyrotechnic Installation Building, where Petrone's team usually installed rockets and their fuel on spacecraft. Now the process was reversed.

Then the examination ramped up, with components being examined under magnification and X-rayed, their physical state recorded and photographed.

While this was going on, spacecraft 014, once scheduled to carry Schirra's crew to orbit, was shipped to the Cape. It was placed in the Pyro Building, where it was disassembled alongside 012, allowing Abbey, Borman, and others to compare damaged components with their pristine twins.

A thousand different items were removed from 012. The investigation quickly eliminated several possible causes for the fire: there was no evidence of leaking batteries, spontaneous ignition, chemical or mechanical failure, and no electrostatic discharge. The only cause that could explain the type of fire that occurred? Electrical ignition. A spark of some kind.

This was an important conclusion. NASA needed to know what caused the fire, in case radical technical changes were required in the upcoming Block II command module. The entire program was on hold until the cause was identified and corrective measures taken.

There was also political pressure. Both houses of Congress had begun hearings into the disaster, and the Senate committee under Sen. Clinton P. Anderson of New Mexico was especially aggressive. One of the committee's members, Sen. Walter Mondale of Minnesota, had gotten hold of the Phillips Report, which was filled with harsh criticism of North American's management and manufacturing. When Mondale asked James Webb about it in an open hearing, Webb denied that any such thing existed. (These notes, as Phillips called them, had not been considered a formal agency report and had not been briefed to Webb. The administrator was quite angry when he did learn what Mondale possessed—and that no one on his team had warned him.)

Reports on the disassembly of 012, with charred wires and panels, only added to the damning picture of North American's assembly procedures.

For example, a socket wrench was discovered nested in the wiring in 012. An examination of a third Block I, spacecraft 017, scheduled for the first unmanned Saturn V launch, turned up even more apparently sloppy work, with skinned wires. A furious and shamefaced Petrone ordered 017 shipped back to Downey for repairs. There, a number of executives begin to wonder if they might face prosecution for criminal negligence.

Joe Shea was at the Cape through all of this, working manically every waking hour, with breaks for frantic rounds of handball. There were no more puns and no more red socks. Shea was also drinking himself to sleep. It was clear to Abbey that his boss had taken the loss of Grissom, White, and Chaffee personally, in spite of a plea by Deke Slayton to remember that test flights frequently led to fatal accidents—that they were no one's "fault." Abbey knew that Shea felt he had ignored several warnings about potential fire danger in the Apollo cockpit.

One Friday in March, as the disassembly was winding down and Abbey and Shea were preparing to finally fly home to Houston for the weekend, Shea called Abbey to his room at the Holiday Inn for a drink. He introduced Abbey to forty-year-old John Yardley, a highly regarded member of the Mercury and Gemini team at McDonnell Aircraft who had been consulting with the review board since the weekend of the fire. "John's going to be my deputy," Shea said. "He'll be flying back to Houston with us."

Abbey had doubts about this move. Yardley was capable, but he had no knowledge of Apollo. This was not the time to try catching up. And as soon became apparent, Yardley was dealing with his own stresses. He had worked with the dead astronauts, especially Grissom, for years, and was, like Shea, obsessed with discovering why his friend was dead. Within a week, in Houston at a meeting of Shea's senior staff, Yardley suffered a breakdown that also had a visible effect on Shea.

Things happened very quickly after that. Abbey learned that George Low and Bob Gilruth wanted Shea to take a vacation. He refused, and, feeling pressured, offered to resign. Nobody at NASA wanted that: Shea was too capable to be discarded lightly and a departure now would also undermine the agency in the ongoing congressional investigation.

Shea then agreed to meet with several psychiatrists—whom Shea completely bamboozled, convincing the shrinks that he had perfectly processed the tragedy and was completely capable of moving forward.

But while Shea could win any contest where he was pitted against one or two people, he couldn't fight all of NASA management. Sure enough, at the end of March, Shea told Abbey that he was being promoted to HQ to become agency deputy for manned flight. "It's more responsibility," he said, but Abbey knew better: Gilruth, Phillips, and Mueller had finally found a place to park him.

To him, Joe Shea was a lightbulb filament that had burned too brightly.

15 | Commitment

FLOYD THOMPSON'S 204 BOARD transmitted its full report to administrator James Webb on April 5, 1967. Two days later, the US House of Representatives held public hearings on the Apollo fire. And Joe Shea's transfer from Houston to HQ—his "promotion" from Apollo spacecraft program office to deputy associate administrator for manned spaceflight—was also announced.

Soon after, Harrison Storms, North American's Apollo boss, was replaced by William Bergen, formerly a senior exec and engineer at Martin Marietta. There would be other changes at North American, in launch operations and manufacturing, and at the very top of the company.

In a spectacular demonstration of bad corporate timing, North American was in the process of being taken over by Rockwell Standard, a Pittsburgh-based manufacturer of industrial machinery, as well as parts for light and heavy vehicles—a truck axle company, some sneered. Rockwell had total annual sales of $636 million dollars to North American's $2.37 billion; the smaller company somehow absorbed the larger.

The acquisition had been engineered by Lee Atwood to diversify North American, and also to provide some cover if NASA blamed the company for the Apollo fire. (Announced in March 1967, the deal would be delayed for months on account of a review by the Department of Justice.)

While prudent, the rush for cover proved to be unnecessary, especially after Frank Borman testified before the House committee on April 17 about agency and especially astronaut confidence in NASA management. "The response we have given is the same because it is the truth. . . . We are trying to tell you that we are confident in our management, in our engineering, and in ourselves. I think the question is really: Are you confident in us?"

The investigation was as complete as it would ever be, since no definitive cause for the fire was ever discovered. Nevertheless, the conditions that made the fire so deadly were obvious in hindsight, and were going to be rectified.

But NASA needed new leadership in the spacecraft team. Gilruth and Low wanted Chuck Matthews—the highly regarded Gemini program director—to replace Shea. But Matthews had just relocated to Washington after two intense years on Gemini. He declined—and said that Low ought to be the new Apollo spacecraft program manager.

Low had already accepted one demotion, moving from HQ to Houston, becoming Gilruth's deputy in February 1964. Replacing Shea would be a second step down. As Abbey had learned, however, titles and rank meant little within NASA. Low was willing to take the job.

But Gilruth, Phillips, and Mueller had yet to inform James Webb of the decision, so Gilruth staged a dramatic "hold the plane" scene the next day at Washington National Airport, allowing Webb to "offer" the job to Low as he was about to fly back to Houston.

George Michael Low was born (as Georg Wilhelm Low) near Vienna, Austria, in 1926; his father was a manufacturer of Jewish descent. The Lows fled the Nazis in 1938, settling in New York State. Young Georg attended Rensselaer Polytechnic, but his college work was interrupted by World War II, when he served in the US Army (and changed his name to the Americanized "George.") He returned to Rensselaer and graduated in 1948. After a brief tour with aircraft manufacturer Convair in Texas, he bounced back to Rensselaer for his master's of arts.

Low joined NACA in 1950, at the Lewis Research Center in Cleveland, Ohio, establishing himself as a premier specialist in aerodynamics, the arcane science of thermodynamics and boundary layers—all areas that would become critical for humans surviving the return from Earth orbit or lunar trajectory.

In 1958 he was one of the NACA specialists who led the transition to the National Aeronautics and Space Administration. That same year he became head of NASA's manned spaceflight office at HQ. He was considered to be brilliant, yes, but unlike many of the Space Task Group engineers, also canny and wise. After moving to Houston, he became famous for his ability to skillfully maneuver inside the NASA bureaucracy.

Of NASA's senior leaders, it was said that Gilruth was loved, Shea was admired, Kraft was feared, but George Low was revered.

Abbey's daughter Joyce BK would say, years later, "My father adored that man."

While these personnel shifts were taking place in Washington, Abbey was busy in Houston, where on April 1 Joyce gave birth to their fourth child, a boy they named James Vincent Andrew, in honor of Abbey's older brothers.

This family time, while busy, allowed Abbey to ponder his own future in Apollo. He had come to NASA with experience as a pilot and instructor, as an engineering student, as an aerospace vehicle developer, but with no particular set of precepts or rules—nothing beyond his own intelligence and common sense.

Two and a half years on Apollo, especially his work with Joe Shea, however, had allowed him to form judgments about the best way to manage large technical programs. *The program must be defined.* You have to know what your goals are, what your plans are, as early as possible. Even though a deadline is essential, *Safety can't be sacrificed for schedule.* And the big lesson from watching Shea . . . *Major programs can't be run by one person.*

When he learned that George Low was going to replace Shea, he immediately asked to see him, and met with the new Apollo spacecraft program manager on Tuesday, April 4. He told Low of his work for Shea, offering to provide the same support to him. Low replied that he knew Abbey's background, and his role throughout 1966, and offered him a job as his technical assistant. (Years later Low would say that he wanted Abbey because "he knew everyone and knew everything about Apollo.")

Low told Abbey that Shea had operated as a lone wolf and that his program office failed to communicate with the center and its engineering and operations teams. He wanted to change that and was open to Abbey's suggestions.

Abbey believed that the solution was to directly involve the center's engineering and operations leaders in Low's decision-making process. He proposed the establishment of a configuration control board (CCB) whose members would include center directors as well as program managers from North American and Grumman. The CCB agenda would review and approve—or disapprove—not only technical changes to the Apollo vehicles but also policy or program-wide matters that could lead to changes. The CCB would meet every week in Houston, and board members would be required to attend in person. All the principles would hear the same reports and make binding decisions then and there. (If you couldn't make a binding decision, you didn't belong in the meeting.) Low liked Abbey's proposal and decided that he would chair the CCB. He was determined to hear all voices and arguments, then make final decisions.

There was one complication, however. Abbey had grown to love Apollo and its goal of man, Moon, decade—it was a way of fulfilling his childhood dreams about spaceflight. But his air force tour in Houston would end in November 1967. He wanted to remain in the service. Military academy graduates with thirteen years of active duty rarely left by choice. The country had invested money in his education and training with the expectation that it would be repaid, in tenure if not in blood. And, in blunt economic terms, it was better to hold on for seven more years to reach the magic twenty that qualified for a pension.

Not that his options in the air force were especially appealing. He would be thirty-five that August, likely too old for acceptance at ARPS, the test pilot school. Most of his contemporaries were either flying missions over Vietnam or training to do so. Always Sam Abbey's son, Abbey would embrace a combat role, but would require at least a year of transitioning and training in one of the fighter jets being flown in the war, meaning that he wouldn't go to Southeast Asia until 1969 at the earliest.

It was also possible that, with his helo background, he would wind up in a rescue unit—again, after a year of requalification and training on the new Hueys (Bell UH-1 Iroquois). Foolish as it would seem in hindsight, in the spring of 1967, few expected the Vietnam conflict to last that much longer.

His NASA experience might also make him attractive to the air force space team in Los Angeles working on the mysterious Manned Orbiting Laboratory program.

But Abbey's sole focus was the Apollo program and its goal of a successful lunar landing before the end of the decade. His feelings had grown stronger after the fire and the loss of Grissom, White, and Chaffee. He realized that his only alternative was to resign from the air force and join NASA, if the agency would have him. Hearing this, Low offered to speak with Gen. Sam Phillips about the matter.

Phillips, Low, and Abbey were all at the Cape the following week. When Phillips learned of Abbey's planned resignation, he asked Low if he might speak with Abbey directly. (Phillips had been the officer who assigned Abbey to Houston in 1964.)

One night, after a meeting that ran into early evening, General Phillips sat down with Major Abbey in an empty office in the operations and checkout building at the Cape.

Abbey had no reason to be nervous about the encounter. He knew that Phillips was anything but a by-the-book martinet on military matters. Phillips immediately put Abbey at ease, telling him that he was pleased that he would be assisting George Low, that such experience would be invaluable for Abbey's future air force career. And he could not promise that air force assignments would provide the same excitement, challenge, and accomplishment as Apollo. His best offer was to extend Abbey's tour in Houston for another year, until the summer of 1968. "That probably won't be long enough for you to be involved in the first landing."

Abbey agreed that mid-1968 wouldn't be enough time. Without an extension through 1969, Abbey admitted that his only choice was to take the NASA position being offered by Low.

Phillips surprised Abbey by admitting that he once faced a similar decision in his military career, when he was "stuck at Wright Field" for six years in what seemed like a dead-end missile development post. He had been offered a job in the civilian world and had seriously considered it. However, he had ultimately turned it down, and "bigger things came along" to keep him in the air force.

"If you can't stay in the air force," Phillips said, "I'm pleased that you're going to work for two of the finest individuals I know, George Low and Robert Gilruth."

The two men shook hands. Abbey returned to Houston and resigned from the air force effective April 30, 1967.

16 From the Earth to the Moon

MAX FAGET WAS LATE.

The wry and wiry chief engineer was rarely punctual, and his NASA bosses had tolerated this—among other quirks. But now he was dealing with George Low.

As head of engineering at MSC, Faget was a principal participant in Low's new configuration control board. But at the CCB's first official meeting on Friday, June 16, 1967, Faget managed to miss the 1:00 PM start time.

By five minutes.

Low had Abbey lock the sixth-floor conference room. Faget started banging on the doors, demanding to be let in. Finally, Low told Abbey to open the door. With some grumbling, the engineer took his place at the large table.

The others present were flight crew operations chief Deke Slayton, NASA medical chief Charles Berry, science director Wilmot Hess, head of flight operations Chris Kraft, manager for Apollo's command and service module Kenny Kleinknecht, and Bill Lee, Kleinknecht's counterpart on the lunar module. Scott Simpkinson was there in his new role as Low's assistant for flight safety. North American Rockwell's Dale Myers and Grumman's Joseph Gavin represented the prime spacecraft contractors.

In chairs around the perimeter were engineers from Owen Maynard's team. There were also North American and Grumman managers for each subsystem on their respective spacecraft, and each of these had a counterpart in Faget's engineering team. Low would insist that each change or issue had to be presented by *both* the contractor and his NASA counterpart.

Low sat at the head, wielding a gavel (shades of NASA Langley in 1960). When Low rapped the gavel, the meeting started.

Low never raised his voice. The same did not hold for the others, especially the NASA veterans trained in the Langley style. Abbey would remember loud arguments between Kraft and Faget; between Faget, Kraft, and Slayton; between Faget, Kraft, and Berry; between all of them and science chief Wilmot Hess; and then between various NASA people and contractors. Low's method was to let the arguments play out, then slam down the gavel and state the resolution.

There were amusing moments. One of the questions later facing the CCB was the inclusion of a television camera in the Apollo command module. Ed Fendell, a flight controller in Kraft's team, made the presentation, starting with his first chart and the statement "Flight Operations has no requirement for television on board the spacecraft."

Kraft immediately roared to his subordinate, "You're crazy as hell. *I* am Flight Operations and I *have* a requirement for television!"

Without blinking, Fendell announced, "What I meant to say is, 'Flight Operations has a firm requirement for television on board the spacecraft.'"

Whatever battles he fought inside the CCB meetings, Max Faget was never late again.

In taking the job—the title was "secretary of the configuration control board"—the newly civilian George Abbey had agreed to develop an agenda for each Friday meeting on the previous Monday. This document would go to Kleinknecht and Lee, and to North American Rockwell and Grumman, giving each entity most of the week to prepare presentations and recommendations.

Abbey moved into a small office adjoining his new boss, with an interconnecting door, inside Low's corner suite.

In those first weeks, Abbey began to know his new boss, who was probably the most universally respected leader in the entire agency. Brilliant and capable, Low was also incredibly self-effacing, freely sharing credit for achievements. He could be direct, cutting through red tape while still managing to maintain good relationships.

Slight in stature, and a bit stooped in posture, the very picture of a college professor, Low was actually fit and athletic. He routinely arose at 5:30 AM to jog, then drove his white Mustang to the center and was at work in his seventh-floor suite in building 2 by 6:30 or 7:00 AM most days. That car, like those of most center staffers, including Abbey, would still be in the lot at seven that night.

He had two secretaries—Marilyn Bockting and Judy Wyatt—who worked in shifts, with Wyatt generally arriving at MSC by 7:15 and Bockting hours later, then staying until 8:00 PM, when Low would still be in his office dictating a daily note to Gilruth. Saturdays were shorter; Low went home in time for dinner.

Low had Abbey create schedule boards for each spacecraft, every major test and associated test hardware, the qualification test for each subsystem, and all major activity, displaying these in a sixth-floor conference room where they were maintained by a team from Boeing.

Boeing had won a contract to do "program integration" for Apollo—integration being one of those aerospace buzzwords that meant whatever the speaker intended. For an engineer, it meant designing and building spacecraft and launch vehicles that could operate as one unit—with common communications, consistent power and guidance, ensuring that a spacecraft wasn't too heavy to be lifted, or too large.

For program management, it meant identifying activities and tests and placing them in a schedule. Nevertheless, Low and Abbey would meet here every morning to record progress—or the lack of it. Abbey made notes that shaped the Friday agenda.

After each CCB meeting, no matter how late it ran (and several ran deep into Friday evening), Low dictated a longer memo for Gilruth, recounting the CCB's decisions and rationales, ensuring that the center director was no longer cut off from Apollo matters, as he had often been with Shea.

At the same time, Abbey would be preparing the minutes of the CCB meeting complete with orders for departments or contractors as well as completion dates. The minutes would be signed by Low on Saturday morning and would be sent out over the weekend. On Monday a new agenda would be sent to the contractors, and to the Apollo spacecraft managers.

During that first week, Abbey was in a meeting with Gilruth and Low when the conversation turned to oversight of engineering and manufacturing at North American—and how to rebuild astronaut confidence in the organization.

He said, "The best way to do that is to make an astronaut head of the oversight team. Frank Borman would be great for that job."

Low noted that Borman would provide "the right kind of leadership." Gilruth agreed with the choice, too. Borman had served as a member of Floyd Thompson's committee investigating the Apollo fire and had distinguished himself in congressional testimony. He also had the engineering skills to lead the team.

As they were leaving, Low turned to Abbey. "George, you just earned your paycheck many times over today."

The first formal meeting of the CCB was the culmination of six weeks of restarts that followed the delivery of the report on the Apollo fire. Much of NASA's work on Apollo and other programs had continued, of course: the Saturn V program was only tangentially affected by the command module problems.

But schedules needed to be set again, now incorporating time for needed technical fixes in all the modules. And new fixes needed to be tested.

Low wrote later:

> We built mockups of the entire spacecraft, and tried to set them on fire. If they burned, we redesigned, rebuilt, and tried again. By vibration we tried to shake things apart; we tested in chambers simulating the vacuum of space, the heat of the Sun, and the cold of the lunar night. We subjected all systems to humidity and salt spray, to the noise of the booster, and the shock of a hard landing. We dropped the command module into water to simulate normal landings and on land to test for emergency landings. . . . We overstressed and overloaded until things broke, and if they broke too soon, we redesigned and rebuilt and tested again.

This attention wasn't all devoted to North American Rockwell's Space Division and the command module: Grumman's lunar module needed work, too. In testing the lander's systems, "We set off small explosive charges inside the burning rocket engines, and to our horror found the all-important LM ascent engine was prone to catastrophic instability—a way of burning that could destroy the engine on takeoff and leave the astronauts stranded on the Moon."

To solve that major problem, NASA forced Grumman to rebid the lunar module ascent engine.

The spacecraft teams, both NASA and contractor, were always asking for more time, but Abbey—backed by Low—refused, feeling that it was more important to bring problems to the CCB promptly. (Abbey's experience had taught him that it was better to have the teams working on a CCB-approved goal rather than charging off in the wrong direction.)

Adding to the challenges, every fourth week the whole CCB show would go on the road, with Low, Abbey, and the NASA principals flying to Long Island, New York, on Wednesday evenings for a daylong Thursday meeting at Grumman.

That night the team would fly to Los Angeles for a Friday meeting at North American Rockwell.

At Low's insistence, the MSC team made the New York–Los Angeles flight in coach, even as the Grumman execs and engineers rode up front in first class.

It didn't take long, however, before Chris Kraft rebelled. After working all day at Grumman, flying most of the night, then facing another all-day session at North American Rockwell, he was going to fly up front where he could get some sleep. He booked a seat in first class. Eventually, Slayton, Berry, and Faget wore Low down, winning the right to fly first class too.

Abbey's job on these trips was to be chauffeur for Low, Faget, Kraft, and sometimes Slayton. This was an unlikely collection of personalities. Slayton was the classic test pilot, not as hard-partying as some, but fascinated by exotic cocktails, upon which he frequently conducted "test runs."

Faget was more abstemious. So was Chris Kraft. The forty-three-year-old Virginia native—his full name was Christopher Columbus Kraft—had a degree in aero engineering from Virginia Tech and had worked at NACA, then NASA. He was a member of the original Space Task Group and had taken the lead in developing ways of controlling spaceflights. When Abbey began to work closely with Kraft, he had already turned the job of flight director over to his protégés, Gene Kranz, John Hodge, and Glynn Lunney.

Brilliant and occasionally hot tempered, Kraft was a mentor to his flight controllers, many of them still in their twenties, insisting on competence and the open admission of error while also giving them tremendous responsibility. His motto, according to Lunney, was "To err is human, but to do so more than once is contrary to flight operations directorate policy."

Slayton, Faget, and Kraft would become Abbey's close friends and coworkers for the next fifteen years or more.

The transition from Shea to Low wasn't without personal pain.

Joe Shea left the agency in July 1967 without any notice to his former associates in Houston. From what Abbey knew, one day Shea was still at HQ, the next he was headed to Boston to work for the Polaroid Corporation.

Maynard's deputy Bob Williams, Abbey's closest friend in the engineering office, had difficulty adapting to Low's managerial style, having worked for Shea on unmanned military spacecraft for many years. And soon he left.

Low had initially been cool to the idea of allowing deputies into the CCB meetings, but Abbey got him to allow them in, provided their presence was justified by the agenda—and that each deputy had the power to make a binding decision for his organization or company.

Abbey was now more visible to the members of the astronaut team, though still fleetingly. Mike Collins would only remember Abbey as "a man who sat in all the meetings and frowned a lot."

Tom Stafford, who was working on Apollo software at MIT during the recovery, realized that Abbey could answer any question asked. "He had a memory like three elephants."

Navy pilot Ken Mattingly arrived at the center in April 1966, fresh from a year at the Aerospace Research Pilot School at Edwards Air Force Base, and found himself baffled by NASA. "They had org charts, they had telephone directories with lists of names, and yet none of those descriptions about who did what and what their titles were seemed to match what they really did in life."

His first impression of George Abbey was dismissive. "He was this nondescript guy. He was called the secretary for the CCB in Apollo, and I had no idea what he did."

But Mattingly soon had a realization:

> The two Georges were a remarkable team in that George Low did everything in public and did all of the formal stuff and wrote memos and gave directions, and everything he did was a matter of record.
>
> Abbey, on the other hand, was intimately involved in every one of those things, every conversation, but he also had this network of working people. He knew all the troops and all the buildings, and he'd wander around and just talk to people and bring all of that stuff back. He knew what George Low was concerned about, the kind of questions, and so he would bring that stuff back, and informally, just no attribution, he would just make sure that George Low was aware of everything going on as perceived from the bottom of the barrel, as well as the reporting structure that officially brought things in.

Abbey's method was simple: take the visiting "troops" from North American Rockwell, Grumman, or other contractors—or other centers—out for beer. Listen to them. Take that information to Low.

This was a classic leadership technique, most famously employed by the legendary Ernest King, the controversial chief of naval operations during World War II. Earlier in his career, whenever his ship was in a liberty port, King would park himself in a local bar and invite every officer who came through the door to spend some time with him—very casually, very relaxed, very well lubricated with beer. Abbey did the same with those who visited Houston for the control board meetings.

This informal contact allowed King—or Abbey—to learn what job each man did, how it was working or not, who was helping or hindering. The more subtle goal was to get a sense of an individual's intelligence, knowledge, and behavior. The method gave Abbey fantastic and unprecedented insight into Apollo's workings.

While Abbey certainly enjoyed a beer, he was, in many ways, not typical of the test pilot and engineer world. For example, he didn't use profanity, a very uncommon trait in the flying community. Gus Grissom and Pete Conrad were famously profane. And it could be said of Deke Slayton—echoing a characterization from the movie *A Christmas Story*—that "he worked in profanity the way other artists might work in oil or clay."

One side benefit of Abbey's new association with Slayton was a resumption of his flying privileges. Upon leaving active duty at the end of April 1967, Abbey lost access to the Texas Air Guard T-33s he had been flying. Colonel McCrory offered him a slot in the guard, and Abbey was planning to take it until he had a conversation with Frank Borman. The air force astronaut, famous for his intelligence and directness, told Abbey, "There's no way you're going to have time to do the job for George Low and still fly for the guard."

Abbey's first weeks with Low confirmed Borman's judgment, so he turned McCrory down.

But he wasn't ready to give up flying. Since he had a private pilot's license, he joined a club at a small airport in League City, eventually logging a few hours in a Cessna, but even that proved difficult to fit in his schedule.

Finally, he was able to convince Deke Slayton, who was in charge of NASA's Ellington fleet, to let him stay current on the T-33—since he was a qualified instructor. Slayton agreed, and he and Abbey flew together more frequently—with Slayton offering compliments on Abbey's flying. When NASA later phased out the T-33, Slayton allowed Abbey to transfer to the sleek supersonic T-38.

Throughout the summer of 1967, Apollo's modules moved toward completion.

Grumman officially delivered the first lunar module at a weeklong customer acceptance and readiness review at the end of June that was attended by Phillips, Rees, Gilruth, Low, Kraft and, of course, Abbey. The delivery was prolonged and testy. LM-1 was still overweight—Grumman was in the midst of a radical weight-reduction program.

Over the next month, however, Low was still finding fault with the work in Downey, and on August 19 fired off a letter to Dale Myers about slips in the production of a test vehicle and spacecraft 101, complaining specifically about managers who apparently did not understand their own production charts.

Nevertheless, there were signs of improvement. The new chief at North American Rockwell (NAR), Bill Bergen, had already hired Bastian "Buz" Hello from Martin Marietta to oversee NAR's team at the Cape, where morale was even lower than it was in Downey.

And in midsummer Bergen instituted a system of spacecraft team managers—appointing a senior, trusted engineer as the point of contact for each vehicle.

For command module 101, it was John Healey, another veteran Martin Marietta engineer, whose first move was to blast both NASA and Downey teams for their poor lines of communication and constant changes, insisting that from this point forward all work had to go through him.

It took several weeks, but before long Low was so pleased with Bergen and Healey's system that he recommended it to Gavin and Tom Kelly at Grumman as well.

NASA added a new group of astronauts in August 1967, eleven scientists who had been recruited to serve as potential crew members for Earth and lunar orbit (and possibly landing) missions in the Apollo Applications Program.

Unfortunately, AAP's robust schedule—one early draft called for two dozen manned missions from 1969 to 1974—had been radically reduced by the fiscal year 1968 budget.

Future flight opportunities for the new astronauts were so grim that on the night of their arrival, Deke Slayton gathered them at the Kings Inn Hotel to tell them they were no longer needed as flight crew members, offering them the chance to withdraw from the program.

No one was willing to take that dramatic step that night, but the group soon dubbed itself the XS-11—"Excess Eleven." In the next five years, four of them would leave NASA without flying in space. The seven who stubbornly persisted would eventually fly, though after waiting more than a decade.

Abbey would come to know three of the group—Bill Lenoir of the Massachusetts Institute of Technology, Robert A. R. Parker of the University of

Wisconsin, and Joe Allen of Yale—quite well. But not for almost two years: all of the XS-11 were headed for air force flight schools to qualify as jet pilots.

Not long after the new scientists arrived, the astronaut office suffered another loss. Clifton Curtis "C.C." Williams Jr. was a Marine Corps test pilot who was selected as an astronaut in 1963. He was a tall—close to the upper limit for astronaut selection—and genial Alabaman. Following a tour as backup pilot for the Gemini X mission in July 1966, he had been assigned as lunar module pilot in a crew commanded by Pete Conrad, with Dick Gordon as command module pilot. They were scheduled as backups to McDivitt, Scott, and Schweickart for the second manned Apollo mission, intended to test the lunar module in Earth orbit for the first time.

On the morning of Thursday, October 5, Williams took off from Cape Canaveral in a new NASA T-38, headed back to Houston. In the skies near Tallahassee, Florida, something went wrong with the aircraft, which went into an unplanned and uncontrolled aileron roll and began a steep dive. Williams had time to radio a Mayday before ejecting.

The aircraft dove straight into a forest and Williams, trying to recover, waited too long to eject. His chute still not deployed, he hit the ground nearby and died instantly.

Williams left a widow, Beth, and a daughter, Catherine, who was just ten months old. Beth, it turned out, was pregnant and would give birth to a second daughter, Jane, in May 1968.

17 | Challenge

BY EARLY NOVEMBER 1967 von Braun's team was ready to launch the Saturn V in the first real test of Apollo hardware since the January fire. The mission was designated Apollo 4 and would carry command module 017 (Block I, though with some improvements, including a better crew hatch) and a dummy lunar module.

Every discussion of the Saturn V led to superlatives. It was 363 feet tall, 58 feet taller than the Statue of Liberty, 33 feet in diameter at its widest. It weighed 6,540,000 pounds—reporters took to comparing the vehicle to a navy destroyer—and had to be stacked in the Vehicle Assembly Building, a structure so large and unique that it had its own list of superlatives: 526 feet tall (forty stories), 726 feet long, and 518 feet wide, covering eight square acres of land and enclosing 3,655,000 cubic meters of space.

Low had traveled to the Cape to watch the launch in person with Phillips, Mueller, and other program managers. Abbey was his representative in mission control, observing Kraft's team under flight director Glynn Lunney as, right on time, seconds before 7:00 AM (EST) on Thursday, November 9, the five giant F-1 engines in the vehicle's first stage ignited. For several seconds, the whole stack sat on the pad as orange and yellow flame erupted at its base.

Then it rose, slowly, even ponderously, shedding flakes and sheets of ice that had condensed on its skin.

The S-I first stage performed beautifully, burning for two minutes and thirty seconds until shutting down and separating.

Then North American Rockwell's troubled S-II stage had its debut, performing as designed for the six-minute burn of its five J-2 engines.

The S-II separated and the third stage, the S-IVB, burned its single J-2 engine for two minutes and twenty-five seconds, shutting down on time.

Apollo command and service module 017 was in a hundred-nautical-mile circular orbit, the same one to be used on flights to the Moon. Two orbits later, the S-IVB reignited, reshaping the orbit to one with an apogee of 9,297 nautical miles and a new perigee of forty-five miles below the Earth's surface—deliberately designed to ensure that the CSM would reenter the Earth's atmosphere.

The CSM separated, and its service propulsion system fired twice, testing its start and restart capability, diving into the atmosphere at high speed to simulate a return from a lunar mission.

The command module survived reentry in good shape, splashing down in the Pacific near Midway Island, within reach of the recovery ship USS *Bennington*.

Originally scheduled to fly in April 1967, Grumman's LM-1 finally had its test flight on January 22, 1968.

After a troubled countdown, Apollo 5 launched just before sunset aboard a Saturn IB—the same launcher originally scheduled to carry Gus Grissom's crew into space. Two orbits into the mission, the legless LM's descent stage fired for the first time, but only for four of the planned thirty-nine seconds. Flight director Gene Kranz's team quickly diagnosed the problem—an incorrect computer setting. As they tried to fix it, they risked losing the mission, since LM-1 would fly out of range of the tracking network in a little over three hours.

Linked to the orbiting LM via a tracking ship off the California coast, the mission control team then successfully ignited the descent engine a second time. Now it fired for a minute and twenty seconds, shut down as ordered, and then restarted.

The next key event was a "fire in the hole," ignition of the ascent engine while the ascent module was still attached to the descent stage. (Astronauts might have to execute the same maneuver in an aborted landing attempt.) That went well too. Half an orbit later, the ascent engine was fired a second time. Kranz would write later, "As a result of the command problems, we had ruptured a control fuel tank, blown a jet nozzle off the LM, tumbled the gyros and expended all ascent rocket fuel. But we had satisfied all objectives in our last-minute maneuver sequence."

The LM ascent and descent stages were left to reenter the atmosphere, where they were destroyed.

Because of the problems, Grumman wanted to fly a second unmanned test with LM-2. But after what Low described as "considerable technical debate," the CCB concluded that the lunar module was ready to carry a crew. LM-2 was canceled just as the vehicle was to be delivered to NASA on February 17, 1968.

In spite of these technical triumphs for Apollo, the early months of 1968 were a difficult time for Abbey and his colleagues at NASA, and for Americans in general.

The conflict in Vietnam continued to drag on, costing the lives of thousands of American soldiers and pilots every week. It was never far from Abbey's mind. "Two Naval Academy roommates and several of my classmates were flying and serving there." He knew that the North Vietnamese had staged an unexpected attack on American interests in Vietnam during the Tet holiday, and that this setback, combined with other challenges, prompted President Lyndon Johnson—the Apollo program's most important supporter—to announce on March 31 that he would not seek the Democratic nomination for reelection in 1968.

There was also growing political unrest in the United States and in Europe, much of it in protest of America's involvement in Vietnam, but not all: some students and activists were calling for civil rights for minorities, others for antipollution laws. In addition, the Soviet Union's dominance of its Warsaw Pact nations was triggering low-level rebellion.

In later years, some members of NASA's Apollo team would admit that they felt "disconnected" from events in the real world—the Vietnam War, politics, even popular music and entertainment. In spite of his hours at MSC, Abbey wasn't one of them. For example, he had always listened to music and was a fan of both traditional performers like Frank Sinatra, and also—thanks to his children—the Beatles. He thought you would have to be "deaf and blind" to miss what was going on outside Houston.

Early on the morning of Tuesday, April 4, 1968, von Braun's team launched the second Saturn V carrying CSM 020 and an LTA (lunar module test article) on a mission scheduled to last ten hours. The S-IVB upper stage would temporarily propel the combined CSM and LTA toward the Moon. Then the CSM's engine would fire, slowing the vehicle and causing it to fall back toward Earth. A second firing would then accelerate the CM into the atmosphere, simulating the speed and heating required for a "direct return" abort.

The first moments of launch were perfect, much like those of the November maiden voyage. But two minutes after liftoff the five first-stage F-1 engines began to experience pogo—an oscillation caused by fluctuations in the thrust of one or more engines. These vibrations began to damage the adapter linking the upper stage to the command and service module. Cameras observed pieces falling off.

The first stage survived to shutdown, but when the S-II stage—the bane of Downey's existence—fired up, one of its five J-2 engines underperformed and was automatically shut down 412 seconds after launch. And then a second one shut down. The three remaining engines burned for fifty-eight additional seconds in an attempt to make up for lost speed.

The fun didn't stop there. Because the three second-stage engines had burned longer than planned, the entire vehicle had remained pitched up. When the third stage, the S-IVB, ignited, its guidance system responded to the unusual attitude by pitching down—back toward the center of the Earth.

The S-IVB's single engine burned for eighty seconds, then pitched up—and because its guidance judged the vehicle to be over the intended speed, the stage kept pitching up even as it continued thrusting. According to Jay Greene, the flight dynamics officer in mission control, "It went into orbit thrusting backward."

That initial orbit was 93 by 194 nautical miles, not the planned hundred-nautical-mile circular orbit.

Low was in mission control, watching from a console with Kraft and Gilruth. Abbey was in the VIP area behind them. There was nothing any of them could say or do to make things better. Abbey would remember thinking, *What a disaster.*

There was a second problem with the S-IVB that required mission controllers to separate the command and service modules and fire the SM's rocket to complete the reentry test. CM 020 splashed down forty-three miles from Hawaii and was recovered by the USS *Okinawa.*

Later that day, civil rights leader Martin Luther King was assassinated at a motel in Memphis. Within hours, riots broke out in African American communities in Washington, DC, Chicago, and Baltimore, spreading to over a hundred cities over the next four days.

Abbey was appalled by the assassination and by the riots, but also all too aware of the causes. He remembered the struggles Jim Wiley faced in his life.

By June 25—ten weeks after the incident—Marshall issued a six-hundred-page report identifying the causes of the Apollo 6 pogo and other problems, and noting that nine of sixteen primary objectives of the mission were accomplished, with six others partially accomplished. Only one, the restart of the S-IVB, was a failure.

As for the structural failure of the adapter, NASA engineer Don Arabian performed some classic engineering detective work and discovered a flaw in its manufacturing, which was corrected.

18 | One Week in August

WITH MARSHALL PROCEEDING WITH repairs to the Saturn V, spacecraft 101 at the Cape and on schedule for launch as Apollo 7 in October, Low finally took a vacation, returning to his family home in Upstate New York. While there, he pondered a looming problem: Setting aside any speculation about Soviet intentions, Low's goal was to achieve President Kennedy's vision of May 25, 1961. A series of Apollo flights, designated by letters, had been laid out leading to a lunar landing attempt. The two unmanned Saturn V flights were the A missions, the unmanned lunar module test was B. The upcoming Apollo 7 flight by Schirra's crew using a CSM but no lunar module was the C mission. The D mission in December 1968—Apollo 8—was to test a CSM and an LM in low Earth orbit. This was to be followed, in early 1969, by an E mission in which a combined CSM and LM would be launched into a high Earth orbit taking them four thousand miles from Earth. The F mission would take a CSM and LM to lunar orbit. The G mission was the landing.

But Grumman's LM-3 was not going to be ready for launch on the D mission in December, meaning that NASA faced a six-month gap in Apollo missions—making any landing attempt before the end of the decade almost impossible.

Low's new notion: Why not take advantage of the LM delay and come up with a new sequence of missions that would lead to a landing before the end of 1969? Assuming the third Saturn V would be ready for flight in December 1968, and Apollo 7 was successful, NASA could launch a manned Apollo 8, command and service module only, to the Moon. That would allow extra time to prepare LM-3 for the D mission in early 1969, followed by the F mission to lunar orbit as a dress rehearsal for a landing. The E mission would be skipped.

Returning to MSC on Monday, August 5, Low told Abbey about the idea. Abbey loved it. Low then spoke to Kraft, who was definitely enthusiastic. Kraft set about trying to find out whether such a mission was even possible in December.

The next morning Low flew to the Cape to meet with Sam Phillips, Kurt Debus (scientist and former director of KSC), and Rocco Petrone about the real status of LM-3 and the Saturn V. They confirmed what Low had suspected: the Saturn V could be ready, but LM-3 was not capable of a flight before February 1, 1969.

Low didn't mention his crazy idea to them. Not yet.

While Low was traveling, Kraft's mission control team conducted its evaluation of the realities of a December flight around the Moon—the trajectory, the guidance and navigation systems, the software. By Thursday they had an answer for Kraft: yes, it was possible, and the best launch date was December 20. When Low, who was now back in Houston, heard this, he telephoned Abbey at home. "We're going to scrap tomorrow's meeting."

The next morning, Friday, August 9, Abbey arrived at Low's office to find his boss already meeting with Kraft about the new lunar mission. Then Low went upstairs to pitch the whole idea to Gilruth, who was immediately enthusiastic.

Kraft and Slayton joined the pair. Phone calls were made—Low to Sam Phillips, who was still at the Cape, and Gilruth to von Braun in Huntsville.

Within hours the Houston team was on a NASA Gulfstream headed for Alabama and a 2:30 PM meeting with von Braun and Rees, as well as Phillips, Debus, and Petrone.

There was universal support for the idea and general agreement to go ahead with serious planning—in secret, Phillips cautioned. Mueller and Webb needed to sign off on any decision this momentous, and both were in Europe attending the UN Conference on the Exploration and Peaceful Uses of Outer Space. The planning would therefore be given the cover name "Sam's Budget Exercise."

In Houston that evening, Low met with Abbey and a handful of others about preparing command module 103 for such a flight. Flight crew operations director Deke Slayton had already decided that the new lunar orbit mission, if flown in December, would be crewed by the E mission team of Borman, James Lovell, and William Anders. Borman was then in Downey leading the Apollo CSM redesign effort. Low asked Slayton to have Borman return to Houston on Saturday, August 10.

Abbey was working in his office when Borman walked into the suite that afternoon. The astronaut was as annoyed as he was curious. "What's this all about?"

Abbey smiled. "I think you're going to like it."

Borman was asked to join Low in his office. Low got right to the point, laying out the LM-3 issue and his plan to insert a new Saturn V mission into the schedule—to send a manned Apollo around the Moon in December without a lunar module.

Borman liked the idea. And when he learned that Slayton wanted to swap the crews for Apollo 8 and 9—that is, leaving McDivitt to fly the LM-3 test and giving Borman the new circumlunar mission—he was even happier. (Slayton had already alerted McDivitt to the upcoming change in mission and vehicles. McDivitt said years later that Slayton made it clear that if McDivitt argued strenuously, he could have kept the lunar mission. But by then McDivitt had devoted two years of his life to the lunar module; he wanted to be the first to fly it.)

There were others in Houston who did not warm to the new concept, believing that it couldn't be supported. Bob Gardner, an engineer in Faget's organization, barged into Low's office that first week, arguing vociferously that the guidance software would never be ready in time for a December launch. Calm as ever, Low thanked Gardner for his thoughts. Gardner stormed out, slamming the door. Low turned to Abbey. "Gardner says it won't be ready, but Chris Kraft says it will be. Who do you believe?"

Abbey said, "Chris."

"Me too."

The idea had yet to be sold to Mueller and Webb. Phillips had returned to Washington and made his presentation to Tom Paine, the deputy administrator. He liked the idea, though he recognized its audacity.

Phillips telephoned Mueller in Vienna on Tuesday, August 13, catching his boss in a bit of a trap of his own making: Mueller didn't respect Gilruth and treated his team's ideas with skepticism. Yet he had been an advocate for all-up testing (in other words, skipping the incremental, one-stage-at-a-time approach favored by rocket designers) and for taking leaps with new systems—which was exactly what Phillips, Gilruth, and Low were proposing.

Abbey, who was privy to these conversations, knew that Mueller heartily disliked Low's idea but couldn't say no. Mueller's only option was to insist that NASA not openly commit to the lunar mission until after Apollo 7 flew, which Low had been stating all along.

This lack of a firm negative left Phillips and Paine with free rein to get James Webb on a conference call. Webb was a political operator, not an engineer. He had always deferred technical judgments to deputy administrators, and to Mueller and Phillips. He was horrified by the risks and the potential political fallout from a lunar mission failure.

But he didn't refuse, either, insisting that Phillips and the whole team continue their work quietly until Apollo 7 flew. When he returned to the United States, Webb was apparently feeling boxed in by circumstance and this apparent rebellion in the ranks.

He had also been head of NASA for seven and a half years, basking in the acclaim of the Mercury and Gemini flights while also suffering the loss of Grissom's crew and the congressional inquisition. His political sense convinced him that he would not be retained as administrator past January, no matter who became president.

With no warning to any of his staff, Webb went to the White House on September 16, ostensibly to brief President Johnson on Apollo and rumors of Soviet activity. He took the opportunity to renew a suggestion he had made earlier, that perhaps it was time for him to resign as administrator. Johnson surprised Webb by suggesting that he resign immediately. And on October 7, 1968, James Webb left the agency and was replaced by Paine. Apollo 7 was scheduled for launch the next week.

Hearing the bulletin, Low, who was at the Cape at the time, turned to a friend and said, "This makes C-prime [Apollo 8] possible."

As if to put more pressure on the decision, in mid-September the Soviets launched the unmanned Zond 5, which became the first vehicle of any kind to loop around the Moon and return to Earth. It carried a biological payload that included tortoises.

American intelligence agencies noted that Zond 5's reentry trajectory—the vehicle splashed down in the Indian Ocean—was too hot and fast for humans. But nevertheless it was a clear first step toward that long-rumored manned mission.

On October 26, the Soviets resumed manned spaceflights, launching cosmonaut Georgy Beregovoy on Soyuz 3. Beregovoy flew his vehicle to a rendezvous with the unmanned Soyuz 2, launched on October 25, but failed to dock—even though the Soviets insisted that the mission had fulfilled all its goals.

In mid-November Zond 6 was launched, repeating Zond 5's loop around the Moon. This time, however, the vehicle executed a skip reentry—literally diving into the atmosphere, then skipping out for a slower second entry—that would have enabled cosmonauts to survive.

Many in Houston and in Washington, DC, believed that the Soviets would conduct a third mission, this one possibly carrying cosmonauts—a launch window opened in early December ahead of Apollo 8's planned date of December 21. *Time* magazine's cover for December 6 depicted two humans in space suits, one American and one Soviet, striding for the Moon.*

The commander of Apollo 7 was Mercury and Gemini veteran Wally Schirra, the same astronaut who had questioned the need for the test that eventually killed Grissom, White, and Chaffee at the Cape on January 27, 1967.

As a result, during training, Schirra and his crewmates, Donn Eisele and Walt Cunningham, were quick to complain about schedules and hardware.

* But there was no launch. Only years later did NASA learn that Zond 6 had suffered a mechanical failure. Even then, it would not have been possible to attempt a manned launch until March 1969—the manned version of Zond, the L-1 spacecraft, was simply not ready.

While he acknowledged the wisdom of Schirra's protests the night before the Apollo fire, Abbey was convinced that by October 1968, Apollo was a much different program, with Low and his managers dedicated to doing things the right way in spite of schedule pressures.

Schirra expressed concern right up to the day of launch, on October 11, protesting the decision to launch when offshore winds briefly gusted to and over the maximum. Flight director Glynn Lunney pressed Schirra to go ahead anyway, and Apollo 7 lifted off beautifully.

Low was at the Cape. Abbey was in the second-floor mission control as his representative, updating his boss by telephone on significant milestones until he could return to Houston.

It was on the second day of the mission that things got ugly again, when the crew was scheduled to make the first live television broadcast from an American spacecraft. Schirra, who had resisted the idea of television in the first place, had developed a head cold, an unpleasant state on Earth, and even more uncomfortable in zero gravity. That and changes to the crew's flight plan (added tasks) caused him to refuse to proceed with the broadcast, getting into a testy exchange with Slayton.

Eisele developed a cold of his own, and Cunningham wasn't feeling great, either, likely due to what would later be called space adaptation syndrome, a form of motion sickness with symptoms ranging from mild nausea, disorientation, and headaches to severe nausea, vomiting, and "intense discomfort."

The broadcast—titled by the crew "The Wally, Walt and Donn Show"—took place on the third day of the mission, and was well received by NASA and the public.

But by then Schirra and Eisele were openly criticizing the mission control team, led by Lunney, for last-minute changes or poorly thought-out exercises. As the mission dragged on—there was no better description—Abbey was witness to this, and the usually genial Low seethed with anger. Slayton was furious, too, since the crew worked for him. Gilruth and Kraft were also present. Kraft was volatile by nature, and the crew's public sniping with Lunney made him especially snappish. Even the placid Gilruth was exasperated.

One final blowup took place as the crew prepared for reentry on October 22, and Schirra, fearing damage to his ears, decided the crew would not don helmets. Slayton tried and failed to argue with Schirra, ultimately telling him, "It's your neck, and I hope you don't break it."

Apollo 7 splashed down safely—Schirra, Eisele and Cunningham in their suits, but with no helmets—after a mission rated by NASA as "101 percent successful."

It cleared the way for the announcement, on October 28, that Apollo 8 "might" attempt a flight around the Moon, pending "a detailed analysis and review of the Apollo 7 mission."

The landing also effectively ended the astronaut careers of Schirra, who had already announced his retirement, as well as Eisele and Cunningham, who were moved off to dead-end assignments and never flew again.

In spite of these issues, Apollo 7 was a technical success and paved the way for more ambitious missions. Without it there would have been no Apollo 8.

Republican Richard Nixon defeated Democrat Hubert Humphrey on November 5. Members of the Apollo team in Houston had wondered what a Nixon presidency would mean for the future of the program—or for Vietnam, for that matter. (Nixon had campaigned on a promise to bring new leadership to the war.) But for the most part, they, like Abbey, were totally engaged in preparing for Apollo 8, and the other vital missions to follow, hopefully culminating in a first lunar landing and return in the next year.

All through the fall of 1968–1969, the CCB met most Fridays, in Houston, New York, and California, though its agenda shifted from schedules and contractor issues to more operational ones, including selection of lunar landing sites.

Program managers and flight directors preferred open, flat plains while the scientific community wanted geologically interesting highlands. Either choice affected launch dates and times—a single day's delay would force a change to a secondary site farther to the lunar east (on account of the angle of the sun, which affected astronaut visibility). This complicated crew training, since the astronauts had to be prepared to land at multiple sites. One of the mission planners, Joseph Loftus, spent "the most strenuous afternoon of my life in front of the CCB" explaining the matter.

Finally, Low rapped his gavel and announced, "That's the way it is."

As the number of issues and decisions piled up—by the end of 1968 the number was over a thousand—Low and Abbey never found a reason to change their methods and schedule.

To Abbey, this period, from October to December 1968, showed NASA at its very best, with every center from Houston to Marshall to the Cape and every department working steadily and obsessively toward one goal: launching the first interplanetary manned spaceflight.

The most active, and crucial—aside from the teams actually preparing the launcher and spacecraft—were the members of Chris Kraft's flight control, who designed the trajectory and fought battles with NASA managers and with Borman himself over the notion of firing Apollo's service propulsion engine to put the spacecraft in actual lunar orbit as opposed to a single loop. (Ten orbits of the Moon would give the crew time to photograph potential landing sites.)

On the predawn darkness of Saturday, December 21, Abbey arrived in mission control and took up his place in the spacecraft analysis support room (SPAN). Low was at the Cape, where astronauts Frank Borman, Jim Lovell, and Bill Anders had finished breakfast with guests Deke Slayton, Robert Gilruth, and George Low and had ridden out to pad 39A to board Apollo 8.

During the final run-up to the launch, they had been hosts to Charles and Anne Morrow Lindbergh—it had been only forty-one years since Lindbergh flew the first solo flight across the Atlantic.

The countdown for the launch was relatively trouble free. At 7:51 (EST) on a beautiful sunny Florida morning the Saturn V ignited and rose from the pad. Following what Gemini veterans Borman and Lovell would later describe as a ride smoother than their Gemini-Titan launch, the stack containing Apollo 8, a dummy lunar module, and the S-IVB upper stage reached orbit.

Flight controllers checked the status of the S-IVB over the first complete revolution of the Earth and found no problems. At two hours and thirty-seven minutes, capcom Mike Collins (who had been removed from Borman's crew because of neck surgery) radioed that Apollo 8 was "Go for TLI"—translunar injection—burning the S-IVB engine for five minutes and eighteen seconds. Abbey heard magic in this dry piece of jargon. It meant that a manned spacecraft was headed for the Moon.

Low returned to Houston, having left the Cape directly after the launch (and radioing Abbey for updates on TLI while en route). When Low arrived, he took a seat right inside mission control while Abbey remained where he was in SPAN, occasionally popping in and out as Low needed him.

At dinner time he would break away to go home for an hour; then he would return, remaining in mission control as long as Low did.

The most notable event of the three-day climb to the Moon, alas, was Frank Borman's illness: twelve hours into the mission, as he was trying to get some sleep, he vomited—and that wasn't the worst of his bodily torments. The crew didn't discuss the commander's situation openly, leaving comments on a private channel that Gilruth, Low, Kraft, and others heard in a separate mission control.

They were concerned, but concluded that Borman was suffering from a twenty-four-hour flu—later research would suggest that it was a case of space adaptation syndrome, a malady that would afflict half of all shuttle astronauts in flight.

At mission-elapsed time of fifty-five hours and forty minutes, Apollo 8 slipped from the Earth's gravitational field to the Moon's. Now the astronauts were falling toward the Moon—which, thanks to Apollo's tail-first orientation, they couldn't actually see.

Monday turned to Tuesday, December 24, Christmas Eve. Around 4:00 AM Houston time Abbey and the mission control team—not to mention millions of television viewers around the world—watched with great attention as, at sixty-eight hours and fifty-eight minutes, Apollo 8 disappeared around the Moon, preparing for a lengthy firing of its service propulsion rocket on the far side.

It was incredibly tense—if the spacecraft emerged around the other limb of the Moon too soon, it would mean the burn had failed. And while the astronauts might still have a chance to get home, it would be challenging. If it were late, something terrible had happened.

Fortunately, Apollo 8 emerged right on schedule, with Lovell calmly radioing numbers on the farside burn, then going on to observe that the Moon was "essentially gray; no color; looks like plaster of Paris or grayish beach sand."

Here was a Buck Rogers moment.

Soon after, Anders managed to snap what would become the most iconic photo of the space age—a blue Earth rising over the surface of the Moon.

Then, several hours later, the astronauts having completed their photographs and preparations, and gotten some rest—as Abbey spent his Christmas Eve in mission control instead of home with Joyce and the children—the television screen lit up with black-and-white imagery of the cratered lunar surface passing below.

"This is Apollo 8, coming to you live from the Moon," Borman radioed to a global audience. Then Lovell, Anders, and he briefly gave their impressions of what they had seen and were experiencing.

"We are now approaching lunar sunrise," Borman said. "And for all the people back on Earth, the crew of Apollo 8 has a message we would like to send to you."

Anders began to read: "In the beginning, God created the heaven and the Earth . . ." Continuing through Genesis to "God divided the light from the darkness."

Lovell took over then. "And God called the light Day and the darkness He called Night."

Borman took over, ending the reading with "And God called the dry land Earth and the gathering of waters called He Seas." Then he said, "And from the crew of Apollo 8, we close with good night, good luck, a Merry Christmas, and God bless all of you . . . all of you on the Good Earth."

It was, for Lutheran church elder George Abbey, the most perfect expression of faith, patriotism, and technical achievement he could have imagined. Proud of his own small role in the Apollo recovery, he made sure to shake hands with Low—the individual most responsible for the Christmas Eve moment.

Apollo 8 made its final loop around the Moon, fired its engine to begin the journey home, Lovell announcing, "Pleased be advised, there is a Santa Claus."

There were other critical milestones to come, especially the high-speed dive back into Earth's atmosphere—and the parachute drop to the Pacific.

As for that famed picture of Earthrise—one of the tasks Low had assigned Abbey was to review photographs from Apollo missions for unusual or notable images. After Apollo 8's return on December 27, Abbey made sure he was at the center photo lab with its chief, John Brinkman—where the two of them, along with the lab tech who processed the image, were the first to see Anders's stunning Earthrise.

It was another reminder of the majesty of the Apollo 8 mission. The upcoming tests of the lunar module and—oh, yes, an actual lunar landing—seemed almost anticlimactic.

At headquarters, buoyed by the triumph of Apollo 8, George Mueller began pressing for a landing attempt by Stafford's crew on Apollo 10—assuming that Apollo 9's test of the lunar module was a success.

In Houston, however, at Low's meetings, this idea was dead on arrival. For one thing, the Apollo 10 lunar module was not configured for a landing and liftoff—its construction took place before Grumman's super weight improvement program.

The other reason was new: analysis of Apollo 8's lunar orbits had found that there were notable deviations from targets. (The unmanned lunar orbiters had been subject to the same effect.) Something was affecting them—not a major problem for an orbiting mission, but potentially disastrous for a landing mission, where a lunar module would have to rendezvous and dock with a command module.

No, if Apollo 9 succeeded in testing the lunar module in Earth orbit, Apollo 10 would have to do the same in lunar orbit. Only if both missions went as planned would Apollo 11 make that first landing attempt.

But which astronauts? Deke Slayton would admit years later that his original plan was to simply recycle Borman's crew to Apollo 11, and McDivitt's to Apollo 12, because he considered these trios to be the best trained.

But in a brief conversation with Borman before Apollo 8, he found that the astronaut was burned out on training and travel and hoped that Apollo 8 would be the triumphant capstone to his career.

Since Slayton couldn't use Borman's crew, it made little sense to use McDivitt's. So he stuck to his rough crew rotation, and while Apollo 8 was in lunar orbit, took Neil Armstrong, the mission's backup commander, aside to tell him he was to be commander of Apollo 11, which might well be the first attempt at a landing. Slayton and Armstrong assembled a crew that included Mike Collins as command module pilot (Collins having lost a coveted seat on Apollo 8) and Edwin "Buzz" Aldrin as lunar module pilot.

Abbey thought they'd do fine, if given the chance to attempt the landing.

On the morning of Sunday, July 20, 1969, Abbey and the family attended House of Prayer Lutheran Church as always. Then, leaving Joyce and the kids to watch daylong television coverage on all three networks, Abbey went "across the street" to mission control.

It was four days after the launch of Apollo 11 with its crew of Neil Armstrong, Mike Collins, and Buzz Aldrin. For twenty-four hours now the crew had been in lunar orbit in the command module *Columbia*; Armstrong and Aldrin were about to fly away in the lunar module *Eagle* to attempt the first manned landing on the Moon.

Neil Armstrong, Abbey's colleague from Dyna-Soar, was about to become the first human to walk on the lunar surface. (That decision not made in the CCB but rather in a smaller meeting in Gilruth's office in early May. There had been some uncertainty between management and the astronauts—and between Armstrong and Aldrin—about who was to take the first step. Gilruth, Low, Kraft, and Slayton unanimously opted for Armstrong, owing to his seniority.)

The past six months had seen one triumph after another, with two incredibly complex Apollo missions fulfilling all goals—Apollo 9 in March, making the first manned tests of the LM and of the lunar space suit and life-support backpack.

Then Apollo 10 in May, in which the LM (and related tracking and communication systems) was tested in lunar orbit, with astronauts Tom Stafford and Gene Cernan flying the vehicle to within forty-seven thousand feet of the surface.

Abbey had been at his post in SPAN for all the critical moments, marveling at the performance of the vehicles—the same command module and lunar module, and Saturn V, that had required so much labor for so long.

It was as if all the NASA centers and organizations, and contractors, had finally come together—a championship football team like, say, the Green Bay Packers.

Abbey knew that the landing was no easy task but, based on what he had seen in the past year, felt confident that if Apollo 11 didn't put an astronaut on the Moon and return him to Earth, Apollo 12 or Apollo 13 would—before the decade was out.

Unless something went wrong.

As Abbey entered mission control in search of George Low, he saw flight director Gene Kranz in his signature white vest, calmly working with his team of thirty.

A quarter of a million miles away, Neil Armstrong and Buzz Aldrin had boarded their lunar module *Eagle* and separated from command module *Columbia*, piloted by Mike Collins.

Everything was going as planned. But Low had a task for Abbey. "Joe is here."

After two years away from NASA, with no contact, Joe Shea had turned up at the center that morning. While driving in, Low had spotted Shea walking, all alone and apparently unrecognized by any other drivers. He had picked up his predecessor, and installed him in the third-floor viewing room.

Fearing some bitterness, Abbey was pleased to find Shea calmly enthusiastic—and taking some justifiable pride in the day. He had not only led the spacecraft office for three years; he had pushed for the lunar orbit rendezvous concept.

And there they sat, surrounded by other center officials, politicians, family members, listening to the air-to-ground conversations between Armstrong and Aldrin and capcom Charlie Duke, their link in mission control.

Abbey felt his heart rate rise when he heard the "go" for powered descent initiation—*Eagle* was at nine miles above the lunar surface. This was the point where Apollo 10 had stopped.

He could picture *Eagle*, on its back, descent motor firing. Armstrong and Aldrin would be looking up at the black sky as they dropped toward the flat, dark Sea of Tranquility.

Then the spacecraft began to pitch over, giving the astronauts their first close-up view of their landing site. It descended—much like a helicopter, Abbey thought—but to a dead gray rocky surface.

Abbey heard "program alarm," and his heart rate rose again. He knew that *Eagle* was close to the lunar surface, that Armstrong was surely looking for a place to set down.

Duke radioed to the crew, "We're go."

Abbey listened as Aldrin counted off the altitude and rate of descent. There were several mentions of "four forward," meaning that Armstrong was thrusting.

Time was running out—*Eagle* only carried a finite amount of fuel. Abbey prayed that the crew didn't have to try a fire-in-the-hole emergency ascent now—

Then, "Contact light."

And a stream of status reports from Aldrin: "ACA out of detent, modes control both auto. 113 is in."

Duke radioed, "We copy you on the ground, *Eagle*."

Finally, Armstrong sent word: "Houston, Tranquility Base here . . . The *Eagle* has landed."

The viewing room erupted into celebration—some blinking back tears.

In the control room, however, Kranz ordered his team to work. *Eagle* had landed, yes, but was it in shape to stay on the Moon? And if so, then it was time to prepare for Armstrong and Aldrin's EVA.

Joyce and the kids—George Jr., Joyce BK, Suzanne, and Jimmy—were planted in front of the black-and-white television in the Abbey living room. Astronauts were familiar figures to all of them, many of them neighbors. Yet George Jr. remembers thinking, *Neil Armstrong came to our Seattle house . . . Now he's on the Moon.*

They were all frustrated by the low-resolution black-and-white images, and struggled to hear Armstrong's first words. "Small step and what else?" Nevertheless, they all knew what an amazing moment this was.

After the EVA, the kids were put to bed. Joyce eagerly waited for her husband to come home.

It was less than twelve years from Abbey's roadside sighting of Sputnik in Montana, less than five since his arrival in Houston. A little more than two since he resigned from the air force in order to help make this moment happen.

Part III

"The Quintessential Staffer"

19 | Mission Accomplished

MAN, MOON, DECADE.

The work of Low's configuration control board ended with Apollo 11, and not just with the accomplishment of the lunar landing and safe return. It had met ninety different times between June 1967 and July 1969, considering 1,697 proposed changes, approving 1,341 and rejecting 346.

With the crew safely onboard the aircraft carrier *Hornet*, splashdown parties erupted throughout the Clear Lake community. Abbey personally drove Low to many of them, listening as his boss reflected on the past two years and the people who had made important contributions, with Chris Kraft high on the list. Low revealed that he was going to return to his prior job as deputy center director under Gilruth, and that he wanted Abbey to come with him as technical assistant to both men.

Low and his wife, Mary, had bought a beautiful lot in Nassau Bay right across from the center; they were building a new home with a dock for their boat and views of the water from every room. But NASA administrator Tom Paine wanted Low back at HQ as deputy administrator, filling the post that opened up when Paine moved to the top job. Paine felt he needed an agency insider to manage the next Apollo flights, the upcoming Apollo Applications Program, and to shepherd the new space shuttle into development.

Low agonized over the offer. He had his dream home under construction and knew that Robert Gilruth wanted him at the center. But Gilruth recognized Low's gifts and ultimately encouraged him to say yes. With great reluctance, Low accepted Paine's offer in September, though he couldn't move to DC right away: the NASA deputy administrator was a political appointee requiring congressional approval, which would not take place for months. Abbey had agreed to serve as technical assistant to Gilruth and the deputy center director prior to Low's acceptance of the headquarters job. Low assured him that he was still needed in Houston. He could also assist Jim McDivitt, Low's successor as Apollo spacecraft program manager and head of the CCB. By the end of November 1969, Abbey had moved to the ninth floor of the headquarters building.

With the accomplishment of the lunar landing, nine more Apollo missions remained to be flown—and at least a pair of Apollo Applications Program flights. Slayton had already judged the current astronaut office to be adequately staffed for these missions. Even with the retirements of Borman and McDivitt and the expected departures of Armstrong, Collins, and Aldrin, there were thirty men in the program, not including members of the XS-11.

Strangely, however, NASA wound up with seven new astronauts the first week in August 1969. They were all military test pilots from the Manned Orbiting Laboratory, which had been summarily canceled in early June 1969, leaving fourteen astronauts with no jobs and—thanks to their exposure to highly classified National Reconnaissance Office programs—limited options in the air force for the near future. Air force chief of staff James Ferguson approached George Mueller about moving all or some of the men to NASA. Mueller, eager to have air force support for the proposed space shuttle, asked Gilruth and Slayton to make the effort.

Thirteen of the pilots traveled to Houston the week Apollo 11 was on the Moon, and they returned for brief interviews with Slayton. Although convinced that he didn't need more astronauts, he agreed to take seven of them—all those under the age of thirty-six, the maximum age for new astronaut selectees.

Slayton was honest with them, telling them they would be waiting until the mid-1970s for flights.*

---

* This turned out to be optimistic. The first to fly in space—Bob Crippen—had to wait twelve years, until April 1981.

One of the MOL orphans was Abbey's friend from the Dyna-Soar program, Lt. Col. Albert Crews, who was in a tough position with the air force. He had been parked in dead-end space programs for seven years and had not attended schools or held jobs that would allow him to rise in the service. Because of security restrictions, he couldn't even be assigned to combat in Vietnam. Fortunately, Slayton saw potential in Crews, and while he couldn't use him as an astronaut, he hired him as a deputy in flight crew ops.

With Apollo going strong, Skylab and soon a space shuttle in development, and all the research and engineering development at the center, the workload of the director and deputy, and their technical assistant, was staggering.

One of the other residents of the ninth floor, John Eggleston, had less traffic—and one of his secretaries, twenty-five-year-old Cheryl Bouillion, who had recently returned to Houston after her husband's military deployment, wandered over to Gilruth's suite and offered to help with any extra work. Gilruth's team was happy to pile it on, and pretty soon Bouillion moved from Eggleston's office to work directly for the technical assistant, George Abbey. Abbey needed the help, since he had been asked by Gilruth to read all his mail and tell him what, if anything, required his input. The only exception was anything from astronaut John Young, currently assigned as backup commander for Apollo 13. "He's the best engineer I've got working for me," the director said. "If he's concerned about something, I'm concerned about it." Abbey was to do the same for Kraft.

Most of this mail was administrative—letters from other NASA personnel, contractors, politicians, and the occasional celebrity. Some, however, came from American schoolchildren. These letters, many handwritten and asking for information or photographs, would wind up on Abbey's desk, and it was up to him to draft a response, or send a photo if it could be sent. He enjoyed this, especially when finding messages from young women asking about possible careers with NASA. He knew that women had made contributions to Apollo, especially in mission planning and flight dynamics, and believed they should have greater opportunities. And when one young woman named Michele Brekke, a student at the University of Minnesota, wrote in, Abbey took the letter directly to Gilruth—a Minnesota alum—for a personal answer. Brekke came to work for NASA in the 1970s, and was later selected as the first woman flight director.

Abbey's primary job was to sit in on all meetings, with the exception of Gilruth's sporadic one-on-ones. These weekly sessions were with the Apollo program manager (Jim McDivitt), the Skylab manager (Kenny Kleinknecht), and the heads of the various directorates at MSC: Faget's engineering, Berry's life sciences, Slayton's flight crew ops. Staff offices, those departments such as legal, facilities, and personnel, or other center operations, met with Gilruth every two weeks.

The third member of the leadership was associate center director Frank Bogart, a 1931 graduate of West Point and MIT, who had served in the army in the Philippines and on diplomatic assignments in Moscow during World War II. (He was a member of President Franklin Roosevelt's staff at the Yalta Conference.) He had later taken part in the creation of the US Air Force, serving as its controller and retiring as a lieutenant general, then worked at NASA HQ under Mueller and Phillips. He was incredibly knowledgeable about the way organizations operated and how money flowed—a few early encounters with Bogart convinced Abbey that he had much to learn.

Since coming to the MSC in November 1964, his focus had been the Apollo program and its spacecraft. Now, working for the center's leadership, Abbey found that while he knew a lot about spacecraft engineering and operations, he knew very little about the MSC—which at that time had nineteen thousand employees, both NASA civil servants and contractors.

In those days, incoming NASA engineers were given the task of tracing the circuitry of a Gemini or Apollo spacecraft. Abbey decided to apply the same technique to the Manned Spacecraft Center. Soon after moving into his ninth-floor office, he created a center map showing which teams were based where, and then he began a systematic series of visits, literally knocking on doors and introducing himself, finding out who was behind each door—from secretary to senior engineer—noting the name and location, then following up, where possible, with an invitation to coffee or beer. Over the next two years he became as familiar with the center and its personnel and operations as he was with the Apollo command module.

One of the other projects that began to consume Abbey's time was NASA's search for a low-cost replacement for Apollo, the space shuttle. For all its virtues, Apollo-Saturn was an incredibly expensive and inefficient spaceflight system. For one thing, most of the giant launch vehicle and the spacecraft were used up and thrown away each mission. The gumdrop-shaped command module was the only piece that returned to Earth—and even that was destined for a museum, not a reflight.

Then there was Apollo's method of returning: splashing down in the ocean. This meant that some percentage of the US Navy needed to be available and on call to NASA for every mission. This was also expensive and, with the Vietnam War making severe demands on the military, increasingly difficult to arrange.

What NASA management—and many of the agency's political supporters—wanted was a reusable vehicle that would take off like a rocket but land on a runway. The vision was nothing new. It went back to World War II, to von Braun's popular concepts of the 1950s, and to the Dyna-Soar that Abbey knew so well.

The new Nixon administration had created a Space Task Group to ponder the next step. Headed by Vice President Spiro Agnew, the Space Task Group had proposed a reusable Earth-to-orbit craft as well as a space tug and a space station, all with the goal of launching human flights to Mars by the mid-1980s.

The price tag was far beyond anything Nixon would accept, however, and this ambitious program was scaled back to the reusable vehicle and a possible space station. Vision was one thing, of course. Making it happen was another.

20 | Apollo 13

"LET ME GET MY briefcase and then we'll go to the Singing Wheel."

Abbey and astronaut Ken Mattingly were sitting together in the third-floor viewing room of mission control. They and the families of the Apollo 13 crew had just watched an in-flight press conference with astronauts Lovell, Swigert, and Haise as they headed for their planned lunar landing. It was just after 7:00 PM on Monday, April 13, 1970. The crew of Apollo 13 had launched on the third lunar landing mission two days earlier.

Four days earlier, Mattingly had been a member of that crew, until exposure to measles prompted skittish flight surgeons to ground him. Given his familiarity with the Apollo 13 flight plan, he had been pressed into service as a capcom for Jim Lovell, Fred Haise, and his replacement, Jack Swigert. But with the conclusion of the broadcast, he had nothing to do. And as a single man, no one to go home to.

Along with astronaut boss Deke Slayton, Abbey had been keeping an eye on Mattingly, who was a team player but obviously frustrated and bitter about what had happened. Hence the invitation to a local watering hole. As Mattingly would say later, "We never got that beer."

It was 9:00 PM as the pair prepared to leave the viewing room. They had just heard the flight controller responsible for electrical, power, and instrumentation aboard the Apollo CSM asking the flight director for a "cryo stir," a routine procedure in which a metal "straw" inside a tank of liquid oxygen in the service module stirred the frozen material inside it. At 9:07, astronaut Jack Lousma, the voice link between mission control and the crew, radioed that request. A minute later, a terse message from Swigert got everyone's attention: "Houston, we've had a problem."

Swigert, the command module pilot in the Apollo 13 crew, had flipped a switch to initiate the cryo stir. Unknown to Swigert, his fellow crew members, or mission control, there was a bare wire inside the tank—its insulation had been burned off during an improperly monitored heating test months back. The wire shorted, sparked, and ignited the pure oxygen, which quickly boiled and expanded, blowing the top off the tank, destroying piping and wiring and ripping a panel off the side of the service module.

None of this was immediately apparent to the mission control team; it was only when Lovell reported seeing "venting" from the service module that controllers realized that they didn't just have a problem or malfunction—they faced a crisis.

With Swigert's first call, Mattingly rushed to the control room to join Lousma at the communications console. Abbey listened to the conversations between the flight director and Apollo 13 commander Lovell, then found the nearest telephone. He called Gilruth and Kraft, catching the center deputy director in the shower. Abbey told both men that there was a major problem aboard the spacecraft and that they needed to come to the center immediately. While awaiting their arrival, Abbey went to SPAN to follow developments as engineers from Max Faget's team and from North American Rockwell struggled to understand the problem. He also noted the arrival of backup astronauts John Young and Gene Cernan, who lent their skills.

He joined Gilruth and Kraft when they arrived, stayed with them as they met with Sig Sjoberg, who had succeeded Kraft as head of flight control, as well as Faget and his engineers, all formulating plans to bring the crew safely home. For Abbey it was the beginning of four long days and nights. He would return home only for a few hours, grabbing sleep whenever he could.

Having moved to the lunar module as the command module died, astronauts Lovell, Swigert, and Haise had enough oxygen to return to Earth, but their trajectory had to be adjusted, new procedures read up to them, power and water managed carefully. Charles Berry and his flight surgeons played an increasingly important role in saving the crew as the astronauts grew tired, cold, and dehydrated, conditions that not only threatened their health but also affected their ability to execute orders from flight control.

On the morning of Friday, April 17, the three Apollo 13 astronauts returned to their dead command module for the first time since evacuating it on Monday. They powered it up, then jettisoned the dead service module followed by their lunar module lifeboat, and reentered the atmosphere. They landed safely southeast of American Samoa less than four miles from the recovery carrier *Iwo Jima*.

President Nixon flew to Houston to award the Presidential Medal of Freedom to the Apollo 13 operations team. Accepting the award on behalf of the team were Sig Sjoberg and four flight directors—Gene Kranz, Glynn Lunney, Gerry Griffin, and Milt Windler. They were honored at a ceremony on the front lawn of the MSC on April 18.

Even before Apollo 13 launched, NASA had canceled Apollo 20, then scheduled for 1974. Its Saturn V would be needed for the new "dry" orbital workshop of the Apollo Applications Program, now named Skylab. In September 1970 the agency canceled two more landings, meaning that the program would end in December 1972 with Apollo 17.

One factor was institutional fear. Many NASA leaders had been shaken by the close call on Apollo 13. What if the astronauts had died? A horrifying public disaster might have ended American manned spaceflight permanently. And the length and complexity of lunar missions was only going to increase; scientist-astronaut Jack Schmitt was already leading a clandestine effort to send the last landing mission to the Moon's farside.

To the public, it seemed as though NASA had rolled the dice with Apollo. It was time to collect the winnings and leave the table. Abbey was less sure. He had faith in Apollo's hardware and wanted to push it as far and as long as it would go.

The main reason for the decision was money. Had Apollo 18 and 19 remained on the schedule, they would have been launched in late 1973 and early 1974, after Skylab—forcing NASA to carry the Apollo workforce for two more years. The future of the program was the upcoming space shuttle. Better to end Apollo in 1972 and redirect money and personnel to the new, reusable vehicle that would be its future for the next generation.

One casualty of the decision was administrator Thomas Paine, who in July 1970 announced his intent to resign. In mid-September, he was replaced on an acting basis by George Low.

21 | Stability

SINCE ABBEY'S ARRIVAL IN the fall of 1964, the area around the MSC had grown explosively. Housing developments had sprouted seemingly within days, certainly weeks. Stores were built. Streets were created. And people moved in. Thousands of them. "Everyone was either NASA or a contractor, or an astronaut," Suzanne Abbey remembers. It looked like a typical suburb, with the exception of the frequent roars of low-flying T-38s and other aircraft. She remembered the community as "like *Leave It to Beaver*," that television series about a happy American family in an idyllic unnamed small town. "We had birthday parties and Easter Egg hunts and parades, and baseball games." The Abbey neighbors were largely part of the space community, too. Engineer Charlie Haines lived close by.

And as 1970 turned into 1971, George Abbey's professional and personal life found a kind of rhythm and stability. George Jr. was thirteen, Joyce BK nine, Suzanne six, James three. Sundays had a routine that recalled Brenta Abbey's of the 1930s and 1940s in spirit if not in detail. There was church in the morning, then afternoon sports on television. Astronaut Jack Schmitt, a bachelor, frequently joined the Abbeys for viewing. Abbey had continued Brenta's ritual of Sunday night family dinners, which would start with Marlin Perkins's *Wild Kingdom* television show, followed by *The Wonderful World of Disney*. Then came one of Abbey's dishes. He was particularly fond of salmon from the Northwest.

There were other family activities, of course, notably the saga of Henry the duck. According to Suzanne, "We rescued this poor duck that had a birth defect—couldn't flap his wings, since they were stuck and upside down. One day he just disappeared from our yard. We were devastated and thought he'd been eaten or stolen." Some weeks later, her father was driving on the center grounds with George Low. Seeing the duck, he shouted, "It's Henry!" Low was startled, looking around for a man by that name. Abbey opened the car door and Henry hopped in. Abbey brought Henry home for a few days, but then the family decided that the center was the best home for a flightless duck, and they returned him to the big pond there.

Family fun aside, in those days Abbey was the classic workaholic. As Apollo 13 showed, he was on call twenty-four hours a day. According to Suzanne, "We knew that when Dad said he was 'going to go across the street' to the center, it meant he could be gone all night.

"It was tough on my Mom, because she was interested in what he did, but when he came home, he was usually too tired to share—he wanted to think about anything but work. So she would hear things elsewhere."

It got so bad that during a gathering at a neighbor's home, James Abbey ran through the living room, passing a man in a dark crew cut, shouting, "Hi, Daddy!" The man was Jack Schmitt, not George Abbey. Abbey and other witnesses were amused.

Abbey made periodic attempts to expose the family to his work. James Abbey remembers going to Ellington one morning to watch astronaut John Young flying the lunar landing training vehicle.

Flying was another commitment that took up his time. Since being given access to the NASA T-38s, he took advantage of every opportunity to fly.

Abbey would join old friends from Boeing on fishing trips to Westport and Neah Bay, small towns on the coast of Washington state. At Westport, Abbey met a decorated World War II vet named Ted Holand, who was a Union Oil distributor for southwest Washington. Abbey and several fishing buddies kept a boat at the Union Oil dock in Westport for several years. Through Holand, Abbey became friends with Bob Bush, another World War II veteran, who had volunteered for service at the age of seventeen—becoming a corpsman because his mother didn't want him in combat—and who had won the Congressional Medal of Honor for combat in Okinawa.

James Fletcher was appointed the fourth NASA administrator and took office on April 27, 1971. Returning to his post as deputy administrator, George Low approached Abbey about coming to Washington to work for him and Fletcher. With Gilruth's encouragement, Abbey flew up to meet with Fletcher. His interview went well. He could certainly work with the new administrator. And Low was there too.

And yet, Apollo was still flying, Skylab was to come, the shuttle was imminent. All that action was in Houston, where he was happily working for Gilruth and Kraft and with Slayton, Berry, and Faget. He was content to be, in George Jeffs's term, "the quintessential staffer." Houston was exciting. Houston was operations. Every day brought something new. There was no better place to be.

22 | Gravity

"A YEAR AGO," Dave Scott told George Abbey, "I was launching to the Moon on Apollo 15. Now look at me."

It was Wednesday, July 26, 1972. Abbey and Scott had just landed a NASA T-38 at Los Angeles International Airport. They had flown from Houston to visit North American Rockwell to discuss the new Apollo-Soyuz Test Project (ASTP) and NAR's role in the project. Two months earlier, US president Richard Nixon and Soviet general secretary Leonid Brezhnev had signed an agreement committing the nations to a joint human spaceflight in July 1975. An American Apollo would dock in orbit with a Soviet Soyuz. In addition to providing the Apollo command and service modules, North American Rockwell would supply a docking module to allow the American craft to link up with its smaller Soviet partner.

It was a busy time at the Manned Spacecraft Center, at NASA Marshall and the Cape, and in Downey. After the Apollo 13 accident and investigation, Apollo returned to flight in February 1971 with Apollo 14 astronauts Al Shepard and Ed Mitchell landing on the Moon at Fra Mauro.

That summer, in June 1971, Apollo 15 became the first extended lunar mission, with Dave Scott and Jim Irwin spending three days—twice the previous duration—at Hadley Rille on the Moon's surface, conducting multiple EVAs and driving to far-off sites with a lunar rover. The mission had gone smoothly, though the lunar activities put a huge strain on the crew. Jim Irwin, in particular, showed signs of heart trouble.

More recently, John Young and Charles Duke had accomplished a second extended lunar mission on Apollo 16, setting their LM down in the Descartes Highlands. Their landing had been delayed by uncertainty about the rocket on their service module, which had suffered an unusual vibration and failure of a backup system during a burn. Mission rules dictated a cancellation of the landing attempt, but program manager Jim McDivitt and the mission control team, looking at the data, chose to go ahead. Abbey's fondest memory of Apollo 16 was a deep space EVA by command module Ken Mattingly; they had become good friends and neighbors since Apollo 13 two years earlier.

That winter, the center had experienced its first change of leaders. Beloved Robert Gilruth stepped down as director in January 1972, with Frank Bogart taking retirement at the same time. Gilruth moved to a different office on the eighth floor, where he would continue to advise Kraft, James Fletcher, and George Low. Kraft in turn brought Minnesota native Sig Sjoberg from flight operations to the ninth floor as his deputy director. At fifty-two, Sjoberg was five years Kraft's senior, but Sjoberg had largely worked under Kraft since Mercury. He was a capable engineer but, more important, also cool and reserved, a calm counterpoint to the mercurial Kraft.

As Dick Truly said, "Chris was brilliant, but Sig was wise."

Abbey agreed. "Kraft began to rise in NASA as soon as he teamed up with Sig."

The new leaders still had the Apollo 17 lunar landing mission scheduled for December 1972, with Abbey's friend Jack Schmitt in the crew.

The year 1973 would see the launch of the Skylab orbital workshop on the last Saturn V, to be followed by three different visiting missions ranging from twenty-eight to fifty-six or more days in duration.

A year earlier, in July 1971, David Scott had commanded Apollo 15, which had proved to be both a spectacular and a popular technical achievement. He seemed to have unlimited potential.

But in the spring of 1972, news reports revealed that the entire Apollo 15 crew—Scott, command module pilot Al Worden, and lunar module pilot Jim Irwin—had carried hundreds of specially printed commemorative postage stamp covers (envelopes) to the Moon. According to Worden, the understanding between the crew and the dealer who supplied the covers was that none would be sold until after the astronauts had left NASA. A dealer put dozens of the covers on the market, however, embarrassing the astronauts, the space agency, and especially officials in Houston.

No one was angrier than Deke Slayton, who had always operated on the assumption that his astronauts knew how to follow the rules. He summarily fired Scott, Worden, and Irwin, who had been assigned as the backup crew for Apollo 17. Irwin had been planning to retire; Slayton told him to "feel free" to move that forward. He told Worden, "Well, Al, the good news is that the air force will take you back." With no desire to return full-time to the service, Worden found a haven at NASA Ames.

Scott was transferred out of the astronaut office and given a desk job as technical assistant to the Apollo spacecraft program manager.

Abbey felt the crew had betrayed Deke Slayton's trust. They purposely did not disclose that they were carrying the covers, nor that they would be sold for profit. Astronaut Pete Conrad felt that Scott had instigated the whole affair and that Worden and Irwin had just followed his lead. He held Scott responsible for constraints that were subsequently placed on other astronauts concerning personal items they could take on spaceflights. He was so incensed that he never spoke to Scott again.

Unbeknownst to Abbey and Scott, NASA had chosen July 26 as the day it would release the name of the company winning the space shuttle contract. For the better part of the year, the two companies that had built the spacecraft that won the Moon race—command module builder Rockwell International and lunar module builder Grumman—had been competing for the $2.6 billion award. Rockwell had just won. As soon as they arrived in Downey, Abbey and Scott found themselves invited to a huge celebratory party.

23 | The Space Shuttle

ABBEY HAD TAKEN PART in shuttle development talks as early as fall 1969, when he first became Gilruth's assistant. At that time NASA hoped to develop a reusable vehicle that would have two stages—a winged orbiter and a manned launcher that would carry it to altitudes as great as two hundred thousand feet before separating and returning to the Cape. To lead this effort, in early 1970 Gilruth selected Robert Thompson, another Langley veteran who had worked on Mercury, Gemini, and Apollo, and most recently as manager of the Skylab program.

Abbey liked the idea of the flyback booster and the advantages of a totally reusable spacecraft and booster.

McDonnell Douglas and Rockwell were given the study contracts, and as the hard numbers emerged by May 1971 it was clear that future NASA budgets would not support the two-stage, fully reusable design, and the agency was forced to consider using expendable components that would reduce development costs. By the end of 1971 the most cost-effective design had evolved—a vertically launched reusable orbiter and its three engines mounted on an expendable external tank carrying liquid oxygen and liquid hydrogen, flanked by booster rockets. By carrying the liquid fuel and oxidizer in the external tank, designers were able to reduce the size of the orbiter. This option also made the propellant tanks simpler and less expensive. The shuttle was intended to support not only NASA missions but also those of the Department of Defense. Two of DOD's requirements had a significant impact on the new design, notably the size and capability of the payload bay (now fifteen by sixty feet and carrying up to sixty-five thousand pounds) and the orbiter itself, which also needed to have a cross range of eleven hundred nautical miles—that is, to be able to maneuver to a landing site eleven hundred nautical miles east or west of its reentry point on a polar mission.

During the first week in January 1972, James Fletcher and George Low flew to the Western White House, President Nixon's home base in California, and won his approval for the space shuttle program. Left open was the question of using pressure-fed liquid-fuel strap-on boosters (preferred by Robert Gilruth and Max Faget) or those using solid fuel. Faget also questioned the need for an eleven hundred nautical mile cross range, since that required the orbiter to have a large delta wing. He lobbied for a straight wing model that had lower landing speeds.

Abbey agreed with Gilruth and Faget, and with Werner von Braun, who frequently said, "Solid-fuel rockets are not suitable for human flight." Liquid-fueled rockets could be tested before launch; they could be shut down if problems developed prior to launch—and even after launch, with the shuttle diverted to an overseas airstrip. Solid-fueled rockets could not be tested or shut down.

In March 1972, based on presumed cost considerations, NASA decided to proceed with a shuttle using solid-fueled boosters. The DOD cross range requirement was also accepted, meaning that the orbiter would be a delta-winged vehicle. The decision made Gilruth, Kraft, and other center leaders unhappy, but it was out of their hands.

The agency then sought a prime contractor, with Grumman, Lockheed, McDonnell Douglas, and Rockwell all joining the bidding. The source selection board, which Thompson chaired, opted for Rockwell.

Albert Crews had served on the NASA technical committee evaluating the Grumman proposal, and he was disappointed with the choice of Rockwell. "The Grumman proposal was much better than the other two." What Crews didn't understand was that technical issues were not the sole reason to select one contractor over another. Yes, Grumman's proposal was the closest to the final design—but only because NASA had asked Grumman to explore that model.

Abbey approved of the choice of Rockwell. After its tragic failures in 1965–1967, the company had proved itself, building a competent technical and managerial team under George Jeffs. Grumman had done well on the LM, of course, but NASA still had lingering doubts about its overall strength.

With Rockwell in place as the prime contractor, Rocketdyne designing and building the main engines, Martin Marietta building the external tank, and Morton Thiokol developing the solid rocket boosters, Thompson assembled his team in Houston and in Huntsville at NASA Marshall.

As in Apollo, where Sam Phillips at headquarters was the primary authority for the program, for the shuttle HQ was designated Level I, meaning that Robert Thompson—and Kraft—made day-to-day decisions but only with Level II authority.

To Abbey, Apollo was simplicity itself compared to the shuttle. The variety of different major hardware developments was proof of that, and so was the need to make the orbiter, engines, and boosters reusable. It wasn't just these challenges. There were also computer hardware and software, and the tricky thermal protection system for the orbiter. This element alone would ultimately delay the first shuttle launch by eighteen months. Abbey took part in all of Kraft and Sjoberg's meetings with Thompson and his team, and quickly developed reservations.

George Low and Max Faget had created a structure of NASA managers for every subsystem inside the Apollo command and service modules. Each manager had been responsible for development, budget, schedule, and production of the subsystem, and Low had held each personally accountable.

A similar structure was put in place for the space shuttle, but NASA subsystem managers found that they had less latitude than their counterparts on Apollo, and the process for implementing changes proved to be lengthy and difficult. There was constant pressure on budget and schedule from NASA headquarters. The 1970s were a time of high inflation, and the money allotted to the shuttle program kept losing purchasing power—but the program management structure also lacked flexibility and timeliness.

Abbey voiced his opinions in private. One lesson he had learned from his air force days and time at NASA: you gain nothing by debating in public, especially after a decision has been made.

24 | The Soviets

AS ABBEY AND SCOTT were paying their visit to Downey and Rockwell, Abbey and the family were adjusting to a new arrival. Joyce had given birth to their fifth child, their third son, on April 1, 1972.

Earlier in the year, on January 20, Abbey's close friend Stu Present and another NASA pilot, Mark Heath, had taken off from Ellington in a T-38. They were evaluating the value of the T-38 as a simulator for shuttle landings. After a number of successful approaches to an airfield at Matagorda Island—a frequent T-38 destination—fog suddenly rolled in, obscuring the runway. The T-38 struck the ground. Present and Heath were killed.

The new son was named Andrew Stuart Richard Abbey—the Stuart in memory of Stu Present. With this addition to the family, Abbey realized it was time to search for a newer, larger house.

Although it was still difficult for Abbey himself to take vacations—family trips were generally weekenders—the children spent several weeks each summer in the Pacific Northwest, not just Seattle but also southern Canada, visiting with their father and grandfather's old friends. There were fewer trips east to see the Widermans, Joyce's parents, largely because they preferred to visit Houston.

Whenever there were family trips, however, Abbey would happily get behind the wheel of his '55 Olds, the same car he had acquired twenty years earlier. He would also don a cowboy hat—as a fan of John Ford's Western movies, he had eagerly adopted this Texas affectation after moving to Houston.

When he was able to leave the ninth floor and head home for dinner, Abbey would stop the Olds at the corner of Hereford Lane, where the house was, and Saxony, the nearest cross street. Suzanne or Jimmy would be waiting, eager to jump in the Olds and "drive" the rest of the way to the house.

Suzanne remembers, "I was kind of a tomboy and I just lived to take trips with my dad. On Saturdays he would take me to the center, though every couple of weeks we would make a stop at Mark & Stricks barbershop. Just walking into building 2 was exciting—you had to go up these steps and make a long walk to the entrance. That building, and all the other NASA buildings of that era, had a specific smell that I can remember to this day . . . and have never experienced anywhere else." Whatever the source, it was at least partly from the cigarette smoke. Abbey did not smoke, but many of his NASA colleagues did. "We would usually stop at the vending machines on the first floor. Dad would buy me peanuts."

Although a cold warrior by training and experience, Abbey had grown intrigued with possible joint US-Soviet space programs as far back as September 1963, when President Kennedy—speaking to the United Nations in the wake of the Cuban missile crisis and a new test ban treaty—first raised the idea of the United States and Soviet Union conducting a joint mission to the Moon. This had been either ignored (by the Soviets) or alternately embraced and condemned in the United States. After JFK's death in 1964, Congress had added a rider to the fiscal year 1965 budget forbidding the expenditure of any NASA funds on a joint lunar landing.

Eventually, however, once the United States beat the Soviets to the Moon, NASA administrator Thomas Paine initiated a private correspondence with Mstislav Keldysh of the Soviet Academy of Sciences to discuss future joint programs in vague terms.

By 1970, largely initiated by George Low, the number and type of contacts had increased, and in late October of that year, a small NASA team led by Robert Gilruth flew to Moscow. By the time the working session ended on October 28, Gilruth's team and the Soviets had taken the first steps toward the development of a joint docking system for possible future space rescues. There were more discussions on the American side, within NASA, and between NASA and the Nixon White House and the Department of State.

Low made a second visit to Moscow in January 1971. Abbey was fascinated by Gilruth and Kraft's reports of their meetings with the Soviets, and the possible development of a common docking system. Follow-on visits by the Soviets to Houston would give Abbey the chance to meet and work with cosmonauts and engineers. He was beginning to believe that any large future space program—a space station, a return to the Moon, a flight to Mars—would be an international effort, just as JFK had suggested in that 1963 speech.

Apollo-Soyuz would be the last NASA manned flight for several years, and throughout most of 1972 one major question remained: Who would fly it? In addition to twenty experienced astronauts, the agency had twenty who had not flown in space. Some veterans were unavailable or not interested, though at least two, Apollo 13's Jack Swigert and Skylab's Jack Lousma, had begun taking private Russian language lessons in hopes of an ASTP assignment. And all the unflowns were eager. Tom Stafford, Slayton's deputy in flight crew operations, received personal visits from most of them, pitching for a seat.

Then there was the case of Slayton himself—grounded for an irregular heartbeat in 1962 and forbidden to fly aircraft without a second pilot, he had begun a regime of vitamin therapy after contracting a cold during the Apollo 13 crisis in April 1970. By late 1971 he noticed that his irregular heartbeat seemed to have vanished. He arranged for tests at the Mayo Clinic, and by early 1972 was able to convince NASA doctors that he was eligible for solo aircraft operations—and thus flight assignment. Kraft restored Slayton to full flight status in March 1972; Slayton promptly recommended himself for assignment to the "next available" mission, which happened to be ASTP. He was eager to command the flight.

Kraft felt strongly that Slayton should have the chance to fly but not as commander; he lacked spaceflight experience. Tom Stafford, on the other hand, was experienced and could work with Slayton as a member of his crew.

Swigert's chance at a second flight vanished about this time. In the wake of the Apollo 15 postage stamp mess, Chris Kraft ordered all astronauts to disclose whether or not they had signed similar deals. Many of the flown Apollo astronauts had indeed—though not on the scale of Apollo 15.

According to Walt Cunningham in his book *The All-American Boys*, Swigert "at first denied having signed the stamps. . . . A few weeks later, Jack had a change of heart or conscience, or maybe someone advised him to clear the air. He went to George Low, the acting administrator of NASA, and admitted that he'd signed, but said he'd given the money to charity." With that information, Low and Kraft realized that they could not consider Swigert for Apollo-Soyuz.

In April 1973 Swigert left JSC to become executive staff director of the House Committee on Science and Astronautics for the US House of Representatives.

In January 1973 Kraft named Tom Stafford, Vance Brand, and Dick Slayton as the prime crew, with Skylab and Apollo veterans Alan Bean, Ron Evans, and Jack Lousma as their backups.

Slayton was disappointed not to be in command, but after ten years of watching his astronauts flying missions, he was happy for any opportunity. He joined Stafford and Brand in Russian language lessons.

At this time, the Manned Spacecraft Center changed its name. Former president Lyndon Johnson died on January 22, 1973. Members of the Texas congressional delegation sought to honor the man who had been one of the earliest supporters of the manned space program and NASA by naming the Houston center after him. President Nixon signed the resolution on February 19. The official dedication would take place on August 27, 1973, what would have been Johnson's sixty-fifth birthday. Lady Bird Johnson, his widow, attended the dedication ceremony in Houston.

Announcement of the Soviet crew for Apollo-Soyuz took place at the Paris Air Show in May—Soyuz would be commanded by veteran space walker Aleksey Leonov, with another veteran, Valery Kubasov, as flight engineer. Since the Soviets promised to have a second Soyuz on standby in case Apollo was delayed, another pair of veterans was revealed: commander Anatoly Filipchenko and flight engineer Nikolay Rukavishnikov.

At the time of the announcement—indeed, during all the Apollo-Soyuz discussions—the Soviet-manned program was essentially on hold. On June 30,

1971—a month before the launch of Scott's Apollo 15—three cosmonauts had died during their return from the world's first space station mission.

Georgy Dobrovolsky, Vladislav Volkov, and Viktor Patsayev had been launched aboard Soyuz 11 on June 6, 1971, rendezvousing and docking with the Salyut space station two days later. This alone was an achievement; the previous mission, Soyuz 10, had accomplished a rendezvous and soft docking on April 12, then had to disengage and make an emergency return to earth without its crew ever entering the station. The Soyuz 11 cosmonauts spent twenty-three days in space, a record at the time, but a leak caused their Soyuz return capsule to depressurize. Due to the cramped interior volume of their vehicle, the cosmonauts were not wearing pressure suits, and they perished.

The flaw in the Soyuz return vehicle—a valve that opened prematurely—was relatively easy to diagnose and repair, and the Soviets were ready to resume manned missions in July 1972. But then Salyut 2 failed to reach orbit. In the spring of 1973 two other space stations also failed for various reasons—leaving Leonov and Kubasov available for ASTP.

The Soviet cosmonauts and their support team arrived in Houston on July 8 to begin serious training, to be followed by a trip to Downey. It was during this session that Abbey began to learn about the technical challenges in the mission plans: Soyuz was limited in its ability to remain in space (six days maximum), its ability to maneuver, and even in its orbital altitude. The active role in the docking would be Apollo's responsibility—as was construction of the docking module, which astronauts and cosmonauts saw in Downey during the summer 1973 trip.

In the fall of 1973, the Soviet docking module team was given an office in building 13 at JSC. They also rented several apartments in a complex in nearby Webster, a "redneck area" according to Abbey. He feared the neighbors' reactions when on October 6, the anniversary of the Russian Revolution, Vladimir Syromyatnikov, the young docking specialist, displayed a Russian flag. That evening Syromyatnikov invited his neighbors, his NASA colleagues and Abbey, to the celebration, with all attendees sampling the limitless supply of vodka and singing Soviet songs.

Abbey and Syromyatnikov would become close friends. The Russian was not only a soccer fan but also a skilled player and years later, during the Shuttle-Mir program, would play matches with Andrew Abbey's team.

25 | Skylab

APOLLO 17, THE LAST lunar landing mission, launched on the evening of December 6, 1972. Jack Schmitt was the lunar module pilot, landing on the Moon near Taurus-Littrow Valley with commander Gene Cernan.

All of the Apollo landing crews after Apollo 12 had undergone extensive field geology training under Gordon Swann of the US Geological Survey. Abbey had gotten to know Swann, Bill Muehlberger, and Lee Silver, geologists who supported lunar surface activities during each Apollo mission and trained the crews, largely in the southwestern United States. When Silver planned a trip to New Mexico's San Pedro Parks Wilderness, he invited Abbey and George Jr. to go along. Abbey would remember it as a fantastic trip for both, and instilled in him a love for New Mexico's beauty, culture, and people.

And, realizing the value of this field training, Abbey supported Silver in pushing for realism in these outings, and more of them. "He knew the astronauts were smart cookies," Silver said. "They could absorb a lot." He judged Abbey to be "better at matching humans with technical challenges than anyone I ever met."

Another phase of astronaut training ended with Apollo 17, though this passed into history with few regrets. In the mid-1960s, the agency had developed the "flying bedstead," a jet-powered lunar landing training vehicle (LLTV) that would give lunar module pilots some experience with a vertical terminal approach to a landing site.

Trouble was, the LLTV—and its early version, the lunar landing research vehicle (LLRV)—failed all too frequently. Neil Armstrong barely escaped death in May 1968, forced to eject from one LLRV after an engine failure. After that accident, Slayton and aircraft ops chief Joe Algranti decided to treat LLTV flights like space missions, appointing a flight director, Charlie Haines. Nevertheless, Algranti had to eject from LLTV 1 in December of that year. And in January 1971 Stu Present had had to eject from LLTV 2.

Apollo commanders Shepard, Scott, Young, and Cernan had continued to use the vehicles, but the moment Cernan made his last flight in late November 1972, Kraft expressed a desire to visit Ellington personally and chain the LLTV to the ground.

Before ASTP could fly, NASA faced the challenge of Skylab. Based on the S-IVB upper stage for Saturn, the orbital workshop was an eighty-five-ton life science laboratory, earth resources facility, and observatory designed to host three different crews for missions lasting up to three months.

When launched on May 14, 1973, atop the last Saturn V, there was an immediate problem: one minute into flight the micrometeorite shield designed to protect the lab on orbit deployed prematurely and tore away. When it did, it took one of the two solar arrays with it. Without that array, Skylab lost much of its electrical power. Without the micrometeorite shield, the station was less protected against intense solar heating. Flight controllers, notably lead flight director Don Puddy, didn't realize these problems for several hours.

The whole series of Skylab missions was at risk. The launch of the first crew—Pete Conrad, Paul Weitz, and Joe Kerwin—was postponed as the mission control team and JSC engineering examined the telemetry. They learned that debris from the lost micrometeorite shield was preventing the deployment of the one surviving solar array, and they began to consider repairs.

Jack Kinzler, chief of the center's technical services, came up with the idea of building a sunshade that the astronauts could deliver to Skylab and, once inside, extend it through a scientific airlock and open it like a parasol to shade the exterior and lower temperatures inside. Kinzler and his team had been responsible for such items as the American flag erected on the Moon by Armstrong and Aldrin and the lunar golf club swung by Al Shepard. "Kinzler and his deputy, Dave McCaw, were old NACA guys who had started out as model builders," Abbey remembered. "They didn't have degrees, but they could design and build anything."

A good friend of Skylab commander Conrad, Kinzler had started thinking about the micrometeorite problem as soon as he learned of it. Abbey joined Kraft and Sjoberg on visits to Kinzler and McCaw, approving their concepts for the parasol and its swift construction. The tools were stuffed into the crammed Apollo command module and the crew equipped with fresh checklists.

Skylab 2 launched on May 25, 1973, ten days behind schedule, making a fast rendezvous with the damaged orbiting station. Conrad's crew performed a soft dock to allow for rest and refreshment before tackling the exhausting work of repairing Skylab. They erected Kinzler's parasol, which did exactly what it was supposed to do, shielding Skylab from the sun and lowering temperatures in the lab.

Two weeks after their arrival, Conrad and Kerwin performed an EVA that freed the stuck solar panel, restoring much of Skylab's power, though there was—as always seemed to be the case with EVAs—a wild moment: Conrad attached a hook with a long rope to the strap that had restrained the solar panel; then he and Kerwin crouched under the rope to apply more pressure on the strap. The rope broke free, flinging Conrad and Kerwin into space. They were, of course, tethered to the station and were eventually able to pull themselves back to handholds. They were rewarded by a beautiful view of the solar panel deploying.

With cooler temperatures and power, Conrad, Kerwin, and Weitz were able to press on with their mission, spending a full twenty-eight days aboard Skylab (a space duration world record) and completing 392 hours of experiments, taking twenty-nine thousand images of the Sun with the Apollo Telescope Mount as well as performing medical tests. They landed in the Pacific on June 22, 1973.

Skylab 3 launched five weeks later, on July 28, 1973. The crew of Al Bean, Owen Garriott, and Jack Lousma docked safely that day, though their service module developed a propellant leak in a reaction control thruster that caused concerns in mission control. The new crew successfully erected an improved sun shade while continuing and expanding on the series of experiments begun by Conrad's crew.

Astronaut Robert Parker served as "science czar," helping the crew with scheduling and priorities—a task that Abbey would later reuse for the International Space Station.

The Skylab 3 crew's mission went so smoothly that, after its bumpy start, its duration was expanded by three days to fifty-nine. Bean, Garriott, and Lousma returned on September 25, 1973. NASA doctors could find no barriers to even longer stays in space.

The third and final crew, Skylab 4's Gerald Carr, Ed Gibson, and Bill Pogue, lifted off on November 16 for a mission scheduled to last at least sixty days, though program managers and the crew hoped to remain in space for as many as eighty-four days.

The crew reached the station safely on the first day, though Pogue was suffering from nausea and fatigue, a condition later to be diagnosed as space adaptation syndrome. The condition was related to motion or seasickness—ironic for Pogue, who had come to NASA after a tour with the air force Thunderbirds' aerobatic team.

Worse yet, the crew—all rookies and perhaps all too aware that they were conducting NASA's last major spaceflight for years to come—hid Pogue's condition from mission control, only to have the information discovered anyway. (Their "private" conversations had been recorded.)

Suitably chastened, Carr, Gibson, and Pogue tried to be good soldiers, diligently adhering to a crowded flight plan until early December. Ed Gibson noted that he felt as though he had been through a "33-day fire drill." Pogue realized that he was unable to complete solar photography assignments. Complicating matters was pressure from the astronomical community as well as NASA itself for photographs of comet Kohoutek, which would only be visible for a limited time.

Finally, on December 28, commander Carr made a statement to mission control, where capcom Dick Truly, JSC director Chris Kraft, and flight crew ops chief Deke Slayton were present. "We need more time to rest," he announced.

"We need a schedule that's not quite so packed. We don't want to exercise after a meal. We need to get the pace of things under control." He requested that the ground teams spend the next loss of signal period mulling these requests.

Twenty minutes later, when Skylab was back in contact, Truly relayed some of mission control's concerns. Carr would say years later that Truly "bent our ear with all of the things that we were doing, including our rigidity that made it difficult for them to have the flexibility to schedule us how they needed to."

Truly made sure to add, however, that Kraft and Slayton were pleased with the crew's performance. And the next morning mission controllers modified the Skylab schedule, promising to add flexibility to any events that did not require specific timing—giving the crew a "shopping list" to be completed when possible—and promising to avoid contact during mealtimes. Further, the crew would not have tasks scheduled after dinner.

The exchange, intended to be kept between the crew and mission control, wound up in the news, where it was labeled a "mutiny." Though perhaps the word *mutiny* was overexaggerated, the crew's problems served as the basis for a Harvard Business School case study.

While Abbey sympathized with the crew, he believed that Skylab was the last such opportunity Americans would have to do long-duration research for the next twenty years. The crew should have made it work.

Carr, Gibson, and Pogue worked through their final five weeks on orbit, ultimately accomplishing all the mission objectives—medical studies, astronomical observations, Earth resources studies, four EVAs—over six thousand astronaut-hours of work. They splashed down in the Pacific off the coast of California on February 8, 1974, after a total mission duration of eighty-four days—a record for the longest spaceflight that would stand until the Soviets broke it in 1977. No American would log that much time in space until 1995.

Just like that, it seemed, Skylab was over—first launch in May, last splashdown in February.

Abbey worked with Kraft and Sjoberg in reviewing astronaut photographs from space. The Skylab 4 imagery presented a special challenge.

All through Gemini, Apollo, and Skylab, crews had been warned not to take pictures of a certain part of central Nevada. Known years later as Area 51, it was home to the Groom Lake air base, testing site for classified American aircraft like the U-2 and SR-71, as well as home to America's collection of captured or stolen Soviet aircraft.

Carr, Gibson, and Pogue forgot, and here were several clear images of the region and its classified air base. NASA could have just deleted them from the release, but each image had a specific sequential frame number—some curious journalist might notice the gap and start asking questions. Given the volume of Skylab photos, there was a better chance that the forbidden image might just go unquestioned. Kraft concluded that that was the best approach. Sure enough, Skylab's Area 51 images remained unnoticed for forty years.

26 The Ace Moving Company

"WE GENERALLY START OUR days at 8:30."

It was a Monday morning in July 1973. George Abbey was calling from JSC headquarters building 1, and astronaut Dick Truly answered the phone in his fourth-floor office in building 4. He realized that Abbey somehow expected him to be on the ninth floor of building 1.

Truly had been in Houston for four years, mainly working on Skylab, including recent duty as capcom. He had first gotten to know Abbey three years earlier. Truly and scientist-astronaut Joe Allen had been tasked with supervising a psychological experiment for the Skylab crews. Having just come from MOL, where astronauts trained for a serious national security mission, Truly found the experiment trivial.

"Let's talk to George Abbey," Allen suggested. Truly only knew Abbey as one of the ninth-floor assistants. They met at a bar in Clear Lake for a few beers, and Truly unloaded on the idiocy of this experiment. Abbey, he noted, smiled but said little. The very next morning Truly and Allen learned that the experiment had been canceled, written off, its team to be redeployed. He surmised that Abbey had more power than he'd realized.

By July 1973 Truly had spent enough time with Abbey that he stopped being surprised by his means and methods. Even so, Truly had never visited the ninth floor. When he arrived he was impressed by the palatial offices for director Kraft, Sjoberg, and even Abbey.

Abbey was brief. "I'm going on leave and Kraft and Sjoberg have selected you to fill in for me while I'm gone." Kraft had decided—with Abbey's prodding—to reach elsewhere in JSC and bring "bright young people" (Kraft's term) up to the ninth floor for a tour. Truly was thrilled. Abbey took him in to see Kraft, who confirmed the plan.

Then Abbey and Truly went back to Abbey's office. "When are you leaving?" Truly asked.

"Today," Abbey said. He was going to Seattle.

Cheryl Bouillion, who had, as usual, not been informed of Abbey's plans or Truly's arrival, happened to glance out the window. Nine floors below, parked in front of building 2, she saw the Abbey family car—with Joyce and the five kids obviously waiting. She would learn later that Kraft had decided that Abbey was working too hard without a break, so he'd granted him a three-week vacation, provided he could find someone to fill in.

Without another word, Abbey departed, leaving Truly standing in the doorway.

"Cheryl, what do I do?"

She sat him down and began talking him through Abbey's daily activities, the meetings, mail, and calls.

Abbey returned in mid-August, in time for the start of the local school year. Truly stayed in the assistant job for five more months, when he was replaced by another "bright young person."

There would be other temporary assistants, eventually to be known around flight operations as "bubbas." Years later Truly would say that he was the first "bubba."

Truly and Abbey teamed up with other JSC pals in a group that eventually became known as the Ace Moving Company. Members included A. J. "Kojak" Roy, Abbey's friend from aircraft ops; Jay Honeycutt, Truly's successor as bubba; astronauts Robert Crippen and Bill Lenoir; and flight director Phil Shaffer.

"We ran, we played, we drank," Crippen would say later. Their playgrounds were the Outpost Tavern, off the NASA Parkway, just across the road from the center, and Griff's Shillelagh Inn, a sports bar downtown owned by Michael "Griff" Griffin. Griff was well connected to the local community. Houston police officers used to frequent his establishment. On home game Sundays for the Oilers football team, Griff would rent a bus, load it with fans, and with a motorcycle police escort, would deliver his patrons to the entrance of the Astrodome. And to this day the bar is the focal point of the annual St. Patrick's Day parade.

Much of the group's play was sports related. The Ace Moving Company fielded a tug-of-war team coached by Jim Casey, one of the famed Casey brothers who had won a tug-of-war championship in their native Ireland in 1932. His experience allowed Casey to instruct the Ace Moving Company in effective tug-of-war techniques, and with the addition of a few stalwarts, and Abbey as captain, the team won the event at the Houston highland games at Rice University in 1975. They defeated the defending champions, a group of Rice University football linemen.

The Abbeys, Trulys, and Crippens became good friends.

One day in November 1973 astronaut Pete Conrad, veteran of Gemini, the Apollo 12 lunar landing, and the first Skylab mission, came up to Chris Kraft's office to talk about his future. Conrad was forty-three, a navy captain who had been out of the naval line for eleven years. He would be welcomed back in the service, but surely in a job where he would be little more than a "potted plant." (This was Wally Schirra's expression, coined when he explored his own return to the navy.) Conrad would have no real prospects for command responsibility.

Besides, Conrad had a plan. "I want to be the new Deke," he told Kraft. Mercury astronaut Deke Slayton was stepping down as director of flight crew operations to concentrate on ASTP training. Kraft would need to replace him, and who better than Conrad?

But Kraft had other ideas. For years, as a flight director, head of mission control, and now center director, he felt he had been on the short end of astronaut arrogance, whether on flights such as Apollo 7—where Schirra and his crew openly rebelled against and mocked Kraft's teams—or in personal situations. While Kraft liked and admired Slayton personally, he felt the astronaut had been too protective of his men, letting them run wild with aircraft, cars, and women, treating flight surgeons with contempt and ignoring the press whenever possible.

"There isn't going to be another Deke," Kraft told Conrad. He thought the Gemini-Apollo-Skylab veteran was better suited to a senior role in the shuttle program, but Conrad said no. If he wasn't going to be Deke, he was leaving the agency.

And it wasn't just about Conrad. With the third and final Skylab launch in November 1973, only one firm piloted mission remained on NASA's schedule—Apollo-Soyuz in July 1975. Shuttle missions might start in 1978, if then.

Kraft had already given Abbey the unenviable task of figuring out how to combine and consolidate flight crew and flight operations, reconciling their two cultures—the take-the-hill mentality of Slayton's pilots and the systems-oriented flight controllers. Relations between the two groups were frequently tense, and it was only the mutual respect of their leaders—Kraft, then Sjoberg, then Bill Tindall with Slayton—that allowed them to work productively.

With the exception of a crew equipment division, which was transferred to Faget's engineering team, flight crew and flight ops were merged under a single leader in February 1974, shortly after the last Skylab crew splashed down.

To head the new combined organization—replacing Slayton—Kraft selected Kenny Kleinknecht, a forty-five-year-old veteran of the original Space Task Group who had risen through Mercury, Gemini, and Apollo program offices to become manager of the Skylab program. Flight director Gene Kranz was appointed as Kleinknecht's deputy.

The astronaut office was also reorganized for the shuttle, with several astronauts assigned to Thompson's program office and the scientist-astronauts to life and sciences. Mercury and Apollo veteran Al Shepard retired and was soon replaced by John Young. Abbey had suggested Young to Kraft, feeling that the Gemini and Apollo veteran (and favorite of Gilruth) was the best choice to represent the astronaut office in shuttle development.

27 | Apollo-Soyuz

ABBEY'S PARTICIPATION IN ASTP working group sessions in Houston led to a growing friendship with astronaut Tom Stafford.

Soviet space failures were a continuing issue as ASTP moved forward. In August 1974 the Soyuz 15 spacecraft was unable to dock with the Salyut 3 station. And in April 1975, another Soyuz failed to reach orbit, its booster shutting down prematurely, forcing its crew to make an emergency landing.

In Houston the mood was near panic. It was one thing to have the iffy Soviet docking system fail, as it had on Soyuz 15, but to have the primary Soyuz booster fail ten weeks before the scheduled ASTP launches—that was a problem. Sen. William Proxmire of Wisconsin, a longtime NASA critic, threatened to end the joint program.

Fortunately, Soyuz program manager Konstantin Bushuyev contacted George Low within days, discussing the booster failure and relaying the first results of a preliminary investigation that revealed that half of the upper stage pyrotechnics had failed to ignite.

As flight preparations entered their final phase, Stafford insisted on visiting the Soviet launch center at Baikonur. He wanted to see the spacecraft his Apollo would be docking with. "I'm not flying in space on a vehicle I've never been inside."

It would be an unprecedented visit—French president Charles de Gaulle had been the only Westerner previously allowed at the secret center. The NASA team was finally cleared for its trip, flying from the Moscow area to Baikonur on April 28.

For the former MOL astronauts working as support crewmen—Truly, Crippen, Robert Overmyer, and Karol "Bo" Bobko—it was a bit of a homecoming. "We had spent hours looking at imagery of that launch site," Truly would admit years later, after declassification of the GAMBIT and HEXAGON spy satellite programs, and their link to MOL. "So it was great fun for me and Crip to stand there and look toward the horizon where we knew there were pads for the Proton and other vehicles."

On May 24—having investigated their launch failure in record time—the Soviets were back in the manned space business, launching cosmonauts Pyotr Klimuk and Vitaly Sevastiyanov on Soyuz 18. They docked with Salyut 4 two days later and commenced on a mission that was expected to double the previous Soviet endurance record—sixty days, as opposed to thirty. The new crew would be in orbit until late July.

The successful launch reassured Congress and NASA to some degree, but a new problem presented itself: could the Soviets exercise proper control of the ASTP mission with two flights at the same time?

One thing NASA learned, as Stafford and Lunney reported back to various officials about their trips to the Soviet Union, was that that nation's flight control was extremely limited: most manned missions were controlled from Baikonur (as the United States had, from Cape Canaveral with Mercury), then from a site in the Crimea that was actually designed for interplanetary and lunar missions—the equivalent of NASA's Goldstone Observatory in California.

During ASTP meetings, the Soviets had revealed the construction of a new mission control dedicated to human spaceflight. But the new center—soon to be known as TsUP, the center for flight control—was barely functioning, and they would be lucky if it were ready for Apollo-Soyuz in mid-July.

Bushuyev assured Low again: the Salyut 4 mission would use the Crimean facility. TsUP in Kaliningrad was for Apollo-Soyuz.

Abbey would follow the ASTP flight from mission control in Houston. He would not be at the Cape for the launch.

George Knutkowski, Abbey's notorious friend from Annapolis, had not only gone on to a flying career but also attended the air force test pilot school, becoming one of Stafford's students. They also became buddies, though Stafford frequently found himself tidying up after mutual misadventures.

Knutkowski had served in Vietnam, retiring from the air force in 1969 to become chief engineer for road and bridge design for the state of Delaware. When Knutkowski had the opportunity to visit the Kennedy Space Center in Florida around the time of the ASTP launch, Abbey arranged for him to receive a NASA VIP invite to the event. Pleased by the news, Knutkowski left word for Stafford in the astronaut crew quarters, letting his old friend know he was at the center.

Stafford, Brand, and Slayton had just arrived at the Cape in their T-38s after struggling to avoid a large thunderstorm in the area. Receiving Knutkowski's message, Stafford immediately telephoned Abbey, half seriously complaining that Knutkoswki was a bringer of bad luck.

On Tuesday, July 15, 1975, Abbey arose early to watch the launch of Soyuz 19 and cosmonauts Leonov and Kubasov on television from Baikonur—the first Soviet launch ever televised live. It went well, clearing the way for the Apollo launch of Stafford, Brand, and Slayton in the early afternoon, by which time Abbey was in mission control. That launch went perfectly, too, with Apollo heading on a more northerly route—to 51.8 degrees rather than the usual Apollo heading of 28 degrees—giving the astronauts a chance to fly over northern Europe, a first for American space travelers.

Two days later, Apollo and Soyuz linked up over Europe, with Stafford flying the final approach and docking. "Soyuz and Apollo are shaking hands now," Leonov reported. For the next two days, the crews visited each other's vehicles, shared meals, and conducted several global broadcasts.

On the nineteenth, it was time for separation. A second docking exercise was scheduled, this time with Slayton at the controls of the command module. As Apollo approached Soyuz, however, Slayton had the vehicles slightly out of alignment, resulting in a noticeable vibration as the docking collar capture latches grappled. There was no obvious damage, but flight controllers at TsUP were horrified—so much so that twenty years later, they would still refer to

Slayton's "crash" into Soyuz. The two vehicles separated, with Soyuz returning safely on July 21.

Apollo remained in orbit until July 24, firing its service propulsion system for reentry and splashdown. During the descent, communications inside Apollo were distorted by a feedback squeal. Stafford didn't hear a reminder to flip a key switch that would shut down the command module's reaction control thrusters, which used toxic fuel. As Apollo's chutes deployed, a valve opened as planned to equalize pressure between the spacecraft interior and the atmosphere outside. The motion of the command module under the chute triggered firings by the thrusters—with toxic fumes sucked into the cockpit.

Nobody in Houston, not flight director Pete Frank, nor Abbey, observing and listening with growing concern, had any idea what was going on.

Inside Apollo the astronauts were coughing and scrambling for oxygen masks as they splashed down. After several harrowing moments, they were able to secure the masks; a recovery team also opened the hatch, flooding the cockpit with fresh air and dissipating the toxic fumes. The astronauts looked shaky as they arrived aboard the carrier USS *New Orleans*.

The crew was lucky to be alive; a planned flight to Hawaii was canceled. Instead, the carrier steamed to port. Aboard ship the next day, Stafford found that his lungs were filled with fluid and he could barely walk. The immediate postflight ceremonies were put off as the crew recovered in Hawaii.

The incident turned out to be a blessing in disguise for Slayton—the extra tests showed that he had an undetected lung lesion that could have been cancerous. It was removed surgically.

Abbey, like Kraft, Lunney, Kleinknecht, and Young, learned most of these details in the postflight debriefs a week after splashdown. They were shaken to realize how close this supposedly simple mission had come to disaster.

By November 1975 Abbey had been assistant to the center director for six years. Apollo, Skylab, and now Apollo-Soyuz had ended, but the shuttle and possibly a space station program loomed, and then back to the Moon and on to Mars. Abbey was part of the leadership, privy to all but the most personal and private discussions and decisions. He controlled the flow of paper into and out of Kraft's and Sjoberg's offices. He knew the center and its offices and personnel as well as any individual could. He was happy right where he was.

Kraft had other ideas. On several occasions he had suggested that Abbey needed to head up a department, to be responsible for a team, not just be the quintessential staffer. Abbey was hypothetically open to the idea, but until the end of 1975 there was no suitable opening.

Flight operations director Kenny Kleinknecht was a superb engineer, trained at Purdue, a private pilot who flew biplanes for fun. In a distinguished career at NASA he had served as program manager for Mercury and deputy program manager for Gemini and had held important positions in Apollo and Skylab. He was accustomed to organizations that operated on a chain of command, and he had had difficulty managing and consolidating the flight crew and flight operations teams, neither of them formally structured. Before long it was clear to Kraft that Kleinknecht had lost the confidence of Kranz and the other division heads. Kraft later admitted that putting Kleinknecht in the flight operations job was "the worst personnel decision I ever made." He and Sjoberg decided it was time for a change.

When Abbey arrived for work on Thursday, January 8, 1976, he was called into Kraft's office, where Kraft and Sjoberg waited.

"You're no longer my assistant," Kraft told Abbey. "Effective today, you're the new director of flight operations."

Part IV

“Ten Consecutive Miracles Followed by an Act of God”

28 | The Right Stuff

"I WANT YOU TO do one thing," Chris Kraft told Abbey.

"Bring the former flight crew operations and flight control teams together in a single organization that works well with life sciences and the rest of the center."

Abbey knew he had the proper experience to do the job and to carry out Kraft's specific wishes. He was an engineer and his work on Low's board had forced him to interact with life sciences, crew systems, and other center units, including the astronauts; he had spent hours in mission control. His six years under Gilruth and Kraft had exposed him to the larger challenges of managing a complex organization. And he was a pilot who had flown out of Ellington with the Texas Air National Guard, so he not only knew the troops there but also shared the pilot mind-set.

Kraft and Sjoberg needed to replace Abbey. He suggested that they consider Henry "Pete" Clements, a former air force communications specialist who had previously worked at the center as assistant to Kraft, where he played a role in the development of the mission control center. Clements was currently at NASA HQ but eager to return to Houston. He accepted the new assignment and soon was working again for Kraft and Sjoberg.

With that, George Abbey was sent off on the most exciting and challenging assignment of his career.

In January 1976 JSC's flight operations directorate consisted of the astronaut office and four divisions: aircraft operations, crew training and procedures, flight control, and payload operations. There were five hundred people in these organizations.

Abbey's deputy was to be veteran flight director Gene Kranz. The forty-two-year-old Kranz had distinguished himself within Kraft's mission control organization since joining NASA in 1960. He was one of the first three flight directors appointed by Kraft in 1965, and led shifts on Gemini and Apollo 5, 7, and 9, and on Apollo 11's lunar landing. He had played a vital role during the Apollo 13 crisis.

Blunt, openly emotional, capable, and highly regarded for his leadership and team-building (he originated the idea of flight director team colors), Kranz had been Kleinknecht's deputy since 1974 and logically expected to be chosen to replace him. Understandably disappointed, he nevertheless agreed to serve as Abbey's deputy.

Kranz had two assistants, Jones "Joe" Roach and Mel Brooks. Roach, an experienced flight controller, had been involved in the Shuttle Training Aircraft development and had frequently worked through Abbey to bring news—usually bad—to Kraft.

Brooks was dealing with planned shuttle payloads, especially the European Space Agency's Spacelab.

For his part, Abbey knew that mission control was a clannish group with its own methods, pride, and jargon. They would regard him as an outsider, only tolerating him because he had been appointed by Kraft himself. So, Abbey resolved to give Kranz and the flight control division chief, Pete Frank, as much independence and leeway as possible.

Abbey would eventually meet with each division chief, but his next priority, after Kranz, was speaking with John Young, chief of the astronaut office.

On this day in January 1976, there were twenty-nine astronauts, fourteen of them with experience in space. Flown or not, most were assigned to shuttle development in two major areas—engineering under Ken Mattingly, and operations/training under Al Bean, who in turn reported to Young.

Short, dark-haired, with a wry, dry sense of humor, Young was highly regarded within NASA for his engineering skills. He was also famous for lengthy memos on technical matters. Gilruth and Low valued these memos; not everyone at JSC did.

Young had had no warning that Abbey was to be his new boss, but welcomed the news. He had not gotten along with Kleinknecht. Abbey knew Young from Apollo, and through Gilruth, but no better than he knew several other astronauts. He and Young would go on to become close friends and colleagues for the next thirty years.

As for the astronauts themselves, Mattingly was surprised to learn that Abbey was now director of flight ops. "I never thought of George as a fellow pilot. He never told aviator stories. Maybe he thought his experiences, while extensive, just didn't match up."

Dick Truly, however, wasn't surprised that Kraft turned to Abbey. What surprised him was Abbey taking the job. "I thought he preferred to work behind the scenes. With the FOD job, though, you couldn't hide."

That same day, Abbey met with his other division chiefs. Jim Bilodeau of crew training was a fifty-two-year-old former World War II fighter pilot who had joined NASA in 1962. Abbey had gotten to know him during Apollo and anticipated no problems.

Charles Harlan, thirty-eight, was the head of payloads, recently transferred from aircraft operations after years as a flight controller. He and Abbey had worked on projects in the past and had a good relationship.

After the astronaut office, aircraft ops (based at Ellington) was the division Abbey knew best, having flown their planes for close to a decade. Joe Algranti, the chief, was a fifty-three-year-old former navy aviator who had joined the NACA's Lewis Flight Propulsion Laboratory in 1951. He had moved to Houston in 1962 to establish the Manned Spacecraft Center's aircraft support team, which now numbered a hundred pilots, mechanics, and administrative people operating dozens of different aircraft.

Abbey simply inherited this team. For now, no one would be transferred.

He only broke the news to his family when he arrived home that night. George Jr. did not recall his father being outwardly excited about it. "It was just a new assignment."

Abbey moved from the ninth floor down to the eighth, to the corner office formerly occupied by Kleinknecht. The office overlooked building 4, where the astronauts were based. Years later, some astronauts speculated that this was Abbey's cunning design, keeping a symbolic eye on their activities, but the truth was that he simply inherited the office.

He brought Cheryl Bouillion with him and also the most recent bubba, Jay Honeycutt.

Honeycutt was thirty-eight, an electrical engineer by training who had spent nine years in mission control. His first memory of Abbey was seeing him entering the control center one night to sit with the late shift for the last Skylab mission.

After serving as one of the "interfaces" with the Soviet team on Apollo-Soyuz, Honeycutt had been selected by Abbey to be one of Kraft's "bright young people," succeeding Truly. Honeycutt moved to building 2 for what was supposed to be six months. He stayed five years.

Abbey immediately set out to repair the relationship between life sciences and the astronaut office. Ken Mattingly described it as "one stop short of open warfare." Flight surgeons were, of course, the natural enemies of pilots, since they had the power to ground them.

Chuck Berry, the director of life sciences, had left his post as Apollo wound down in the fall of 1972, replaced by Dick Johnston. Johnston remained in the post through Skylab and Apollo-Soyuz, then left the agency.

When Abbey became director of flight operations, there was no official head of life sciences. Searching for a suitable candidate to fill the job, Sig Sjoberg contacted Johnston, who was working in Chicago. He offered a name: "Mine." He returned to JSC and served as head of life sciences until December 1979, in a far more collegial environment.

The shuttle program was moving forward, though not all elements were proceeding at the same pace. The first orbiter—*Enterprise**—was being built at Rockwell's plant 42 in Lancaster, California, not far from Edwards Air Force Base. Four other orbiters were to follow.

* Originally intended to be called *Constitution*, it was renamed at the urging of fans of the *Star Trek* television series, who conducted a letter-writing campaign.

The space shuttle main engines were under development not far away at Rocketdyne in the San Fernando Valley. The critical design review for the high-powered reusable units was scheduled for late summer 1976.

In Houston teams were developing simulators and training procedures, creating software, and integrating potential payloads from ESA, the Department of Defense, the scientific community, and commercial satellite builders.

One essential training tool was the Shuttle Training Aircraft (STA). On returning to Earth and landing, the shuttle orbiter would be what astronauts called "a flying brick." The landing speed would be close to two hundred knots, more than twice that of any aircraft.

Kraft and Abbey realized that pilots needed real-life training, and went in search of an aircraft that could be modified to mimic at least some of the orbiter's characteristics. Various aircraft were considered before the Grumman Gulfstream II corporate jet was selected. Modifications would allow the Gulfstream STA to dive at 1.8 Gs with reverse thrust, approximating the steep approach of an orbiter, then flare out just above the runway. It would also have to have a tandem cockpit, so an astronaut could fly with an instructor.

Begun in late 1972 and managed out of aircraft operations, the STA was originally supposed to cost $3 million. The STA quickly proved to be too much—the Ellington team was used to flying and maintaining planes, not designing and building them. Charles Harlan and Charles Haines of JSC engineering took it over. One constant problem, according to Harlan, was that the shuttle design "was a moving target."

"Right after we got the project, they changed the shuttle cockpit," which required the team to change the left-hand cockpit of the STA. "So that cost us a million dollars right there." Additional digging by Harlan and Haines turned up more overruns, now close to $4 million.

Every visit to the ninth floor—and now they were almost daily—resulted in words of displeasure from Kraft and Sjoberg. Haines said, "God, George Abbey was mad at us every day."

Knowing that most of these issues were due to failures by the contractor, Grumman, Harlan finally stood up and fired back at Kraft, saying "Why are you so mad at us?"

"Because you aren't in here pounding my desk enough!" Kraft snapped.

Sure enough, eventually Kraft aimed his complaints at Joe Gavin, the former Apollo Lunar Module manager, now president of Grumman, in a blistering tirade that shocked Harlan and Haines.

Once that passed, all sides began to concentrate on the real problems—notably, a pair of major technical issues with the STA. Abbey brought in more NASA and Grumman technical people to address the problems; one solution actually came from Kraft himself.

With those issues resolved, the STA quickly reached readiness, with John Young and aircraft ops pilot Dave Griggs conducting the first test fight in June 1976. The aircraft proved to be an outstanding trainer for the next thirty-five years.

As it turned out, the NASA program paid dividends to Grumman's entire G-II fleet. Monitored extensively, the STA endured a severe punishment inflicted by the training regime, with thousands of landing gear cycles compared to normal G-II operations. This extreme usage revealed problems that Grumman engineers could correct with their commercial fleet.

The whole space shuttle concept had been sold politically on two tenets, one economic, one technical, that Abbey did not believe. The economic rationale stated that the more the shuttle flew, the cheaper each flight would be. Reusability was the most frequently cited factor. After all, in Apollo-Saturn, every bit of the vehicle was discarded except for the command module, and that was never used again. Surely reusing the expensive, complex orbiter and the expensive, powerful solid rocket boosters saved money. In theory, yes, though no one in NASA or the contractor teams had actually refurbished a vehicle for a second flight into space. In early 1976 the costs and schedule of this process were largely still theoretical.

The early traffic models for the shuttle program predicted as many as fifty flights a year in the program even though the combined number of American military and scientific launches in 1973, for example, was only half that. Many in Abbey's mission control team believed that the shuttle would be flying two dozen times a year, or even more, and were planning budgets, designing facilities, and hiring personnel based on that figure. Abbey and Young anticipated eight to ten missions a year at best. They realized that the challenge to higher flight rates wasn't just refurbishing the orbiter but preparing the operational team. Instructors and their simulators, flight planners and their procedures, engineering and contractor teams, payloads and experiments—all had to be ready to support each mission. But few in Houston—and no one at HQ—were prepared to accept any challenge to their assumptions.

The technical issue was shuttle safety. JSC's shuttle safety and mission assurance office had performed calculations showing that there was only a "one in a hundred thousand" chance of a catastrophic failure. But Abbey and Young knew that every phase of a shuttle's flight regime—launch, on-orbit operations, and reentry—presented unique challenges to safety. There was no escape system that would protect a crew through all three. During launch, for example, a solid rocket booster could fail, an event that would destroy the vehicle. There was talk of giving each SRB a thrust termination system, which shuttle program manager Bob Thompson judged to be as dangerous as any perceived SRB failure. In essence, if the SRB showed signs of failing (itself a tricky subject and one that required a decision within a handful of seconds), its nose would be blown off—and the same with its twin; otherwise the entire shuttle vehicle would tumble uncontrollably.

"So you've got this cylinder with all this solid propellant burning. You can't snuff it out, can't turn it off. So you've got a bunch of hot exhaust coming out this end," Thompson said. "I could never convince myself you were that much better off to blow the nose off and have exhaust coming out both ends in kind of an uncontrolled fashion."

The three high-powered main engines were another potential danger source. Since they were liquid fueled, they could be shut down if a potential problem were detected in time. In certain circumstances, after the separation of the solid rocket booster, the orbiter might be able to accomplish a "return to launch site" (RTLS)—to shut down a failing engine, shed the external tank, then to pitch over and fly back to the Cape. Young described the successful execution of an RTLS as "ten consecutive miracles followed by an act of God." At a certain point in a troubled ascent, approximately four minutes after liftoff, it was possible for the shuttle to do a transatlantic abort, flying to a predetermined location in Africa or Western Europe.*

* For launches out of Vandenberg, where the shuttle would be flying south from California over the wide Pacific, the abort sites were limited to . . . Easter Island.

But if a main engine blew up, the orbiter was finished.

To give the crew some chance of escape in certain limited emergencies, Abbey and Young fought for and won the installation of two ejection seats. But they were only usable below certain speeds, and some astronauts—notably Mattingly—thought they were a waste of time. And the orbiter only had room for ejection seats for the commander and pilot, meaning that they were only for the ALT and orbital flight test programs.

The plan remained: make the whole stack as safe as possible—"as safe as a Boeing 737."

Astronaut Story Musgrave would remember odd moments where Young, addressing new groups of astronauts, would say, "Y'all are going to be filling up graves at Arlington." He and Abbey could rail about the lack of a crew escape system, but they couldn't get the orbiter redesigned. By 1976 it was far too late for that.

Beginning in 1961, NASA flight crew assignments for the Mercury program were made by Robert Gilruth himself. It was he who chose Alan Shepard to be the first American in space, and John Glenn to be the first to orbit the Earth. However, in the summer of 1962, after Deke Slayton was grounded and given the job "coordinator of astronaut activities," Gilruth ceded the power to him. He reasoned that Slayton knew his colleagues' strengths and weaknesses better than any manager or outsider.

Beginning with Gordo Cooper on the last Mercury, and through the last Skylab crew, flight assignments were Slayton's job. His choices were generally successful, even though some astronauts and program managers grumbled.

Kraft, after consulting with Sjoberg, Kleinknecht, and Abbey, had determined who would fly Apollo-Soyuz. But he felt that it was logical for Abbey, his handpicked director of flight operations, to hold the power to make assignments. Abbey would consult with Young, then present crews for Kraft's approval—with John Yardley at HQ having the ultimate authority.

In his first attempt to assign astronauts, for the shuttle's approach and landing tests, Abbey ran into an immediate bureaucratic roadblock. Astronaut Dave Scott had gone to the NASA Flight Research Center at Edwards as its director in 1973, and felt strongly that the ALT flights should be flown by test pilots from his center.

But Abbey and Kraft countered: it would be JSC astronauts flying the shuttle so it made sense to have them involved from the beginning. And they won the argument. (The whole dispute was complicated by the outright animosity between Scott and Slayton, whom shuttle program manager Thompson had appointed as manager for the ALT flights.)

As commander of the first crew, Abbey chose Fred Haise, who before his flight on Apollo 13 was a highly respected NASA test pilot at the NASA Flight Research Center. Haise had been working in the shuttle project office for the last three years. As his pilot, Abbey assigned Gordon Fullerton, one of the Manned Orbiting Laboratory astronauts who had worked in Apollo support after coming to Houston, then on shuttle. Fullerton had specialized in the orbiter's cockpit layout.

The second crew would be led by Joe Engle, the only NASA astronaut to have earned astronaut wings before coming to the agency, flying the X-15 to an altitude greater than fifty miles. Former MOL astronaut Dick Truly was named Engle's pilot.

Abbey informed each of the astronauts separately, calling them to his office to deliver the good news. This would become his signature method, the sly undersell: "I'm thinking of assigning a crew to this. Are you interested?"

A predictable side effect, of course, was disappointment. Hank Hartsfield was especially upset to be passed over for an ALT assignment. "*I* was the specialist in the Shuttle flight control system."

29 | The Next Generation

IN ADDITION TO THESE ongoing and developing challenges, Abbey faced a completely new one. NASA was about to conduct its first astronaut selection since 1967. Kraft and Abbey assumed that the current group of pilot astronauts was sufficient to crew the approach and landing tests, orbital flight tests, and probably the first few operational flights of the shuttle. Scientist-astronauts could easily provide the mission the specialists needed for the first operational missions too.

Given the projected launch schedule as it existed in January 1976, NASA wouldn't be flying any new astronauts until 1982. However, Abbey judged that it took two years to train an astronaut for a specific mission—so that brought the selection date forward to 1980. And, wary of the mistakes made with scientist-astronauts in 1965 and 1967, where some proved unsuitable, Abbey and Kraft decided that new selectees would go through a two-year probationary period, undergoing general training and familiarization as "astronaut candidates."

This new generation were supposed to be arriving at JSC in 1978. Which meant that selection—receipt of applications, review, background checks, and interviews—should take place in 1977. This recruitment, the largest NASA had ever conducted, would search not only for pilots but also for mission specialists, astronauts who were not professional pilots but rather engineers, scientists, physicians, or members of a nonflying branch of the military.* Kraft

* A third class of astronaut was planned—shuttle payload specialists. But these individuals would not be full-time NASA astronauts, nor would they play any role in operating the vehicle. They would be selected on an as-needed basis by agencies flying payloads on the shuttle.

and Abbey were determined that this selection, planned to number forty, would also include women and minorities.

According to Abbey, some members of the astronaut office "did not feel a selection was needed at all. They viewed any additional astronauts as more competition, be they male or female. And of course there were others in the office that did not feel women were qualified for the position."

Other than secretaries, Abbey had not worked directly with many women at NASA. However, his exposure to Gilruth and Kraft's letters, many of them from young women; his appreciation of the intelligence and competence of Carolyn Huntoon, a thirty-four-year-old NASA physician and physiologist; and his role as the father of two daughters convinced him that he was right. But to bring women into the astronaut office he had to overcome seventeen years of history.

Back in November 1958, one month after NASA set up shop as a new agency, several medical consultants working for Gilruth's Space Task Group under Charles Donlan formulated requirements for future space travelers as part of "Project Astronaut." The applicants would have to be between the ages of twenty-five and forty, under five feet eleven in height, under 180 pounds in weight, with a bachelor's degree in physical, mathematical, biological, or psychological sciences. They would have to have demonstrated "(a) willingness to accept hazards comparable to those encountered in modern research airplane flight; (b) capacity to tolerate rigorous and severe environmental conditions; and (c) ability to react adequately under conditions of stress or emergency."

The announcement added:

> These three characteristics may have been demonstrated in connection with certain professional occupations such as test pilot, crew member of experimental submarine or arctic or Antarctic explorer. Or they may have been demonstrated during wartime combat or military training. Parachute jumping or mountain climbing or deep sea diving (including SCUBA) whether as occupation or sport, may have provided opportunities for demonstrating these characteristics, depending upon heights or depths obtained, frequency and dura-

tion, temperature and other environment conditions, and emergency episodes encountered.

These requirements—which could have allowed the recruitment of women—were published as a draft civil service solicitation on December 22, 1958.

And were promptly ignored.

The resistance came from the Space Task Group itself, starting with Gilruth and Low, both leery of the time it would take to recruit this team of adventurers. Describing himself as an "old aviation person," Gilruth wrote, "I thought there was nothing quite like test pilots who were used to flying a vehicle with wings instead of behind a rocket. They are used to altitude, the need for oxygen, bends and acceleration. They are used to high discipline and taking risks. . . . They also had the technical knowledge to understand the ins and outs of the space capsule and the rockets and navigation."

Gilruth, Low, and Donlan revised the proposed qualifications: candidates must now be graduates of a military test pilot school; holders of a bachelor's degree or have equivalent experience; have logged a minimum of 1,500 hours of flying time; and be qualified in jets. They still had to be under five feet eleven and weigh less than 180 pounds. There was no longer any chance NASA would select women astronauts, since there were no female graduates of military test pilot schools.* (Soon after this, in the spring of 1959, the Soviets would take their first steps toward the creation of a team of potential space travelers—cosmonauts. They decided to limit the selection to operational air force pilots, though, not test pilots.)

Two groups of prospective astronauts were briefed on the Mercury program in late January and early February 1959. Thirty-two volunteered for the program and went on to (later infamous) physical examinations at the Lovelace Respiratory Research Institute in Albuquerque. Two were dropped at this stage; thirty others went on to psychological exams at Wright-Patterson Air Force Base in Ohio (where young Captain George Abbey met several of them). Eighteen were judged to be worthy of selection, and after days of further deliberation Donlan and his team recommended seven—the famed Mercury

---

* Abbey realized that the proposed mission specialist applicants could have come from Donlan's original pool.

astronauts Carpenter, Cooper, Glenn, Grissom, Schirra, Shepard, and Slayton who were introduced to the public on April 9, 1959.

After the May 1961 go-ahead for Apollo and a manned lunar landing, NASA could justify the selection of additional astronauts who would work through the end of the decade. On April 18, 1962, the agency announced that it would be selecting five to ten new astronauts. The requirements were similar to those for Mercury, though civilians were now eligible—and the maximum age had been reduced to thirty-six. Women were still not eligible.

Slayton, a highly regarded air force test pilot before coming to NASA, had been scheduled to fly America's second orbital mission, Mercury-Atlas 7, in May 1962, but was grounded for having an irregular heartbeat in March. Knowing that NASA was about to select additional astronauts, and that the agency was searching for a "retired general or admiral" to be in charge of selection and training, three of Slayton's fellow Mercury astronauts, Shepard, Schirra, and Grissom, asked Gilruth to give Slayton the job, which he accepted with energy and enthusiasm. He jumped into the new selection, which had resulted in 250 applications by the deadline of June 1. Thirty-two candidates were invited for medical screening, but now at Brooks Air Force Base—the air force's medical center—not the Lovelace clinic. All but one, judged to be too tall, passed, and were invited to Houston for interviews with Slayton and the board the week of August 12.

Nine new astronauts were announced on September 17—civilians Neil Armstrong and Elliot See; air force officers Frank Borman, Jim McDivitt, Tom Stafford, and Ed White; and navy pilots Pete Conrad, Jim Lovell, and John Young.

Based on his own experience, personal contacts and recommendations, and lingering resentment over his treatment by NASA doctors, Slayton had firm ideas about the makeup of the new astronaut group. It was no coincidence that two of the nine finally selected were Pete Conrad and Jim Lovell—the two Mercury candidates who had been disqualified for what Slayton judged to be minor medical matters.

He also sought recommendations from ARPS commandant Yeager and honored X-15 test pilot Robert White. Slayton would later say that he had "pretty much handpicked" the group. And he would indeed dominate the selection process—and flight assignments—for the next decade.

The test pilot requirement was dropped for the 1963 group because Slayton felt that the agency had largely exhausted the pool of military test pilots who were qualified, willing, and available. Now operational military pilots, those with a thousand hours in jets, could be selected.

The requirement that astronaut applicants be pilots was dropped for the 1965 group, which consisted of physicians and scientists. Six were selected. Two of them already were qualified jet pilots; the other four were required to qualify, and three did so.

In 1966, under orders from George Mueller to "man up" for the ambitious Apollo applications schedule, Slayton and his team selected nineteen more pilot astronauts, using the same criteria as for the 1963 group. (This was the one Major George Abbey applied for.)

A second group of scientists came aboard in 1967—eleven of them.

The final addition to the NASA group came in August 1969, when seven test pilots from the air force Manned Orbiting Lab program came aboard. This group included Robert Crippen and Richard Truly.

In 1960, Randolph Lovelace, head of the private clinic where the Mercury astronauts underwent their medical examinations, invited Jerrie Cobb, an experienced woman pilot, to endure the same set of tests, and she passed easily. Lovelace and Cobb sought public support from Jackie Cochran, another well-known woman pilot, and from Jane Hart, a pilot who was married to Sen. Philip Hart of Michigan.

During the summer of 1961, eighteen other women took the tests at Lovelace, and thirteen of them passed. Cobb called them her "First Lady Astronaut Trainees," or FLATs. Cobb was introduced to NASA administrator James Webb, who appointed her a special consultant to the agency for a year.

What Cobb was planning was a more rigorous series of tests for the FLATs at the navy medical facility at Pensacola. But the navy needed a letter from NASA justifying the expense of the tests. After a considerable amount of back and forth between Cobb and various NASA officials, in June 1961 these navy tests were canceled. NASA decided that while it was possible that women might someday be selected as astronauts, at the present the agency had no information on future needs, so it could not devote time and money to the matter.

Jerrie Cobb flew to Washington, DC, to try to have the testing program resumed. She and Jane Hart wrote to President Kennedy and visited Vice President Johnson, who listened politely but took no action.

Months later, on July 17–18, 1962, Rep. Victor Anfuso (R-NY) convened a session of a special subcommittee of the House Committee on Science and Astronautics and invited Cobb, Hart, and Cochran to testify.

Cobb and Hart spoke about the benefits of Lovelace's private project. Cochran was less supportive.

NASA was represented by George Low and astronauts John Glenn and Scott Carpenter. In their testimony, all three fell back on the catch-22 that since NASA required all astronauts to be graduates of military jet test piloting programs and have engineering degrees, and that no women could meet these requirements, there was no possible justification to include them in the program.

Although the subcommittee was sympathetic to the women's arguments, no action resulted and the issue largely faded from the news.

On June 14, 1963—a month after the final Mercury mission by Gordon Cooper—the Soviet Union launched its sixth human space mission, Vostok 6. The cosmonaut onboard was a woman, Valentina Tereshkova. A twenty-six-year-old parachutist—not a pilot—she remained in orbit for three days, longer than all NASA's male astronauts combined to that point, appearing on Soviet television, and returning safely on June 19. Tereshkova's flight revived the discussion over the lack of women in America's program, but nothing changed.

(No one in the West knew it at the time, but Tereshkova was one of six female parachutists selected to train for spaceflight in March 1962. Members of that group were considered for additional flights, even an all-woman Voskhod mission, but none of them flew. The group was disbanded in October 1969. The Soviet Union would not select women as cosmonauts again until 1980.)

In January 1976, almost thirteen years after Tereshkova's flight, there were still no women in the NASA astronaut team. Nevertheless, it was a new era. In 1975 President Ford signed Public Law 94-106, requiring US military academies to admit women—which they did the next year. The military naval, air force, and coast guard academies saw 327 women enroll as members of the class of 1980.

The lack of African American and other male minority astronauts was a different matter. African Americans—such as Jim Wiley, Abbey's senior officer on Dyna-Soar—were allowed to be pilots and even test pilots. But none appeared on the lists of candidates for 1959 and 1962.

There was a prominent minority male candidate—Capt. Edward Dwight—who entered the Aerospace Research Pilot School in September 1962. His enrollment at ARPS had been directly encouraged by President Kennedy's brother Robert, responding to a request from Whitney Young, director of the Urban League: "Why isn't there a Negro astronaut?"

Dwight's presence at ARPS—and the publicity surrounding him—was resented by ARPS commandant Chuck Yeager, who judged Dwight to be "a good pilot but not a great one." Dwight's flying experience had largely been in bombers at a time when bomber or multiengine pilots were not commonly selected for test pilot school. Yeager claimed that his response to political pressure was to take Dwight, "but only if I'm allowed to take all the other candidates who ranked ahead of him."

Dwight completed the first phase of ARPS in May 1963, ranking exactly in the middle of the class, then entered phase II, the so-called space school, where his classmates included future astronauts Ted Freeman, Jim Irwin, and Dave Scott, as well as MOL selectees Mike Adams and Lachlan Macleay.

Dwight was submitted to NASA by the air force for the 1963 selection, one of three dozen, but failed to pass the preliminary screening. (NASA received 271 applications that year, seventy-one from the air force, navy, and marine corps, and two hundred from civilians.)

Charles Berry was a member of the 1963 board. He remembered a phone call from Robert Kennedy himself, saying, "We want to assure that there is an

African American in the next selection." Berry told the Attorney General that NASA already had a list of finalists and that there were "two, I think, African Americans" on it. But that he couldn't guarantee that one would be selected.

Kennedy said, "You did not hear what I said. I said we want to assure—and you will assure—that there will be an African American."

Berry then learned that Slayton had also been contacted.

Slayton, Berry, and the board examined the applications, using their standard point system: ranking each candidate by flying time, types of aircraft flown, combat experience, test pilot or operational experience, academic degrees and grades—adding weight for outstanding performance (top ranking in the test pilot school, for example) and for strong references from commanding officers and colleagues. They also gave points for candidates like Michael Collins or Richard Gordon, who had applied in 1962 but were deemed to require more experience.

In this "blind" scoring, Dwight ranked well outside the list of thirty-four finalists who would be given medical examinations at Brooks Air Force Base beginning July 31, and interviewed in Houston the week of September 2.

Fourteen white male pilots—seven from the air force (including David Scott, Michael Collins, and Edwin "Buzz" Aldrin), four from the navy (including Roger Chaffee and Dick Gordon), two civilians, and one marine—were announced on October 17, 1963.

President Kennedy was assassinated five weeks later, on November 22, 1963, ending his administration's involvement in astronaut selection.

Dwight himself had "no complaint with NASA" for not selecting him as an astronaut. He did, however, have many complaints about Yeager and the air force. After graduating from ARPS, he was assigned to bomber testing at Wright-Patterson Air Force Base, where he and his family faced discrimination and domestic trouble. Planning to leave the service—and get divorced—Dwight was not submitted to NASA by the air force in 1965.

However, that 1965 selection for the air force's Manned Orbiting Laboratory did produce an African American astronaut, Maj. Robert H. Lawrence Jr. But he was killed in an accident at Edwards in December 1967.

Nine years later, there was no directive from NASA HQ, nor was the outreach dictated by law. The 1964 Civil Rights Act prohibited discrimination, but it didn't require inclusion—especially in a specialized institution like NASA.

Abbey's first involvement with the subject came in 1972, when working for Kraft. Jeffrey Miller, director of the federal Office of Civil Rights Evaluation, had written to Ruth Bates Harris, director of equal employment opportunity at NASA HQ, noting that all American astronauts had been white and what was the agency going to do about it?

Kraft's response, drafted by Abbey, noted that NASA had never categorized its astronauts by ethnicity, so could truthfully provide no data on the subject. It was technically true, but only highlighted the problem.

To Abbey and Kraft, it was a new era at NASA and women and members of minority groups were going to be part of it.

Another question to be answered was, How many new astronauts?

On the day George Abbey became director of flight operations, there were twenty-nine active astronauts assigned to various offices at JSC from shuttle development to life sciences or engineering.

Four other astronauts were detailed to government positions (Bill Anders, Rusty Schweickart, Joe Allen, and Jack Swigert). Allen was likely to return to JSC to fly, but none of the others showed interest.

Of the twenty-nine at JSC, the oldest was Deke Slayton, who was fifty-one. The youngest was scientist-astronaut William Lenoir, thirty-seven. Assuming the shuttle began flying in 1979, none of the current astronauts would be under forty and most would be half a dozen years older.

Pete Clements, now Kraft and Sjoberg's technical assistant, had been given the task of evaluating astronaut office staffing. Sjoberg started by telling him, "We're in a down time now. Skylab is over, ASTP is over and we've got all these astronauts sitting over there and they're not doing a heck of a lot." Clements's first mission was to see if NASA could shed some astronauts or assign them elsewhere.

Clements worked with Abbey's bubba, Jay Honeycutt, and astronaut Bob Crippen, and immediately realized, "Come shuttle, you don't have anybody trained on the shuttle, but you're going to have to train them all, and you've got X number of years before you get there, and then you've got X number of years of training, and you don't know what the astronauts are going to do on the other side of the actual flight. I mean, are they going to want one flight and go home, or they want ten flights, or they're now fifty-five years old?"

Clements and team concluded that NASA couldn't afford to lose astronauts and, in fact, needed to hire *more*, and fast.

Sjoberg was unhappy. "I give you one little job and you come in with the wrong answer!"

But Kraft and Abbey agreed with Clements. Looking at the existing team, Abbey said, "The pool of veterans was actually limited. You had Young, Mattingly, Weitz, Lousma. Engle hadn't flown in space, but he'd earned astronaut wings on the X-15, and making an X-15 landing was the closest anybody had come to landing a vehicle like the shuttle. He was probably our most experienced pilot, looking at it that way.

"Many of those who hadn't were scientist-astronauts, and even the test pilots were far removed from real flight test work—they had no experience with heads-up displays and other new technology."

While Abbey would feel an obligation to offer every scientist-astronaut at least one flight, to honor their commitment to the program, his real goal was to create a cadre of next-generation astronauts who would not only fly multiple missions but also would go on to contribute to the program as managers and engineers into the 1980s and beyond.

Clements and Honeycutt were sent to HQ to brief John Yardley on their study; it was followed soon by a memo from Kraft dated March 8, 1976, stating that JSC had begun the astronaut selection process and formed a board.

Forty new astronauts would be selected, twenty pilots and twenty mission specialists. And forty would not be the end. Abbey and Kraft envisioned additional selections every year beyond 1978.

Abbey's selection board consisted of two panels, one evaluating pilot applicants, the other dealing with potential mission specialists.

Board members on both were chief astronaut John Young, Jay Honeyutt (recorder), JSC personnel representative Duane Ross (who would serve on selection panels for the next forty years), Joseph Atkinson from JSC's minority affairs (the first African American to take part in the selection process), as well as Jack Lister from JSC personnel.

For the first time, there would also be a woman—Carolyn Huntoon, who had worked on the Lunar Receiving Laboratory (where early Apollo landing crews were quarantined), Skylab, and Apollo-Soyuz, before her tour with Kraft. Kraft had actually asked Huntoon if she wanted to become an astronaut herself. Told no, he had Abbey add her to both panels.

Vance Brand would be on the pilot panel.

Also present was Martin Raines, the head of JSC's Safety & Quality Assurance team. As a young army officer assigned to Huntsville, he had taken part in the original discussions about astronaut selection for Mercury.

The mission specialist panel would include astronauts Joe Kerwin, Robert Parker, and Ed Gibson, as well as Robert Piland from JSC engineering development and James Trainor from NASA Goddard.

One JSC veteran would *not* be part of the process. At the board's first meeting on Friday, March 24, 1976, Kraft and Sjoberg emphasized the importance of considering qualified women and members of minority groups. Deke Slayton questioned the availability of such candidates, noting their absence from previous selections, which largely drew from the community of test pilots.

But Kraft cited the shuttle's need for mission specialist astronauts as well as pilots. He felt that qualified women and minority candidates were available and insisted that they would be considered. Further, NASA would reach out to those communities.

Slayton had always believed that potential astronaut candidates were self-motivated and definitely didn't require outreach of any kind, and that such efforts would cause NASA to relax its selection criteria. Then Slayton stood up, announced "I want no part of this," and walked out. He never returned.

After that awkward moment, the board meeting resumed. Under Abbey's guidance, general qualifications for pilot and mission specialists—which had been under discussion for three years by this time—were approved.

Common requirements for both pilot and mission specialists were few: all had to be citizens of the United States. The preliminary age limit—thirty-five—was set aside.

There would be no mention of gender or race. And—here Abbey was remembering his bitter experience with the 1966 selection—there would be no added or extra requirements.

Candidates for shuttle pilot had to have:

A bachelor's degree in engineering, physical sciences, or mathematics—or have completed one by December 31, 1977. An advanced degree or equivalent experience was preferred.

A minimum of a thousand hours of "first pilot" time, with two thousand preferred. High-performance aircraft and flight test experience was preferred.

Ability to pass a NASA Class I physical.

Height between sixty-four and seventy-six inches.

Some officials at JSC had lobbied for a minimum master's or even doctorate for mission specialist applicants.

Yardley opposed this, and so, tellingly, did Abbey. "I also insisted on the only degree requirement be for a bachelor's degree in engineering or science, and advanced degrees be considered to be desirable. I insisted the requirements be based on what was needed to do the job and not written to exclude individuals that could become well-qualified astronauts. A number of excellent astronauts were selected that had only received a bachelor's degree."

Candidates for shuttle mission specialist were required to have:

A bachelor's degree in engineering, biological or physical science, or mathematics, or have completed same by December 31, 1977. Advanced degree or equivalent experience desired.

Ability to pass a NASA Class II physical—much like the Class I though with relaxed vision and hearing requirements.

(Apollo-era astronauts had had to meet Class I requirements in order to fly as first pilot on the T-38 and also to endure high g-forces.)

Because of the shuttle's relatively benign launch and landing profiles, astronauts would not be subjected to loads greater than three times the force of gravity.

A public announcement of the new astronaut selection was made as a press release on July 8, 1976. Applications had to be postmarked no later than June 30, 1977.

During the spring of 1976, Abbey and his board put together a recruitment plan that would reach out to universities, including minority universities with a well-regarded engineering and scientific curriculum. Joseph Atkinson went on the road to make presentations.

The agency also contacted technical and scientific societies.

In early 1977, NASA headquarters did its own minority outreach, hiring *Star Trek*'s Uhura—actress Nichelle Nichols—to make presentations about the space shuttle and astronaut selection at schools and other community organizations. The move wasn't popular in some JSC circles. Abbey said, "People complained, 'She's not a real astronaut.'" He didn't mind, though he was never able to confirm that Nichols's efforts directly resulted in an application.

More routine efforts were made with the military services, with announcements going directly to the Air Force Test Pilot School at Edwards Air Force Base, and the Navy Test Pilot School at Patuxent River Naval Air Station, and to the respective services flight test activities there and at Eglin Air Force Base, Florida; China Lake NAS, California; and elsewhere.

Captain Loren Shriver, an F-15 test pilot at Edwards, was one of those who saw the announcement, and after talking things over with his wife, sent in his application "along with about sixty percent of Edwards."

Capt. Guion "Guy" Bluford was at Wright-Patterson pondering a return to the cockpit (after a tour doing research) when he spotted the NASA announcement in the base newspaper.

Navy test pilot Rick Hauck was back in the fleet, assigned to Whidbey Island in Carrier Wing 14 when he saw the announcement. He applied and so did fellow pilot Dan Brandenstein.

Abbey said, "We also made a conscious effort to reach out to the army and the coast guard. None of this was done as a result of headquarters' direction."

Former marine fighter pilot Norman Thagard, now at the Medical University of South Carolina, had missed the announcements until alerted by his wife, Rex Kirby, who had heard about them on the radio and sent away for an application for her husband.

Air force major Fred Gregory was detailed to NASA already, at the Langley Center in Virginia. Because his background was unusual—trained in helicopters (like Abbey) for a Vietnam tour, then cross-trained to fighters after attending the Naval (not Air Force) Test Pilot School—he knew he had no chance to be approved by an air force selection board.

He was also African American.

On June 30, the last day applications could be postmarked, he applied as a civilian.

Six weeks later his phone rang; Maj. Gen. Tom Stafford, now commandant at Edwards Air Force Base, was on the line. "Who are you?" Stafford asked.

Gregory identified himself as an air force major. Stafford knew that, of course, but said he had been telephoned by John Young asking if Stafford knew why an air force officer was applying as a civilian. Gregory walked Stafford through his reasons, and Stafford laughed, "Okay." That was all Gregory heard until November, when called for an interview.

Sally Ride, a twenty-six-year-old female physicist at Stanford University, learned of the recruitment from an article in the school paper.

James van Hoften, a professor at the University of Houston who had flown combat missions with the navy during the Vietnam conflict, came to meet with Abbey at the recommendation of John Cox, a flight controller who was one of van Hoften's engineering students.

And on it went, not only in the United States but at remote locations too. Duane Ross noted an instance in which he received a call at home one Sunday night at 11:00 PM. "When I answered the phone this fellow said, 'My name is Big John, and I'm a ham operator in Lockquoshena,* California, and I have someone on the line who has called in by ham from Antarctica who wants to talk to you about the astronaut program." The caller was an American scientist on an exchange tour at the Russian Vostok station. "If that guy could find out about it," Ross said, "I would think anybody could find out about it."

There were some strange, and ultimately unproductive applications. Suzanne Abbey remembered, "Some people actually put résumés in our mailbox—and it wasn't one of those on the street, it was a slot in our door! They were pretty crazy too."

Kraft had said, and Abbey hoped, that Americans were still interested in becoming astronauts. That certainly proved to be the case: there were as many

* Probably La Crescenta.

as a thousand requests for the application package in a single day. JSC mailed out over twenty-four thousand.

Ultimately the agency received 8,079 applications postmarked by the deadline.

Knowing they faced a deluge at the deadline, the board reviewed applications as they arrived throughout the spring and summer of 1977. It was a smart decision, since almost all applications from the air force, navy, marines, army, and coast guard arrived in a single package just days before June 30.

The panel's first pass eliminated applicants who didn't actually meet the stated qualifications. This reduced the eight thousand plus to 5,680, still an impressive number.

Then the real work began, with panelists using a scoring system to rate the applicants.

For example, pilots were scored on number of first pilot flying hours in all aircraft (two points for each five hundred), first pilot hours in jets (two per two hundred hours), being a graduate of a test pilot school (add five points), whether or not the applicant had served as a test pilot (three points per year, maximum of nine), and so on.

Combat tours earned points. Number of aircraft flown. Grade point averages in college and, if available, graduate school.

The maximum pilot score was eighty-five.

Mission specialists were graded on level of degree achieved (four points each for bachelor's, master's, doctoral, twelve points maximum), grade point averages as an undergraduate and graduate (twelve possible points in each category), then relatable work experience (one point per year, up to ten) or flight experience (one point per two hundred hours, up to five points).

The maximum mission specialist score was sixty-five.

This work was reviewed in long meetings where panelists might read several hundred applications—and scores—in a single day.

In the next phase, each applicant's references would be contacted. In the case of Rick Hauck, he found himself asked about a fellow pilot named Dan Brandenstein—who was being asked about Hauck at the same time.

For civilians, this was a new and sometimes fraught process. Some had protested the Vietnam War, and to have government officials showing up on doorsteps to interview friends, family, coworkers, and lovers was nerve-racking. The reference check did indeed eliminate some applicants.

Ultimately, Abbey and his team identified 208 subjects they wanted to interview. The plan was to bring them to JSC in groups of twenty for a week of medical tests, orientation, and a one-hour interview with the board.

The first group of twenty—all potential pilot astronauts—arrived the weekend of July 30 and 31, 1977, and were boarded at the venerable Kings Inn Ramada, just outside the JSC main gate.

Abbey, Young, Ross, and NASA physician Sam Pool met with the applicants—all test pilots—to tell them what they would face during the week, and to assign each of them an essay explaining why each wanted to become an astronaut.

During the week, each applicant would endure a session involving two different psychiatrists.

Navy lieutenant commander John Creighton was the first interview subject, at 8 AM Monday, August 1.

The interview was not deliberately designed to unsettle the applicant, though it did. The tables and chairs were arranged in the classic *Meet the Press* model, with Abbey sitting next to the interviewee and both facing the board members.

The questions could be technical, personal, or topical. The primary purpose was to see each applicant's public manner: Articulate or not? Humorous? Personable? Someone who could be trusted to represent the agency with the public?

In every session there would be one trick question—which changed, since Abbey knew that the applicants would be briefing each other. For example, early on the panelists asked subjects their opinion on the Panama Canal—then in the process of being handed back to Panama by the United States.

A day later, the question was about the Suez Canal. There were applicants who answered about the current transfer to Panama, likely doing themselves no good. And one interviewee asked, "Aren't you supposed to ask about the Panama Canal?"

MIT astronomer Jeff Hoffman was at JSC with the sixth group of applicants in the first week of October. On his application, Hoffman had admitted to violating British maritime laws—sailing from England to Norway without a certified ship's captain. The offense had resulted in a trip to crown court and a fine. When he entered the interview, someone piped up, "Here comes the criminal!"

Some applicants were satisfied with their performance on the interviews, others not. Air force test pilot Brewster Shaw was horrified when Abbey offered him a postinterview ride to Shaw's next appointment. He told Abbey, "Jeez, that didn't go very well."

But, Shaw remembered, Abbey only smiled.

The psychiatric session followed the classic "good cop, bad cop" model, with Dr. Eddie Harris trying to disorient or provoke the applicant while Dr. Terry McGuire took on a more supportive and welcoming manner. (Harris had experience at Folsom Prison in California.)

Applicants would be given an extensive medical examination as well as briefings on the shuttle program. They would have some kind of test—blood sampling, treadmill run, the infamous proctological exam, vision tests, and so on—every day.

Other hours would be devoted to JSC personnel—applicants would meet members of the astronaut office for "casual" chats.

And there would be tours of facilities and exposure to hardware.

One test involved a proposed "personal rescue enclosure" for shuttle crews. This piece of hardware—never flown—was intended to allow a suited astronaut to transfer an astronaut without a space suit through a vacuum from one vehicle to another. The sphere was a fabric balloon large enough—thirty-six inches in diameter—to hold one person wearing an oxygen mask curled up in a partial fetal position. Each applicant was zipped into the sphere for a brief time, usually about fifteen minutes, ostensibly to test for claustrophobia. Several reported getting drowsy; others noted the "gym locker" odor. Whatever the goals or results, the sphere test was later dropped.

Kathy Sullivan, who was in the eighth and final group, recalled:

"It's one appointment after another. Drinking from a fire hose—deliberately inscrutable in some respects, the whole sort of 'good cop, bad cop' game with the psychological interviews. One or two folks deliberately planting a sense that everything about you through this whole week is part of the interview. The implication was if you're strolling aimlessly across the campus versus marching purposefully across the campus, someone will note that and write it down.

"It's just all those things to try to figure out how do you react if you can't tell what matters, but you suspect everything matters, and you suspect big consequences hang in the balance. How do you behave? Do you go bonkers? Do you settle down? That was just all a little surreal."

An informal but, to Abbey, vital part of the screening was a barbeque dinner Wednesday night at Pe-Te's, the revered Cajun restaurant near Ellington's main gate. Astronauts and others were invited to the dinner to meet the applicants in order to get a sense of how they would act in social situations, knowing they were under scrutiny.

In addition to the medical tests and personal interviews, there was a third requirement for that week in Houston. Abbey arranged for astronauts and others to speak with the candidates about the true nature of the job.

It didn't just involve flying in space. There would be tedious hours of design reviews, vehicle tests, work on science experiments developed by others, a great deal of public relations activity, and long absences from home. Several astronauts from the 1960s had failed to grasp this and had left the program.

Upon hearing this, some of the 208 finalists decided NASA was not for them. But Abbey was determined to turn every candidate into a full-fledged astronaut and fly that astronaut several times.

Military officers would be detailed to NASA, as they had been during the Apollo years. Tours were expected to be seven years long, double those of normal military assignments.

Civilians would become NASA civil servants under a "special appointing authority," since the forty positions were open to the public at large. The new hires would be paid by the government service schedule, GS-7 to GS-14. Not that anyone was asking about salary.*

* Anna Fisher, an emergency room physician in the Los Angeles area, was taken aback when she realized that her potential astronaut salary would be a quarter of what she was making.

Two years of astronaut candidate (ASCAN) training would be followed by a minimum of five years as an active astronaut. Any ASCANs who failed to qualify would be placed in jobs elsewhere in NASA.

A second group followed on August 15, a third on August 29, and every two weeks until the last applicants departed on November 20, 1977.

Some of the applicants faced unique challenges in making it to the interview. Double applicants Anna Sims and Bill Fisher had planned to get married the week Ross wanted Anna for an interview, August 29. (She and her husband-to-be simply moved everything forward a week. Obviously, it was a fairly casual affair as weddings went.)

Bill Fisher was interviewed the week of November 7.

There was one exception to the two visits with shrinks. A young astronomer from Minnesota named George Nelson, who was at JSC the week of October 3, said of McGuire, "Halfway through the interview, he got up and walked across the room and packed up his briefcase and walked to the door and opened the door, and I said, 'Well, I guess we're done?' And I thought, *Oh, boy, what have I said here?*"

McGuire said, "Yeah, we're done, because you've got a softball game in half an hour out at the field, so Mr. Abbey said you had to get out early."

Nelson had been a star athlete at Minnesota's Willmar Senior High School in both football and baseball, and had been scouted by the Minnesota Twins. (A football injury ended his hopes of an athletic career.) He arrived at the field, was handed a glove, and learned that he was to play for the astronaut office team, coached by Abbey with Jay Honeycutt on the mound. According to Steve Hawley, Nelson hit two home runs. "His future was assured."

Nelson agreed. "I figured that was my real interview."

With the interviews completed, the board turned to final evaluations, hoping to find the perfect forty out of the 208 finalists. Abbey's goal was not to select the most experienced pilots or mission specialists, or those with operational experience over academics, or NASA space veterans over those from industry. He wanted someone from every category—a balanced, integrated astronaut team.

That meant that he tended to support the candidacies of applicants in their late twenties, like George Nelson, Sally Ride, and Steven Hawley, who lacked any operational or flying experience but had proven themselves academically and as members of research teams—and who also seemed to possess athletic prowess and artistic ability.

Abbey and Honeycutt had promised the applicants that the final selections would be announced by the end of December, and asked for Christmas vacation contact information in order to reach everyone.

A list of twenty pilots and twenty mission specialists was complete by early December; Kraft signed off on it.

But NASA administrator Robert Frosch, and Alan Lovelace, the head of spaceflight, wondered whether the agency needed so many new pilots, given that there were still twenty or more in the office. Abbey went back to the list and cut five pilots, all of whom would be selected later.

It was time-consuming. Robert Piland would remember, "We were all tucked away for months."

30 | Approach and Landing

AS THE ASTRONAUT INTERVIEWS were taking place, the shuttle began its approach and landing tests at Edwards using the *Enterprise*.

There had been five inert (orbiter unpowered and uncrewed) flights of the *Enterprise* and Shuttle Carrier Aircraft combination between February 18 and March 2, 1977, with durations ranging from an hour and a half to over three hours. These validated the basic aerodynamics.

The Haise-Fullerton and Engle-Truly teams had been aboard a series of three active flights between June 18 and July 26, in which *Enterprise* was powered up. On these the crews rehearsed the precise profile for releasing *Enterprise*.

Now it was time to turn the orbiter loose—to determine if the calculations and computer simulations were correct.

From the time he became chief of astronaut activities in 1962, Deke Slayton personally accompanied every NASA flight crew to the pad. The only exception was ASTP in 1975, when Slayton was a member of the crew and management was represented by chief astronaut John Young.

Abbey planned to follow Deke's example, even though the ALT flights were atmospheric drop tests, not rocket launches. He and Young joined the four ALT pilots for an early breakfast on Friday, August 12—bacon and eggs prepared by the cafeteria at the flight center and delivered in Styrofoam boxes.

On Friday, August 12, the SCA took off with *Enterprise* and Haise-Fullerton. At an altitude of 24,100 feet and a speed of 310 miles an hour, *Enterprise*—sporting a giant white tailcone for additional lift—separated from the

SCA and began its dive toward the lake bed at Edwards. Haise and Fullerton tested control surfaces and monitored the operation of the on-board IBM computers running the vital flight control software.

It took less than five minutes to reach the lake bed, with *Enterprise* at the relatively high (for aircraft) speed of 213 miles an hour. Haise was concerned about two possible surprises, neither good. One was a "vacuum sweep," in which the ground effect (the buildup of air under the orbiter's giant delta wing) would suck the vehicle into the ground. The other was "ballooning," which would keep it from reaching the runway.

Neither took place. The orbiter landed as designed, and the shuttle program reached another key milestone.

Abbey was back at Edwards a month later, hosting Joe Engle and Dick Truly at another breakfast, this one catered by the Lancaster Inn on genuine china. The second tailcone-on test went well too.

After the third tailcone flight ten days later, also a success, ALT manager Deke Slayton elected to move to the final phase, tailcone off. That attempt was scheduled for October 12.

With Engle and Truly in the cockpit, and simulated main engine rocket nozzles installed on the back of the orbiter, *Enterprise* separated from the carrier aircraft. As predicted, without the tailcone providing lift, the orbiter's glide slope toward the lake bed was steeper. But the astronauts landed safely and smoothly.

The second free flight with the tailcone off, and Haise and Fullerton in the cockpit, took place on October 26, and was to be the first to land on the concrete runway at Edwards.

On final approach, Haise made a slight adjustment in the orbiter's attitude to ensure that he would land at the end of the runway. The next thing he and Fullerton knew, they were bouncing back into the air and wobbling, a case of "pilot-induced oscillation."

Enterprise quickly settled down, landed and rolled out, but it was not pretty. Tom Stafford, then a two-star commander at Edwards, was watching this with air force chief of staff David Jones, who snapped, "What the hell was that?"

Postflight investigation showed that there was a slight lag in the orbiter's flight control system—fixable, yes, but a sign of the complexity of the shuttle system, where one minor software glitch could come close to ruining everyone's day.

The original target launch date for the first shuttle had been October 1977. That had slipped to the right, but as the ALT program ended, Kraft, Thompson, and others were still hoping for a launch in October 1979.

Two years into his tenure as director of flight operations, Abbey was a happy man. He loved this job, eagerly getting to his office between 8 and 8:30 each morning. He was always dressed in a jacket and tie, no matter the weather. Cheryl Bouillion remembered Abbey lunching quietly by himself most days, his lunch a bowl of soup. And moving through his schedule of meetings and phone calls calmly—"I never saw him frustrated or angry."

There was one day early on, however, when he alerted her by saying, "I might be a bit terse." When Bouillion didn't recognize the word, he said, "Look it up." It turned out he was having to make some reassignments that day.

Abbey could be oblivious to the activity around him. "I do remember going into his office one day and he was working through some paperwork, mumbling to himself, and I tried to get his attention and failed. Finally, I just said, 'My desk is on fire!' And walked out, assuming he had paid no attention at all.

"A few days later he was passing my desk when he said, 'Now *my* desk is on fire.'"

Abbey would often be accused of mumbling or speaking at a barely audible level, which he could not deny. At times it was a strategy—according to Boullion, "usually when he was having a conversation he didn't want to have." It was also just a personal tic. According to Suzanne Abbey, her father's unwillingness to use profanity also tied him in verbal knots. He would call a slow driver "Dumpy Joe" or "Donkey" or, when dealing with the children, would find unique combinations of their names, such as "Jandy" or "Su Joyce." To them his biggest threat was "you could be grounded for life."

At this time, age forty-five, Abbey had almost never been ill. Suzanne credited "his large consumption of tabasco sauce." His eyesight was still good—unlike many of the other pilots, including astronauts like John Young, Abbey did not require reading glasses.

Not that there weren't occasional rough spots, especially working for Chris Kraft. Abbey would recall several occasions where the director would give him an order, then stare at Abbey's face, saying, "You're not actually going to do what I said."

"I didn't say anything."

"But you're not going to do it!"

The only possible response then was to pledge to do better and get out of Kraft's office.

Kraft was largely happy with his appointment of Abbey, finding that he only had to issue "corrections" on a few occasions. "Sometimes George was too secretive."

That was undeniable. Abbey was fond of seeking consensus in private. And, like all the Abbeys, he was fond of pranks and surprises. And there was no better surprise than sudden good news.

31 | Thirty-Five New Guys

ON THE MORNING OF Monday, January 15, 1978, Bouillion arrived at the ninth floor to find, to her surprise, Abbey on the phone—speaking very clearly.

He was calling the thirty-five astronaut applicants to give them the official word that they had been selected. News had leaked from HQ that this was the day of the announcements—in fact, the release was scheduled to go out at noon. The task of delivering the bad news to 170-odd other applicants was shared by the other board members.

The first call was to earth scientist Kathy Sullivan in Halifax, Nova Scotia. Other East Coast recipients were Bell Labs engineer Terry Hart in Pennsylvania, and air force pilot Fred Gregory at the Air Command and Staff College a few miles east, in Norfolk, Virginia.

Navy pilot Dave Walker was at sea, deployed to the Mediterranean Sea on the carrier USS *America*. Marine flight test engineer Jim Buchli was at the Naval Air Test Center in Maryland, as were flight engineer Dale Gardner and Robert "Hoot" Gibson. Navy test pilot John Creighton was assigned to Pax but was on assignment at Marine Corps Air Station El Toro in California; Abbey had to track him down too.

Physician Norm Thagard was at the University of North Carolina medical school. Abbey reached air force test pilots Dick Covey and Brewster Shaw at Eglin Air Force Base in Florida. Flight engineer Mike Mullane was on temporary duty in Idaho—Abbey left an urgent message with Mullane's wife, Donna, to have her husband call, which he did, though not until late in the day.

Then it was on to the Midwest. Abbey connected with Guy Bluford at Abbey's former base, Wright-Patterson in Ohio. Physician Rhea Seddon was doing a rotation at the VA hospital in Memphis. Chemist Shannon Lucid was at work at the Oklahoma Medical Research Foundation in Oklahoma City.

There were two locals, in the Houston area, Professor James van Hoften and aircraft ops pilot Dave Griggs.

In the mountain standard time zone, astronomer George Nelson (frantically trying to complete his thesis) in Boulder, Colorado, and a bit south, Maj. John Fabian at the Air Force Academy in Colorado Springs.

Jeff Hoffman should have been in Cambridge, Massachusetts, but he was on a family ski vacation in Sun Valley, Idaho, where Abbey reached him around 7:00 AM. The furthest reach was to the offices of the Inter-American Observatory in Chile—for astronomer Steven Hawley.

On to the West Coast: Abbey made a half dozen calls to Edwards Air Force Base alone—to flight test engineer Ellison "El" Onizuka, to pilots Steve Nagel, Dick Scobee, Brewster Shaw, Loren Shriver, and army test pilot Bob Stewart.

Navy pilot Don Williams was in his quarters at the Lemoore, California, naval air base with his face covered with shaving cream when he was told Abbey was on the line. Jon McBride was at Point Mugu. Mike Coats was in Monterey at the naval postgraduate school.

Rick Hauck was at the naval air station on Whidbey Island when Abbey gave him the good news. A mile away, fellow pilot Dan Brandenstein was home in the shower when the phone rang.

Physicist Sally Ride was at Stanford University, while Xerox engineer Judy Resnik was in the same time zone, four hundred miles farther south, in Los Angeles.

Physicist Ron McNair was at Hughes Research Labs in Malibu.

Also in the Los Angeles area was Anna Fisher (née Sims). Fisher's husband, Bill, another applicant, learned from Abbey that he wasn't selected in the same call, the only time Abbey delivered a rejection. Anna, Bill, and Judy Resnik, who became friends with Bill during the interview phase, would go out to dinner that night to celebrate.

Abbey's manner was understated and relaxed, frequently leading with some bit of small talk. For example, Jon McBride was awakened in a motel room at Point Mugu. "How's the weather out there today?" Abbey asked.

"I don't know, I haven't been out yet," McBride said.

"Well," Abbey said, "whatever it is, it'll be a nice day. I want you to come work for us."

With Sally Ride, the opening was "Remember that job you applied for? You still interested in that?"

Most of the selectees were amused—they were so happy to be chosen that they would have responded to anything—but not all. Norm Thagard found Abbey's delivery "weird." And Dan Brandenstein was vaguely irritated when Abbey "started asking about the weather."

Mike Coats, who received Abbey's call on his thirty-second birthday, was in Monterey, California. Because of flight schedules, he had to head for Houston several days ahead of the February 1 press conference, so he arrived before the other thirty-four.

After a day of orientation, all thirty-five were introduced to the press in the Teague Auditorium on Tuesday, January 31, with most of the attention focused on the six women and four members of minority groups. Chris Kraft made the formal welcome, and then NASA public relations introduced the candidates one by one.

They would spend the next week in mundane administrative tasks—getting official portrait photographs taken, for example.

Then they were sent home to make moving plans.

The selection and announcement of thirty-five new ASCANs left 170 other finalists disappointed. Abbey hoped that many of those would try again. He had directed John Young to personally telephone promising pilots who had missed the cut, to let them know that the door was still open for the next selection.

It took three months from the rejection for air force flight test engineer Jerry Ross to summon the courage to telephone Abbey directly. Hearing that one of the interviewees was on the line, Abbey left a meeting to speak with him. Ross wondered if Abbey and the board had seen anything that would preclude him from future selection. Not a thing, Abbey told him. Ross had been a promising candidate.

The air force officer was weighing several possible future assignments—did Abbey have any thoughts on which might be beneficial for a future application to NASA?

Abbey had a better idea. "Why don't you come down here and work for us?"

Ross was stunned and pleased, and accepted the invitation, though it took him a year to get a transfer through the air force.

At least three other potential mission specialists—Bonnie Dunbar, John Michael "Mike" Lounge, and Jerry Jost—did the same thing, calling for career advice in hopes of improving their chances of becoming an astronaut. Abbey would assign all four to payload operations.

Shortly after the introduction of the new ASCANs, and with the first of four shuttle orbital flight tests scheduled in eighteen months, Abbey settled on four crews.

John Young was the most experienced astronaut in the office—in the world, for that matter—with two Gemini flights and two Apollo missions, including a lunar landing. He was the obvious choice to command the challenging first flight of the shuttle.

The pilot would be Robert Crippen, highly regarded as the astronaut office's specialist in shuttle software.

Abbey found the perfect time and place to give Crippen the news. On the morning of Friday, March 10, 1978, the orbiter *Enterprise*—fresh from the ALT tests in California—was being ferried to the Cape. The Shuttle Carrier Aircraft stopped at Ellington for the night. Abbey phoned Crippen in his building 4 office and said, "Why don't we go out there and take a look?" Reaching the airfield in Crippen's beat-up pickup truck, the two men walked around the giant carrier aircraft with the orbiter mounted atop it, and Abbey casually said, "How'd you like to fly the first one?"

The second crew would be carried over intact from the ALT program. Joe Engle and Dick Truly learned of their new assignment when called to Kraft's office early the next week to meet with the director, Sjoberg, Abbey, and Clements. They were told that Young and Crippen were flying STS-1 with Engle and Truly as backups.

Engle and Truly would then fly STS-2, with Young and Crippen as their backups.

NASA headquarters and shuttle managers hoped that STS-3 would be launched in time to conduct a rendezvous with the unmanned Skylab, attaching a tele-operator stage that would raise the space station to a safer orbit. Abbey didn't believe that Skylab would remain in orbit long enough for such a mission. But he assigned a crew of Fred Haise—Young's only real competitor for commander of the first mission—and Skylab veteran Jack Lousma to the reboost effort.

The fourth crew would consist of ASTP veteran Vance Brand and Haise's ALT pilot Gordon Fullerton.

Several veterans, especially Ken Mattingly, were disappointed by the assignments. Mattingly was so incensed upon learning that Young—not he—was taking the first command that he snubbed new ASCAN Mike Coats, who had dropped by to introduce himself. (He later apologized.)

Equally important was the selection of flight directors for the orbital flight tests. Abbey largely left this to Kranz. Once mission operations left Apollo behind and concentrated on the shuttle, there was no immediate need for new flight directors. Don Puddy had served as flight director for the ALT missions in 1977, and Kranz decided to retain him, Chuck Lewis, and Neil Hutchinson. It was this team that would spend three years enduring the upgrades and sims for the orbital flight test program.

As STS-1 loomed, in 1979 Kranz added Tommy Holloway and Harold Draughon, two veterans of mission control.

Told that their first day of work would be Monday, July 10, the new astronaut candidates had begun moving to Houston, beginning with Judy Resnik, who arrived in April.

There was one additional arrival—veteran scientist-astronaut Joe Allen, who had been on a detail at NASA HQ but who had chosen to return to Houston and resume flight training. Allen had jokingly specified that his reporting date had to be "one day prior" to that of the new group. He "didn't want to be considered junior." (On the reporting day, when the new selectees were introduced to the rest of the office, Abbey would joke, "I told you there were thirty-five; there are thirty-six. This is Joe Allen.")

Their first meeting as a real group—not posing for photographs and answering questions from the press—took place in the big conference room, #966, in building 1. Abbey and Young spoke briefly about where the ASCANs would go first—classroom work, and a tour of NASA facilities around the United States.

In *Riding Rockets*, his raucous memoir of the 1978 astronaut group, Mike Mullane would write that Young "welcomed us with a few forgettable words," and that he was clearly uncomfortable addressing large groups. He finished with this message to the new arrivals, "Don't talk about nothing you know nothing about."

The first lecture came from Max Faget.

A few days later, all of the new ASCANs were at Ellington being fitted for NASA helmets and flight suits. Then they were off to a tour of other NASA centers.

The new group was so large that it had to be split into teams for classwork and travel. Abbey created a red team under Rick Hauck and a blue one under John Fabian, who were the senior navy and air force officers. One team would attend a class in the morning while the other would be elsewhere; after lunch they would swap.

"There were no responsibilities" with the assignment, Fabian said later. "Somebody had to be the person that George called if there was a problem."

Fabian's life and career profile were much like Abbey's—from Washington State, educated at AFIT, trained as a pilot but not a test pilot. He had grown interested in becoming an astronaut during Apollo but knew that at six feet four he was too tall. The closest he'd come to the space program was having his air force flight training at Williams Air Force Base postponed by six months so a civilian scientist-astronaut named Owen Garriott could qualify on the NASA T-38.

During these first weeks, the new astronauts got to know each other, measuring strengths and weaknesses. Exploring differences.

They were a diverse bunch. Twenty-nine were male, and six were female. There were three African Americans (Bluford, Gregory, McNair). One Asian American (Onizuka). Hoffman and Resnik were the first Jewish NASA astronauts. Onizuka was Buddhist. Most were Christian, some Evangelical. Several professed no religion at all. Ages ranged from thirty-nine (Fabian and Scobee) to not-yet-twenty-seven (Sullivan and Hawley). Fabian, Scobee, Hauck, and Griggs were old enough to be contemporaries of veteran astronauts like Bruce McCandless, Dick Truly, Robert Crippen, and Bo Bobko. In fact, Hauck, as a young naval officer studying for a master's in physics, had attempted to apply for the 1965 scientist-astronaut selection.

Twenty-one were active-duty military; the other fourteen were civilians, though several (Griggs, Hart, Thagard, and van Hoften) had military experience.

For those with military backgrounds, life and work at NASA wasn't radically different from what they'd known. They were familiar with the forms, the furniture, the travel. For the civilian postdocs, especially the women, who had emerged from universities, hospitals, and research centers, NASA was a greater challenge. For one thing, they were pioneers, the first women wherever they went. No matter how NASA officials tried to protect them, the press and others sought them out at every public function, and quite a few private ones.

In *Riding Rockets*, candidate Mike Mullane cited a political figure "questioning several of us men at a party. He was totally focused on our comments until Judy [Resnik] walked by in her flight suit. Then he said, 'Excuse me,' and hurried to catch up to Judy. We were abandoned like the out-of-state voters we were."

On the other hand, Kathy Sullivan irked some future shuttle pilots by stating that they would be "taxi drivers" while the mission specialists "did all the interesting work."

One bonding moment came when the new candidates agreed on a name for themselves. Many of the new ASCANs were veterans of the Vietnam War, where new arrivals in combat units were referred to as "the fucking new guys" (TFNG). Someone—and fingers pointed to Rick Hauck or Hoot Gibson—

adapted it to NASA. Judy Resnik even did a sketch that wound up on T-shirts, though the name evolved to "Thirty-Five New Guys." None of the women objected to being called "guys."

Some of the men were just uncomfortable with women in this environment. They also learned, for example, that Sally Ride was quick to register sexist comments and gestures, even as Rhea Seddon and Judy Resnik would let them go.

Physician Norman Thagard's apprehension began in his October 1977 interview with the selection board. Abbey, as often happened, asked a question from deep into Thagard's resume. "I see that you made a C in ballroom dancing. Why was that?"

According to Thagard, he answered, "Well, our instructor was a woman who liked to lead. . . . I found that very difficult to learn to dance with someone who was leading."

But then the next question he was asked was "Well, what do you have against women?"

At that point, Carolyn Huntoon, "the only female member on the thing, had gotten up and left." Which left Thagard fuming. "First of all, they've drawn this thing out, which to me, I thought was an innocent enough response, but now they're making a big deal out of it. Now this woman is obviously a feminist and offended that I've said this, and so she's left."

Fast forward to January 1978, when Thagard and his family arrived in Houston for the public introduction. Huntoon took charge of Thagard's children, and Thagard took this opportunity to ask her why she'd walked out on his interview.

"I had to get up to leave because my babysitter had to go home."

New astronauts from the military had to make other adjustments. Mullane noted, not entirely jokingly, that he had been in uniform for thirty-two years. "I had gone to Catholic schools for twelve years and worn the uniforms of that system. In four years at West Point I never had a piece of civilian clothing in my closet. The air force also told me what to wear."

More seriously, there was potential tension over politics, specifically the Vietnam War. Most of the TFNGs from the military had served in the conflict.

The veterans were understandably proud of their service, and suspected that some of the postdocs had been antiwar protesters. But as far as Abbey knew, the issue never caused any problems. Nor had he anticipated any: "We made sure to select people who were prepared to work for a common goal and not get distracted by other issues."

And the dividing line wasn't rigidly military-civilian. Civilians van Hoften and Thagard had served in the war. Yet air force pilot Brewster Shaw would reveal in 2002 NASA oral history that after his Vietnam tour flying F-4s, he had returned to the United States in 1973 "very much disillusioned with the war effort and the military," so much so that he applied to law school. It was only acceptance at the Air Force Test Pilot School that changed his mind.

Another factor: for years the all-male astronaut group had generally referred to each other by last names, or some variation of a last name. Al Bean was "Bean-O," for example. Since it struck most as odd to call, for example, Rhea Seddon by her last name, first names became the rule, for women and men.

As with any group of intelligent, fit, outgoing humans, the TFNGs soon found their personal behavior tested. Mike Mullane said "the gold bands on the fingers of us married TFNGs were no deterrent" to many female groupies. "It was easy to see who was taking advantage of the situation. During the head count on the bus to return to the hotel, some MIAs would be noted."

There had been similar incidents in the Mercury, Gemini, and Apollo days, in an era when it was easier to cover up. Then, of course, the sinners and observers were all male pilots. How would these activities play with a more diverse group?

At one point, the group's unofficial leader, Hauck, reminded the others, essentially, that what happened on these trips stayed on these trips. Abbey had traveled with the Apollo astronauts in California in the 1960s and 1970s. Astronauts and groupies were nothing new.

He had never discovered a correlation between private personal behavior and technical competence. Some of the most active womanizers in the early days were the most capable astronauts. (So were several of the most faithful husbands.)

He had the power to inflict punishment, of course—to withhold or delay a coveted technical job or flight assignment. So when he issued a warning, he knew the listener would heed it. For how long, and how attentively, he couldn't predict. He did assume that the offender would more careful.

Some TFNGs were single or, like Ride, involved but not married. Ride dated Hoot Gibson, then Steve Hawley. Gibson later married Rhea Seddon.

Al Bean was one of the veterans of the astronaut team, having landed on the Moon with Pete Conrad during Apollo 12, then spent fifty-nine days in space aboard the second manned Skylab. Following a tour as backup commander for Apollo-Soyuz, he had worked in shuttle development, concentrating on cockpits and layouts.

In June 1978, as John Young began to concentrate on training for STS-1, he found himself with a new assignment. "Abbey said, 'You're going to be the acting chief of the astronaut office.'" Bean immediately found himself leading the Monday morning pilots meetings, something Young had never enjoyed.

As the weeks passed, Bean began to enjoy the job, though it had its challenges. He found the astronauts to be "very ambitious and imaginative and energetic. They want to do everything in the world." But "when you get a whole bunch of them together, then they step on each other's toes."

He also had to learn to deal with Abbey's managerial style, especially compared to that of his famed predecessor, Deke Slayton, who Bean claimed "wasn't trying to tell you every day [that] he was in charge. You knew he was in charge, but he left you alone unless he had to come and make a decision."

Abbey, he says, came to exemplify what he called "a power guy." That is, "I like to have power so I can do what I want." Abbey "wants to do what he wants, but then he wants you to know you can't do anything unless he says so."

Bean was also reacting to Kraft's specific order to Abbey to get the astronaut office under control—and make sure the astronauts knew it.

Abbey had worked out a training schedule with Bilodeaux's people, and with Young and Bean. Phase I started with JSC orientation for both pilots and mission specialists.

Then it was time for the T-38.

NASA had been operating a squadron's worth since the mid-1960s, when the sleek supersonic jet was the air force's new training aircraft. Astronauts needed the vehicles to retain flying proficiency, to log their minimum hours for flight pay, and to commute between far-flung training locations—Long Island, Florida, California. Flying was inherently dangerous. Four astronauts had died in T-38 accidents between 1966 and 1967.

Several TFNG mission specialists were or had been jet pilots—James van Hoften, Guy Bluford, Norm Thagard—and they would qualify to fly as pilots in command: the front seat. For them, and for military pilots like Shriver, Hauck, Brandenstein, and others, coming directly to NASA from flight test centers or operational squadrons, basic ground school was not needed.

But the others were shipped off to Homestead Air Force Base in Florida for survival and ejection training—in case any should have to bail out of a T-38.

Upon returning to Houston, they were put through the T-38 syllabus, beginning with such basics as radio use and protocols. Aircraft operations deputy Bud Ream was their primary instructor here.

All astronauts, whether professional pilots or new mission specialists, were required to log fifteen hours in the cockpit each month. This worked out to three to four hours every week. That goal was easily met, for example, when astronauts flew from Houston to Florida to spend a week supporting shuttle operations—two hours down on Monday morning, two hours back on Friday evening. In other weeks, it meant getting up early, or flying on weekends, between academic studies and later technical assignments.

But Abbey knew that it was valuable, not just in keeping professional pilots proficient, or exposing nonpilots to life-and-death situations. The flights also promoted team-building between pilots and mission specialists—and was also efficient. No empty seats were allowed.

There were incidents. Brewster Shaw allowed nonpilot (though experienced F-111 backseater) Mike Mullane to land a T-38. By itself, this action wasn't unprecedented. Shaw had flown two-seat F-4 Phantoms during Vietnam, and one of his first goals as a pilot was to make sure his backseater could land the aircraft if he were incapacitated.

So far so good. But the T-38's nosewheel got damaged. Shaw took the blame. He was, after all, the pilot in command of the aircraft. But aircraft ops deputy Dave Walker, another TFNG, confronted Shaw with the evidence—dramatically dumping a shredded tire on the main table at a Monday morning pilots meeting. Shaw admitted his complicity and guilt, and endured one chewing out from Young, and another from Abbey. "That was dumb," Abbey told him.

The incident had no real impact on Shaw's career—he was one of the first TFNG pilots assigned to a shuttle crew . . . with Young as commander.

An astronaut's most regular contact was the astronaut office scheduler. Each Monday after the pilots meeting, every astronaut, veteran or candidate, would find a printed schedule for the coming week—classes or meetings, travel, and the all-important flying times.

Outside work there were more informal events created by Abbey himself—softball games, chili cook-offs, and picnics. Attendance was never mandatory, but widely considered to be a good idea, especially by the TFNGs.

Abbey also frequently held court at the Outpost or some other local watering hole on Friday evenings. These were just an extension of information gathering that had proved so valuable when he worked for George Low, and so much fun with Crippen, Truly, and others.

Some TFNGs, sharing Abbey's fondness for "beer call," were eager participants. Others resented these gatherings, thought them a waste of time that they would rather spend with their families. They feared that valuable information was being shared—or key personal impressions made, that those who hung out with Abbey (and who became known as "bubbas") would have first crack at choice assignments.

While Abbey certainly welcomed the insights he would gain from long nights at the Outpost, they had little effect on major personnel decisions. Those were guided by input from aircraft operations and trainers.

By fall 1978 the TFNGs were in phase II—technical training on shuttle systems, flight operations (aerodynamics and orbital mechanics), and on mission operations (flight planning, ground support, etc.), then manned spaceflight concepts and sciences. The candidates were exposed to a shuttle simulator, and to basic flight equipment like the shuttle's robot arm, and EVA suits.

Much of this information was conveyed the old-fashioned way, in lectures, though there were no tests and no grades. Classes were, of course, familiar turf to academics like Hoffman, Nelson, and Ride. The military TFNGs found them dull at times. One thing they all shared were complaints about NASA jargon. Even such basic terms as commander and pilot were rendered as "CDR" and "PLT."

Some teaching moments took place outside the classroom. During the mission operations section, the candidates were taken to mission control for a talk by Gene Kranz that ended with playback of the horrific last moments of the Apollo 1 astronauts, killed in the January 1967 fire—a reminder of the ever-present danger of their work.

By this time, the fifteen pilot candidates had been checked out on the Shuttle Training Aircraft as well as other vehicles, like the NASA KC-135 transport, the so-called Vomit Comet.

The first flight-capable orbiter, vehicle 102, now named *Columbia*, was still undergoing assembly and checkout at Rockwell's plant 42 in Palmdale. There were delays, notably with the orbiter's thermal protection system.

But John Yardley, the associate administrator for spaceflight at headquarters, was pressuring Rockwell and Robert Thompson's program office to get *Columbia* to the Cape.

Abbey and Young flew to California in late February 1979 to check on the assembly in person and were dismayed at the state of the vehicle. Systems were just incomplete, nowhere near ready for tests, much less flight. The spacecraft was declared "complete" on March 5. Three days later the orbiter was trucked overland up the road to the research center (which in 1976 had been officially renamed the NASA Hugh L. Dryden Flight Research Center), where it was erected on its carrier aircraft.

Because hundreds of its thermal protection tiles were still temporary, a short test flight took place on the afternoon of March 10. It was a disaster, with thousands of tiles, close to five thousand temporary and a hundred permanent, peeling off the orbiter even before it left the ground. Over the next week, ground teams scrambled to apply more temporary tiles.

Columbia finally took to the air on its cross-country voyage on March 20, stopping the first night at Biggs Army Airfield at El Paso. The next leg was to San Antonio; the following day *Columbia* reached Eglin Air Force Base in Florida.

Finally, on March 24, it arrived at the Kennedy Space Center. A day later, de-mated from the SCA, it was rolled into the Orbiter Processing Facility for tests aimed at qualifying the vehicle for its official target launch date of November 9, 1979.

It was obvious to Abbey and many others that *Columbia* was in no condition for launch on that date, and likely not for months afterward. Walt Kapryan, NASA's veteran checkout lead at the Cape, said so—Yardley transferred Kapryan to another post.

Because of the premature move of *Columbia*, Rockwell had to transfer hundreds of people to the Cape for three years to complete the work that hadn't been done in California.

The tile installation would eventually be supervised by Kenny Kleinknecht, Abbey's predecessor as flight operations director. Abbey would assign astronaut Robert Overmyer as an assistant to Kleinknecht, giving him eyes and ears on the operation.

The shuttle was far more complex than Apollo—logically enough, since it was conceived as a space launcher, orbital spacecraft, science laboratory, satellite deployment platform, and aerospace plane.

The orbiter had an airframe with wings and elevons, flight controls, a propulsion system, steering rockets, onboard power, communications, and an environmental control system. The mid-deck was for habitation and would include storage lockers and a space toilet. There was a payload bay capable of carrying satellites and their upper stages, and a remote manipulator system

(RMS; the shuttle's robot arm) to remove payloads from the bay. It would also carry an airlock and eventually a docking system. It would be controlled by five general purpose computers that required specialized code and programming.

All of this had to be strong enough to survive the brutal vibrations of launch by the solid rocket boosters. And capable of surviving reentry into the earth's atmosphere at hypersonic speed and temperatures of two thousand degrees Fahrenheit, and landing on a runway (landing gear, tires, brakes). It would carry three main engines, each developing 375,000 pounds of thrust at sea level, each throttleable, each designed to be returned with the orbiter for reuse. Fuel for the space shuttle main engines (SSMEs) would be carried in the giant aluminum external tank 153.8 feet in length and 27.6 feet in diameter. Its inert weight was 66,800 pounds; when filled with liquid hydrogen and oxygen, it weighed over 1,655,000 pounds. Although designed to be ejected from the orbiter on ascent and ultimately destroyed on reentry, it nevertheless required its own thermal protection system to prevent localized heating.

Then there were the SRBs, monster firecrackers (149.16 feet tall, 12.17 feet in diameter) designed to burn for the first two minutes of launch, delivering around three million pounds of thrust each. They would be separated from the orbiter-ET (external tank) combination at an altitude of 220,000 feet, then parachute back to the Atlantic, to be recovered and refurbished for later use.

This vehicle had to be assembled and transported from plant 42 or the landing site at Edwards/Dryden to the Cape . . . there it had to be maintained and refurbished between flights, transported to the pad. It required its own constellation of communications satellites (the Tracking and Data Relay Satellite System, or TDRSS) and downlink stations.

It had to be controlled, so it required computer software, telemetry, and teams trained to use those systems. And there were dozens of potential payloads, ranging from a large Spacelab to satellites for military, commercial, and scientific communities to small "getaway specials." All of these needed to be integrated with the shuttle.

And on and on.

All through shuttle development, veteran members of the astronaut office had worked on these issues. Some had become true specialists on certain matters. Story Musgrave had spent untold hours on EVA, and Bruce McCandless was the office expert on EVA and the Manned Maneuvering Unit backpack.

In early 1979 the office had an orbital flight test group, which included the four assigned crews, and also an ascent/entry group under Mattingly, an on-orbit group under Kerwin, as well as other branches.

And when the TFNGs classroom training officially ended on February 27, they were deployed here, not only to learn but also to serve as the astronaut eyes and ears on those varied shuttle systems.

Not all of the new technical assignments would be equal, of course. A handful of the TFNGs, those with recent flight test or operational experience (Hauck, Brandenstein, Buchli, Shriver), were given support roles for STS-1.

Air force flight test engineer Mike Mullane was stunned to find himself assigned to Spacelab support along with test pilot Dick Scobee and others. On the other hand, civilian postdocs like Anna Fisher and George Nelson were given assignments dealing with key hardware—the RMS for Fisher and EVA suits for Nelson.

It was, of course, part of Abbey's plan. "These technical assignments were deliberately intended—with malice aforethought—to put people where they would be uncomfortable," Abbey said. "We put scientists in operational jobs, military test pilots in science and so on. They had to be generalists, and they had to be able to cope with assignments that went beyond their experience. Some were outstanding. Steve Hawley, who was an astronomer, really proved himself on the shuttle computer systems. So did several of the test pilots. There were some who didn't perform well in those first assignments and wound up needing more time to become generalists."

The Shuttle Avionics Integration Laboratory (SAIL) was something new for flight crews and astronauts—a facility to test the orbiter's software. Located in building 16, it was physically unimpressive—a rudimentary shuttle "cockpit" connected to a bank of computers and consoles. It had been conceived and developed by Max Faget's engineering directorate in partnership with Rockwell, the orbiter contractor. Rockwell engineers would largely staff SAIL, and the facility's first "commander" was Jim Westom, a former air force pilot now working for Rockwell. But when Abbey assigned Brewster Shaw and John Creighton to SAIL, Westom found himself nudged aside.

According to Hoot Gibson, part of the second wave of astronauts assigned to SAIL, "George Abbey would just get that camel's nose under the tent and then take over the whole thing." This was Abbey's Apollo experience at work. Even though computer software in the 1960s was rudimentary compared to 1979, astronauts performed the tests. And now, Gibson said, "George wanted the astronauts to be the ones that flew the SAIL simulators and flew the simulated missions and verified all the software. George was relentless, and there was a lot of resistance to it, and George prevailed, as he usually does."

For Abbey, it wasn't really a power play, just logic. "The astronauts had a greater need to learn the software than contractors did."

Another key assignment was as deputy director for aircraft operations at Ellington, working for Joe Algranti. In 1979, Bob Overmyer had the job, then it was given to TFNG Dave Walker when Overmyer was assigned to the Cape. The assignment gave an astronaut valuable experience in leadership, but Hoot Gibson thought Abbey had another motive: "Joe Algranti was extremely talented and extremely sharp, so wise and knowledgeable in how to maintain a fleet of aircraft and do everything safely, but he could be tough on his people.

"The reason George put an astronaut out there as the Deputy Chief is because an astronaut was immune to him. He couldn't kill him. Joe Algranti couldn't kill an astronaut who was his deputy but he would be able to kill any average pilot who was his deputy."

The only astronauts assigned there were navy or marine pilots. According to Hoot Gibson, who did two tours there (1982–1983 and 1984–1985), Abbey told him, "I have got to have a navy officer out there as the deputy. The air force guys don't actually have the right kind of background and training from the way the air force runs things operationally and maintenance-wise."

Hoot Gibson thought it was a bad deal, that the astronauts who would get flight assignments would be those who were in the astronaut office every day, working directly on shuttle matters—not five miles down the road at the "air patch."

From Mickey Moore, from Joe Shea, from George Low, from Chris Kraft, George Abbey had learned lessons, good and bad, in leading—and managing. His opinions were also colored by his original mentor, Sam Abbey.

What he concluded was that astronauts ought to be well-rounded individuals, not just professional generalists but also artists as well as technicians, athletes as well as intellectuals. In short, engaged citizens.

This was, in fact, not far from the naval academy's goal of producing "officers and gentlemen," ship handlers and warriors who could also dance at a ball or carry on informed conversation.

With that goal in mind, in addition to technical assignments, he also gave the TFNGs one-of-a-kind tasks in which the steps weren't always obvious. This served several purposes: in one case, the subject would find the logical solution to a problem and present it. Another possibility was that the subject would come up with a better set of solutions. There was also the chance that the subject would fail. That happened with several people, so Abbey and Young either gave the subjcct a new task and different guidance—or judged the subject to be less flexible than the others. This wasn't always a bad thing. But it might dictate the type of work the astronaut would be asked to perform in the future.

Just before their departure from Skylab on February 8, 1974, astronauts Jerry Carr, Ed Gibson, and Bill Pogue had fired their Apollo service module thrusters for three minutes, raising the station's orbit by seven miles to one that ranged between 270 and 283 miles.

Then they departed.

Although repaired by astronauts, its mission salvaged, Skylab was past its sell-by date, with ailing or failed control gyroscopes, fading power supplies, and other issues. Controllers placed it in a gravity-stabilized attitude—the airlock adapter pointing directly away from the Earth—and put it out of their minds.

NASA hoped Skylab would simply stay in orbit, drifting ever closer to Earth with each passing year, for its predicted nine-year lifespan, ideally until 1982 or 1983. That goal had driven the planning for a shuttle rendezvous with the station on STS-3, with the attachment of a small booster stage that would boost it again for possible future use.

Abbey was skeptical of this date from the beginning. He had performed his own calculations of the fuel available for the final boost and knew that it could have been much more powerful, leaving Skylab in a higher orbit. But he was also aware of the unpredictability of orbital lifetimes for such large vehicles.

Solar activity and its effect on Earth's upper atmosphere caused Skylab's orbit to deteriorate far faster than NASA had predicted. By early 1979 it was obvious that the station was going to reenter soon, as early as summer.

Thanks to the uncontrolled reentry of a Soviet military satellite, Cosmos-954, which spewed radioactive material over northern Canada in January 1978, the public and Congress grew concerned about Skylab's looming reentry. Nobody wanted it to break up over a populated area.

By summer 1978, a team had been formed in mission control to recontact and, if possible, take control of Skylab. Now, finally, on the night of July 11, 1979, Abbey joined Kraft, flight director Don Puddy, Charlie Harlan, and others in mission control for the death watch.

Shortly after 3:00 AM, as the station passed over North America for the last time, its altitude down to seventy-five miles, flight controllers uplinked a final command to Skylab, sending it into a tumble.

The huge station broke up over the Indian Ocean, with most debris falling into the water. A slight miscalculation by mission control—easy to do, given the lack of hard data on atmospheric density—meant that the station broke up further east than expected, and some debris did rain down on unpopulated areas of Australia.

With the end of Skylab, and its occasional use of facilities, the mission control teams began to concentrate solely on the shuttle, running simulations every Tuesday and Thursday.

One bit of collateral career damage: shortly before the reentry, with the Skylab reboost a dead issue and all shuttle test flights still in abeyance, STS-3 commander Fred Haise withdrew from his assignment, leaving NASA for a job at Grumman.

Abbey reshuffled the crews, promoting Haise's pilot, Jack Lousma, to commander of the STS-3 crew and moving Gordon Fullerton, an ALT veteran, to STS-3.

32 | Problems

THE PARTNERSHIP BETWEEN KRAFT and his deputy, Sigurd Sjoberg, was close and mutually beneficial, though contradictory. Kraft was blunt, excitable, decisive, while Sjoberg remained calm, amiable, and quiet. Sjoberg was an avid golfer and introduced Kraft to the game, which Kraft, with his typical passion, embraced, eventually becoming a ten handicap. Years later he would write, tallying up the money he had spent on fees, equipment, and lessons, "I didn't know whether to thank him or damn him."

In early 1979, as the dire rumors about the shuttle's future circulated, Kraft told Sjoberg that he was retiring. Having no wish to continue at NASA under someone else, especially in these circumstances, Sjoberg put in his own request for retirement—and accepted a job with the Orbiting Astronomical Observatory Corporation that would begin in August.

According to Abbey, when Sjoberg went in to confirm his plans with Kraft, the conversation went like this:

> Sjoberg: I'm retiring at the end of May.
> Kraft: Why the hell are you doing that?
> Sjoberg: You told me you were retiring.
> Kraft: I'm only retiring technically, becoming an annuitant. I'm still going to be here!
> Sjoberg: I didn't know that.
> Kraft: You can do the same thing. Go change it.
> Sjoberg: I can't. I already gave them my word.

And so Sjoberg left, with Kraft scrambling to replace his trusted associate. In August 1979 Kraft appointed Cliff Charlesworth, a former flight director, as his deputy.

In 1979 George Abbey and his wife, Joyce, separated, with Joyce moving out and leaving Abbey in the Davon Street house.

It was not an easy time. George Jr. and Joyce BK were off at college, but that left three children at home. Suzanne was in her first year of high school. Jimmy was twelve and Andy was just seven.

"He couldn't just go off to work and assume that things would be taken care of," Suzanne remembers. "Now he had to make sure lunches were made, homework was done."

Suzanne also found herself weighed down with greater responsibility. "I became the lady of the house for a while.

"We hired a housekeeper, but that turned out to be a bad idea. For one thing, she was really talkative—so chatty that Dad had a hard time getting out of the house in the morning. She was also just sitting around all day watching TV. I finally told him I would get rid of her if he would pay *me* what he was paying her."

When the TFNGs arrived in Houston in July 1978, they expected to spend two years as "candidates," not full-fledged NASA astronauts.

However, on the afternoon of Friday, August 31, 1979, Kraft and Abbey called an all-hands meeting of the TFNGs to tell them that they could drop the "candidate" from their job titles. They had performed beyond expectations in their first technical assignments.

And Abbey was already searching for reinforcements. On August 1 he and the agency had announced a new ASCAN selection, the first of a hoped-for annual series. The next application period would open on October 1 and close on December 1.

33 | Those Other Astronauts

FROM THE FIRST TIME the concept emerged in 1974, Abbey had embraced the idea of flying payload specialists—noncareer astronauts—on the shuttle. Flying on the orbiter made few unusual physical demands on a crew member: any individual in decent health could endure the experience.

A basic shuttle flight would be flown by a commander and pilot. "We had crews of two for the orbital flight tests," Abbey said. "They could handle the basic operations."

In early planning, NASA assumed a single mission specialist—the flight engineer—would join the commander and pilot for operational missions, where satellite deployments and EVA were scheduled.

It was hoped that the fourth seat on the orbiter's flight deck would be filled by payload specialists, individuals from science or industry who had been selected to go into space with their experiments or satellites.

Hoping to take advantage of this opportunity, in August 1973 the European Space Research Organization (later to become the European Space Agency, or ESA) signed an agreement with NASA to design and build a laboratory module—later known as Spacelab—that could be carried in the shuttle's payload bay and flown multiple times in different configurations. Lab module 1 would be provided to NASA free of charge in exchange for flight opportunities by ESA astronauts.

With this goal in mind, on March 28, 1977, ESA announced a recruitment for scientists to fly as Spacelab payload specialists, ultimately receiving two thousand applications.

Four scientists were selected on December 22 of that year, and three were submitted to NASA for Spacelab 1 in May 1978—Ulf Merbold of West Germany, Claude Nicollier of Switzerland, and Wubbo Ockels of the Netherlands.

ESA officials believed that their Spacelab mission would be crewed by three NASA astronauts—the commander, pilot, and flight engineer—with three other positions filled by Merbold, Nicollier, and Ockels.

But by 1979, with planned vehicle operations growing more mature, it was clear to Abbey that the basic operational shuttle crew would be two pilots and three mission specialists. "It also became obvious that a mission specialist could easily do the tasks that payload specialists might do. So in most cases that extra crew member wasn't required."

But two American payload specialists were assigned to Spacelab 1 in June 1978, and NASA announced two mission specialist astronauts two months later. One of the American payload specialists would be a backup, but both mission specialists would fly, meaning that ESA wasn't getting three seats on the mission—just one.

ESA was not happy. Mel Brooks, now the NASA representative to Europe, lobbied Kraft and Abbey to allow ESA astronauts Nicollier and Ockels to go through ASCAN training.

When ESA agreed to pay for the training, Abbey couldn't say no. Nicollier and Ockels would join the 1980 ASCAN group.

A trickier battle awaited Abbey with the air force.

With the shuttle designated the Space Transportation System, NASA, and especially JSC, had to learn to accommodate the requirements of the US Air Force, which had bought (at a terrific discount) a quarter of the manifest.

Some of the adjustments were disruptive and expensive: a secure control room for highly secret national security flights. And just adding a layer of secrecy and compartmentalization to operations, travel, and personnel deployment was inimical to NASA's charter as a public civilian organization.

Military payloads could not be identified, even those like the Defense Communications Satellite Program, which were largely unclassified. Launch dates and especially launch times became national security issues.

The public payloads—communications, weather, navigation—were developed and operated by the Space Division, the Los Angeles–based successor to Bernard Schriever's Space Systems Division.

The secret payloads—imaging reconnaissance systems with names like GAMBIT, HEXAGON, KENNEN, and LACROS; signals intelligence platforms like AQUACADE, JUMPSEAT, and WHITE CLOUD; as well as technology developments and experiments—were under the purview of the National Reconnaissance Office (NRO).

Jointly staffed by the air force and CIA, NRO had been created in 1961 to operate the vital CORONA spy satellite and other early space intel programs. By 1979 NRO had major offices in California and Virginia as well as tracking and data download facilities in Virginia, Colorado, Europe, and Australia. Its director usually held the public title of undersecretary of the air force for space; his staff—the Office of Space Systems—was located behind a blank door inside the Pentagon, room 400C.

Abbey and JSC found themselves dealing with NRO's seven-hundred-person unit in Los Angeles, which operated under the unclassified name Secretary of the Air Force Office of Special Projects. SAFSP's commander from 1975 to 1983 was Maj. Gen. John Kulpa.

Hans Mark, former director of the NASA Ames Research Center, had been appointed as director of NRO in early 1977. Mark was a strong supporter of military use of the shuttle and forced a reluctant air force and NRO to give up expensive unmanned vehicles like the Titan III, and to redesign critical payloads to take advantage of the orbiter's big payload bay.

Kulpa was also a shuttle supporter; indeed, Mark had reappointed the general to a second tour as head of SAFSP for that very reason.

As the shuttle made its first flights, even the name NRO was still secret. And while many in the space industry knew of its existence—contractors like TRW, whose space officials included a former NASA engineer named Dan Goldin, built civilian satellites as well as NRO birds—NASA had to adhere to information blackouts.

These military payloads presented Abbey with a serious challenge when it came to assigning crews: he was cleared for general knowledge of what would be flying on a mission—was it a satellite going to geosynchronous orbit (GEO) atop an upper stage, or would it propel itself? Did the mission require an RMS operator? Or EVA? But few specifics beyond that.

Equally troubling was the air force's initial insistence that only active duty military officers be named to shuttle crews flying with its payloads. (Abbey killed that idea early on.) The service also wanted to assign a military payload specialist to every one of its shuttle missions.

In early 1979, at Mark's directive, the air force had recruited a cadre of thirteen potential payload specialists. Given the title "manned spaceflight engineers," these potential payload specialists were all drawn from Special Projects, the NRO mission control at Sunnyvale, and the NRO launch office at Cape Canaveral.

Brett Watterson, selected from the imaging world, felt that he understood the MSE role. "Our job was to integrate DOD payloads onto the shuttle." Given that the air force and NRO had been ordered to phase out the Titan III expendable launch vehicles, and would be relying solely on the shuttle in years to come, this was an important job.

Several of the young MSEs saw the assignment as a holy mission—"We were scouts sent out from a fort" said Jeff DeTroye, a former flight controller from Sunnyvale. "We figured it was our job to represent the air force to NASA, and to define that relationship for the future."

That meant pushing for opportunities that were far beyond those of the typical commercial or scientific payload specialist: some MSEs were convinced that they would operate the orbiter's RMS and do EVAs.

Among other disagreements, this aggressiveness was typified by an incident in the WET-F (Weightless Environment Training Facility). A crew systems engineer spoke with MSE Keith Wright about the design of an EVA tool for a DOD mission. Deciding there was no better time to test it than now, Wright, who was a qualified diver, grabbed a mask and tanks, and the tool, and splashed into the huge pool.

At the same time, however, two shuttle astronauts were doing EVA training. Their test conductor, seeing unscheduled and unknown activity in the water, went ballistic. Wright and the poor engineer were ejected from the facility.

News of the incident shot up the chain of command at NASA like, well, a rocket—not just to Abbey but also to Chris Kraft, who took to his telephone to express his unhappiness to General Kulpa. Wright effectively became persona non grata around JSC.

Watterson would say later that several of his colleagues "had more air speed than vector," that they seemed to fail to understand that they were "more like the principal investigators for a Spacelab mission. Our first job was to integrate our payload. Flying was secondary."

Interviews for the 1980 group began on Monday, February 25, and continued every other week through May 2—six groups in all.

On the morning of Wednesday, May 28, Abbey made the good news telephone calls, as he had two years earlier, starting with those on the East Coast: Franklin Chang-Díaz at MIT in Boston; Jim Bagian at the University of Pennsylvania; Capt. Charles Bolden and Lt. Cdr. Dick Richards at Patuxent River, Maryland; Capt. Bob Springer at Fleet Marine Force/Atlantic HQ in Norfolk, Virginia; navy lieutenant commander Mike Smith with Attack Squadron 75; air force lieutenant colonel John Blaha; and marine corps major Bryan O'Connor at the Pentagon.

Moving to mountain standard and Pacific time zones, Mary Cleave at Utah State University; Maj. Roy Bridges at Nellis Air Force Base, Nevada (though he was flying with the mysterious Detachment 3 at Area 51); marine captain Dave Hilmers at El Toro in Orange County; navy lieutenant commander Dave Leestma at Point Mugu.

Marine major Woody Spring happened to be in Tucson, Arizona, that morning. He was just returning from an assignment in South Korea and on his way to a new station at Fort Monmouth, New Jersey. Abbey made only one good news call to Edwards, to air force major Ron Grabe.

Among those at JSC learning of their selections were Bonnie Dunbar, Jerry Ross, and Mike Lounge. Anna Fisher's husband, Bill, was also given the good news.

Dunbar's office was in building 4, on the same floor as the astronaut office. Abbey's way of telling her she had been selected was to ask, "How would you like to move your office down the hall?"

One later phone call was to Capt. Guy Gardner, an air force test pilot stationed at Clark Air Base in the Philippines.

Sixteen of the nineteen were present at JSC on Monday, July 7, to be welcomed by NASA administrator Frosch, Abbey, and Kraft. (Gardner was still relocating from the Philippines, and Leestma and Grabe were also in the process of moving from California.) Both O'Connor and Ross would remember Kraft's words as harsh and mystifying. After first telling the new astronaut candidates that they were "not omnipotent," Kraft added, "Remember, we don't need any of you. So don't screw up."

One of the new ASCANs, doing the math and reflecting Kraft's blunt words, dubbed them the "Needless Nineteen, Plus Two." The plus two were ESA astronauts Nicollier and Ockels.

But Abbey needed them, and so did veterans in the astronaut office. Ken Mattingly, who was head of shuttle engineering and development, felt that the Needless Nineteen "came aboard much quicker and came up to speed because we practiced on the first group and kind of learned how to do it."

He compared the TFNGs and Needless Nineteen to "first and second children" in their differing personalities, with second children feeling less pressure and scrutiny. "That's an all-star cast. There's more superb talent in that than any single group. It was just really, really impressive."

(Two members of the group, Jerry Ross and Franklin Chang-Díaz, would go on to fly seven times each—a record unlikely to be broken before the middle of the twenty-first century.)

As for Kraft's harsh greeting, Abbey would simply say, "That was Kraft being Kraft."

As the new ASCANs were reporting, there was another huge shuttle management meeting, this time at the Cape. The goal, according to Richard Nygren, a member of the vehicle integration team, was "to settle on a realistic launch date."

In the meeting, Abbey turned to astronaut Bob Overmyer, who was leading the tile repair—the long pole in the schedule now. Overmyer reported that at the current rate of installation, it appeared that *Columbia* would be ready in November. But then Nygren stood up and said, "March, which made everyone unhappy."

Nygren turned out to be optimistic by a couple of weeks. But it was clear by late summer 1980 that, by God, *Columbia* was going to fly within the next six months.

Not all astronauts were willing to wait. Astronaut Ed Gibson was one of the first scientists selected as astronauts in 1965. He had flown the third Skylab mission in 1973–1974, then left Houston and the astronaut office to take a job with the Aerospace Corporation.

As the shuttle program ramped up, however, he heard from Joe Allen that a space station was finally in the works. After a meeting with Kraft, who assured him that a "space station was right around the corner," Gibson rejoined NASA in September 1977, then went off to Europe for a year to work with ERNO, a German-based company that was involved in Spacelab.

Abbey valued Gibson's experience and intelligence, and as soon as the 1978 astronaut candidates arrived, brought him back to Houston to serve as head of training for the mission specialist candidates. He was then assigned as ascent capcom for STS-1, after which he would likely have gone into mission training for Spacelab 2.

But while Gibson knew that "flying in the back of the orbiter would be interesting," a Spacelab mission of seven to ten days wouldn't compare with three months living, working, and more important, doing science as he had on Skylab. The Carter administration had zeroed out space station funding, and Gibson knew it would take years to restore it.

Gibson did a calculation, adding eight years to his age, and came up with a number that was over fifty. Uninspired by the idea of a Spacelab mission and a long wait for a theoretical space station, he went to see Abbey and told him he was leaving.

Abbey assigned TFNG Dan Brandenstein, Gibson's backup, to replace him.

By October 1980, George and Joyce Abbey were in court, arguing the terms of their divorce, which was finalized on October 21.

34 | STS-1

AS 1980 TURNED INTO 1981 the pace of preparation for the first shuttle launch increased. At the same time, a presidential election was in progress, and even before Ronald Reagan defeated incumbent Jimmy Carter on November 4, NASA administrator Frosch had announced his resignation.

Alan Lovelace took over as acting administrator. John Yardley literally moved from Washington, DC, to the Cape, to personally oversee the work.

On November 7, 1980, the solid rocket boosters and the external tank completed attachment to the orbiter in the Vehicle Assembly Building. On December 2 the fourth and final preliminary certification of the SSMEs was complete. The next day there was a final major test of the shuttle's integrated flight system at Marshall's operations support center.

Later that month, the fully stacked shuttle—*Columbia*, ET, SRBs—was rolled to pad 39A at the Kennedy Space Center.

On January 17, 1981, the SSMEs successfully completed their twelfth and final test firing—ten minutes and twenty-nine seconds, the longest yet—at the Mississippi Test Site.

At the Cape, on February 20, *Columbia*'s three main engines were fired on the pad for twenty seconds, the first readiness firing.

Finally, on March 19, John Young and Robert Crippen, now wearing their orange-colored pressure suits and accompanied by Abbey, boarded *Columbia* for the dry countdown demonstration test. (The pad crews started a day earlier; the whole test ran thirty-three hours.)

It went well; the astronauts returned to the crew quarters and were, according to Crippen, "patting each other on the back," noting that they were getting pretty close to launch.

One of the lingering issues from February's flight readiness firing was a suspected hydrogen gas leak in *Columbia*. For the dry run in March, the aft compartment of the orbiter was flooded with nitrogen, an inert gas, to prevent a possible fire or explosion. A team of Rockwell pad techs entered the compartment prematurely. Two died. Investigators found that the nitrogen purge had not been flagged as hazardous—and that the scheduled duration of the purge was not properly communicated to the pad team.

Abbey, Young, and Crippen learned of the tragedy during their flight back to Houston. Abbey was shocked; he'd experienced loss before, the deaths of fellow pilots and astronauts, either on the pad or in aircraft accidents. But this fatal accident was completely unexpected. There was nothing he could do, since the investigation would fall to the team at the Cape, but he made sure that Young and Crippen would publicly acknowledge the lost men during their flight.

At long last, all the flight readiness reviews, the planning, and the training were complete. A firm launch date loomed.

On Wednesday, April 8, 1981, Young and Crippen and Engle and Truly flew their T-38s to the Cape.

Abbey followed the next day on the NASA Gulfstream.

Shortly after 2:00 AM on Thursday, April 10, 1981, Abbey awoke and dressed. He checked on weather, then breakfast preparations, then went off to wake Young and Crippen.

At 2:45, as TFNG Loren Shriver worked in the *Columbia* cockpit setting up the switches, Young, Crippen, Engle, and Truly had breakfast. After the meal, Young proposed a card game, Possum's Fargo, establishing a prelaunch tradition.

Astronaut Paul Weitz took off in the Shuttle Training Aircraft, flying approaches to the Cape. Weitz's reports on wind, weather, and visibility would be vital if Young and Crippen had to execute an RTLS abort, the maneuver requiring "ten miracles."

Young and Crippen donned their pressure suits—not true space suits but a NASA version of the suits worn by SR-71 spy plane pilots.

Then, at 4:30 AM, in a ritual that Abbey would replicate for the next several years, the crew went downstairs and walked out to the waiting Astrovan, waving to well-wishers and the press. Abbey and the astronauts boarded it, then rode toward pad 39A five miles away.

The chat was almost nonexistent, nothing more than exchanges about the weather reports and countdown status.

The Astrovan stopped at the launch control center. Abbey shook hands with Young and Crippen and wished them a good, safe flight. Then he exited. The van then continued to the launch pad, where the astronauts entered the elevator that would carry them thirty stories up to the white room that gave access to the orbiter's hatch.

In the launch control center, Abbey took his place at a console and listened to the air-to-ground chatter, the launch conductor at KSC, and astronaut Dan Brandenstein at mission control in Houston.

As the count approached the last few minutes, Abbey felt his heart rate rising. He was acutely aware that for the first time in spaceflight history, human beings were being launched on a vehicle that had never been flown before. Abbey had faith in the engineering, but for a moment he pondered the chain of confident assumptions that led to this moment.

T minus one minute . . . as the count ticked down to ignition of the shuttle main engines, Abbey heard a hold—twenty minutes before the scheduled launch.

It was a computer problem, some glitch between one of *Columbia*'s five onboard general purpose computers and ground systems. Abbey knew the moment he heard it that there would be no launch today.

It took close to an hour, but eventually flight director Don Puddy's team agreed.

Abbey, Young, and Crippen had feared that the computer problem would force a delay of weeks, if not months. But the team at JSC resolved the issue within the day. STS-1 was rescheduled for 7:00 AM on Saturday, April 12, 1981—which happened to be the twentieth anniversary of the launch of Yury Gagarin on Vostok 1.

This time the new rituals of launch seemed familiar. Young and Crippen had done this several times by now. They and Abbey all wondered if today would finally be the day.

At 7:00, the three SSMEs lit up, one by one. The entire shuttle stack twanged, its nose swinging backward momentarily. Then, as it returned to vertical, the two SRBs ignited. And up she went with a shattering roar.

Abbey watched the television feed and listened to the air-to-ground between Young and Brandenstein, silently counting down to each major milestone. Go at throttle up. SRB separation, a huge moment.

Press to main engine cutoff.

Wait. Abbey tried to imagine Young and Crippen in the cockpit, experiencing what should now be a smooth ride.

Then—main engine cutoff, followed moments later by another vital step, separation of the huge external tank. The whole unlikely combination of main engines, orbiter, solid boosters—it had all worked.

Columbia was in orbit. The United States had returned to human spaceflight almost six years after Apollo-Soyuz.

Abbey flew back to Houston in a T-38 with Joe Engle to take up a position in mission control. Upon arriving, he learned the disturbing news that the first images from *Columbia* showed that several thermal protection tiles were missing from the pods holding the maneuvering rockets at the rear of the spacecraft.

Had tiles fallen off the bottom as well? Abbey had a hard time believing that, since the whole protection system had been tested and retested. Over the next several hours, there were hushed conversations above his pay grade, between program managers and HQ, about getting some help from the air force and the intelligence community. The air force operated a pair of ground-based telescopes—TEAL BLUE in Maui, Hawaii, and TEAL AMBER at Malabar,

Florida, not far from Cape Canaveral—that routinely tracked orbiting Soviet satellites and might produce useful images of *Columbia*.*

There was a third telescope atop Anderson Peak near Big Sur, California, primarily used to track missile launches out of Vandenberg Air Force Base to the south. NASA had, in fact, planned to use Anderson Peak to track *Columbia* on its descent from orbit into Edwards.

The intel community operated spy satellites that might also obtain imagery from orbit. Hans Mark, the incoming deputy administrator, and former director of NRO, knew that there were four imaging birds in orbit—a low-altitude GAMBIT and higher-altitude HEXAGON, which used film that had to be returned to Earth for processing, and two KENNEN vehicles. These used charged-couple devices and were capable of proving near-real-time imagery.

The problem was orbital geometry. The KENNENs were in polar orbits—*Columbia*'s was largely equatorial.

Of course, the unspoken issue was . . . if images showed serious damage to the underside, Young and Crippen were doomed: they had no ability to make repairs.

Whatever was going on, and something was definitely going on, Abbey had no direct involvement. And he had his doubts about the utility of any such systems. "Photos like that never really show you much."

All he could do was concentrate on the flight plan for the two-and-a-half-day mission. Whatever their concerns about the tile issue, Young and Crippen were generally upbeat, remarking on the views as they put the orbiter through its maneuvers.

There were minor problems. *Columbia* was equipped with data recorders for its test flights. One of them failed, and when Crippen tried to open a plate covering the device, he made no progress in loosening the screws. Young failed too. (They would learn later that an overcareful tech had sprayed a substance called Loctite on the screws to make sure they didn't come loose.) And the new space toilet didn't work.

Nevertheless, the two days went well, and on the evening of Monday, April 13, Abbey flew to Edwards Air Force Base.

---

* These telescopes and satellites were still classified in 1981.

The launch of *Columbia* using three main liquid-fuel engines and two giant solid rockets was a major milestone. Equally challenging was the orbiter's return to Earth.

Bright and early on a beautiful, sunny Tuesday morning, Abbey was on the lakebed landing strip at NASA Dryden.

Also present was a crowd of onlookers estimated at a quarter of a million.

Beginning around 10:00 AM, Abbey could hear Young and capcom Joe Allen, with their astonishing callouts. "We're at 265,000 feet doing Mach 18."

Months earlier, Abbey had assigned navy astronaut Jon McBride as lead for the chase team. Now McBride was in the air in a T-38 with TFNG George Nelson in the backseat taking pictures. The goal was to engage *Columbia* at forty thousand feet, just as the orbiter was going subsonic, and escort it to the lake bed runway.

Columbia signaled its arrival with a sharp pair of sonic booms.

Spectators could see a tiny black arrow in the sky that grew larger. On the lakebed, near the Astrovan, Abbey shaded his eyes and scanned the sky.

Almost before he knew it, *Columbia* and McBride's T-38 were on approach. The orbiter's landing speed was over two hundred miles an hour, twice that of a normal airliner.

Its main gear touched down . . . *Columbia* rolled forward, gently settling on its nose gear. Then it stopped, mission complete.

Abbey was waiting at the foot of the stairs as Young, still in his pressure suit, emerged, grinning. A quick handshake, then the veteran commander began to walk around *Columbia*, fists pumping.

"I was happy. John was *giddy*."

The first shuttle flight was a success. Now it just had to be repeated.

In the next week, Abbey led Young and Crippen through a lengthy debrief about their experience. There were nagging problems, but these could or would be addressed by Thompson's team, understandably energized with the success of STS-1 after years of delay.

At the press conference, the crew for STS-2, Joe Engle and Richard Truly, were handed a giant key, for a "used spacecraft."

Part V

“Go at Throttle Up”

35 | The Used Spaceship

AS THE SHUTTLE PROGRAM and George Abbey basked in the afterglow of a successful STS-1, a new group of agency leaders took over.

James Beggs, a senior executive from General Dynamics, was sworn in as the new administrator on July 10, 1981. So was Hans Mark, former director of the National Reconnaissance Office and before that, head of NASA Ames.

John Yardley resigned as head of spaceflight at HQ; Houston-based shuttle program director Robert Thompson decided that ten years was enough, too. Glynn Lunney, former manager of the Apollo-Soyuz Test Project, ultimately took over the shuttle in Houston.

Eager to reorganize and consolidate shuttle leadership at HQ, Beggs recruited Maj. Gen. James "Abe" Abrahamson, an air force test pilot and program manager whom he knew from the F-16 program. Abrahamson had also worked briefly with Mark. He had been an astronaut in the Manned Orbiting Lab program from 1967 to 1969, and been quite disappointed when judged to be too old for transfer to NASA.

As these leadership changes took place, the Needless Nineteen, Plus Two were well into their first round of technical assignments, and about to be certified as full-fledged astronauts.

Deputy chief astronaut Al Bean's understanding with Abbey was that he would remain as acting chief astronaut until after STS-1. Abbey hoped then to assign Bean as commander of a Spacelab mission.

Bean had other plans. In the spring and summer of 1981, Bean began to fly less and less, to "minimize using NASA assets." Abbey noticed and called him in. "You haven't been flying the T-38s lately." An astronaut had to meet certain minimums in order to remain eligible for flight assignment. Bean was effectively taking himself out of consideration and told Abbey that he planned to leave "in a few months."

According to Bean, Abbey was surprised. "Where are you going? What are you going to do?"

Bean's answer was, "I'm going to be an artist."

He says that Abbey was so shocked by this that he pushed his chair back from his desk so quickly that he bumped into the window behind him. "Can you earn a living at that?"

(Abbey remembered his reaction as less dramatic, but surprised nevertheless. "I knew Al had been painting for years. He even encouraged me to give it a try. But it seemed that he was talking about going from paint by numbers to gallery showings.")

Bean said he was prepared to work at Jack in the Box if he had to, to have time to pursue his dream.

There might have been another reason. Astronaut Walt Cunningham, in his salty 1977 memoir about his time at NASA, *The All-American Boys*, quotes an unnamed "friend"—clearly Bean, who was Cunningham's closest confidant during Apollo—as saying, "I look at the latest spacecraft, and I'm just not sure I want to fly it."

At any rate, Bean left NASA in August 1981 and did indeed become an acclaimed painter working in the style of Robert McCall—Abbey's favorite space artist.

Bean's departure allowed Abbey to give Crippen a leadership role; he appointed him as deputy chief of the astronaut office.

He also revised the crewing for the orbital flight tests. He had originally planned to have Young and Crippen turn around and serve as backups to Engle and Truly on STS-2, since the missions were so similar.

But as STS-1's launch slipped from its original date of October 1979 to April 1981, Young and Crippen had logged hundreds of hours in the STA and simulators. Abbey felt that they could do with a break, and he thought it was time to give other astronauts mission-specific training.

He created a new crew with Apollo veteran Ken Mattingly as commander, assigning Hank Hartsfield as pilot. He also augmented the roster of flight directors, appointing three from mission control or from the JSC staff: John Cox, Gary Coen, and Jay Greene. According to Greene, he didn't apply for the job—indeed, there was no actual recruitment or training program. "You'd sit around your office, and one day George Abbey would walk in and say, 'Guess what you're going to do.'"

As Greene remembered, it "wasn't a request."

Abbey and Gene Kranz had been scouting for new flight directors. "I encouraged some and discouraged others." He and Kranz, of course.

After five and a half years on the job, using the same methods that had served him so well in Apollo, and in his first year with Gilruth, Abbey felt that he had a good sense of the talent at JSC, especially in flight operations.

Which isn't to say that everyone else recognized this, or appreciated it. Given their natures—as pilots, leaders, pioneers, with good social skills, great intelligence, and obvious self-confidence—astronauts as a group were tough to manage.

By now the newer astronauts had realized that it wasn't John Young who made decisions about their careers—it was Abbey, the mysterious figure from the eighth floor of the headquarters building, the mover behind chili cook-offs and beer calls and highland games and who knew what else.

The astronaut office divided itself into three groups. First were those perceived, fairly or unfairly, to be Abbey's. Those included astronauts whose technical assignments included aircraft operations, technical assistant to Abbey, or flight support and capcom.

One example was Mike Smith. The North Carolina native was a graduate of the Naval Academy who had flown combat missions over Vietnam, then become a test pilot. His first technical assignment had been in the Shuttle Avionics Integration Laboratory. Then, in early 1982, he moved up to the ninth floor to take Joe Allen's place as Abbey's technical assistant. Partly because of their shared Annapolis background, but also thanks to Smith's relaxed outgoing manner, he and Abbey became good friends.

Nevertheless, when it came time to begin assigning members of the 1980 class as shuttle pilots, Smith was nowhere near the first to be chosen.

Then there were those who were on the outs with Abbey, and this group—small in number—consisted of holdovers from the Apollo days, scientist-astronauts like Musgrave and Don Lind. Perhaps they still thought of Abbey, as many in mission control did, as a faceless "horse-holder" who had somehow magically worked his way into a powerful job.

It was also because they suspected, correctly, that they really didn't figure in Abbey's plans for the future. Musgrave believed that Abbey was willing to fly the Apollo holdovers once in the shuttle program, "a thank you" for their years of service, then good-bye.

He was correct, at least as far as the scientist-astronauts were concerned. There were talented astronauts in that group, but few (Kerwin, Garriott, Gibson) had flight experience. Age was an issue: in 1982, Karl Henize was fifty-six, Kathryn Thornton fifty-three, and Lind fifty-two, though Tony England, born in 1942, was actually younger than several of the TFNGs.

Abbey's rationale was that, because of their age, the holdovers couldn't realistically be expected to have long careers. In order to create a cadre of crew members to support a dozen or twenty flights a year, Abbey felt that he would be relying on the 1978 and 1980 groups. He was willing to consider multiple flights for these veterans "if they did a good job."

Most astronauts would have been classed simply as "solid citizens," those who showed up, did their work, and assumed that that would lead to a flight assignment.

Sometimes those who worked for Abbey found ways to get his attention.

On Friday, August 21—his forty-ninth birthday—Abbey was in his eighth-floor office when he heard and saw something outside.

It was Superman, though not flying as he had in television shows or comic books—he was dangling from a rope, pounding on the glass and singing "Happy birthday to you . . . !" Then Superman dropped out of sight, rappelling to the ground and disappearing.

Soon, though, word reached Abbey from a very upset Kraft that two culprits, dressed in overalls, had talked their way into building 1 and up to the ninth floor pretending to be window washers.

There they had found an unoccupied office and window overlooking the northwest corner, then attached rope to stanchions mounted on the exterior.

One of the pair took off the overalls, revealing a Superman costume, and then rappelled off the building to complete the prank.

As Superman fled, the accomplice packed up the gear and left the building.

JSC security soon identified the culprits, however—Superman was Jim Bagian and the accomplice was Guy Gardner. Both were astronauts from the 1980 group.

Cliff Charlesworth, the center's deputy director, was not amused and convinced Kraft to order Abbey to order John Young to deliver appropriate written punishment.

Young met with the pair—showed them the reprimand—then balled it up. "I thought it was shit-hot."

At the other end of the spectrum, there was the mysterious portrait of Abbey that found its way aboard *Columbia*. El Onizuka, astronaut support for STS-2, and definitely in the bubba group, connived with Steve Hawley and Kathy Sullivan, two other TFNGs assigned to support, to place a portrait of Abbey on the middeck with the caption, possibly ironic, "To George W. S. Abbey, our leader and inspiration."

It would fly on STS-3, 4 and 5 . . . then disappear.

STS-2 had originally been scheduled for launch on October 9, but that had slipped when a nitrogen tetroxide leak from *Columbia*'s forward reaction control jets had damaged several thermal tiles on the vehicle's nose. Three hundred had to be removed, refurbished, and replaced.

Fortunately, the location of the tiles allowed repairs to be completed while *Columbia* remained on pad 39A, saving a time-consuming trip back to the Orbiter Processing Facility.

Engle and Truly were aboard *Columbia* on the morning of November 4, ready for liftoff, when two of the orbiter's three auxiliary power units began showing high oil pressures, a sign of a larger problem. The launch was postponed again while repairs were made.

Columbia finally lifted off on November 12, 1981, on what was planned to be a five-day mission in which the Canadian robot arm would be tested and the astronauts would operate several Earth resources experiments mounted in the orbiter's payload bay.

The first hours of the mission went fine. And on day two, President Reagan visited JSC mission control and invited Engle and Truly to the White House, once STS-2 was over. But then one of the orbiter's three fuel cells failed. Mission rules dictated a "minimum mission," two and a half days, rather than the scheduled five.

James Beggs, overseeing his first shuttle mission, insisted on taking part in the high-level telecons that included deputy administrator Hans Mark, shuttle program director Thompson, Abbey, and others, including Kraft.

When the fuel cell issue was raised, Beggs wanted to continue the mission.

Kraft just snapped, "You don't know what the hell you're talking about." Glynn Lunney, who was present, said that Kraft "was absolutely correct."

But it was not the kind of thing you say to your boss, not in public, and to Abbey's mind, it was exactly the kind of outburst that Sig Sjoberg would have prevented. He wondered how Beggs would strike back.

On operational matters, Engle and Truly did test the RMS, and operated the Earth resources platforms, and made a more challenging landing at Edwards.

At home, Abbey continued on as a single father. George Jr., now twenty-four, was out of the house and doing additional college work at the University of Minnesota. He was planning to return to Houston in the coming months. Joyce BK was in her last year at Stephen F. Austin and about to return to Houston, "boyfriend in tow." That still left three minors in the house, from Suzanne at seventeen to Jimmy at fourteen and Andy at nine.

Abbey was increasingly reliant on Suzanne, who was entering her junior year in high school. To make her job of family driver somewhat easier, he bought her a new car. "Which only made me nervous," she said.

Jimmy was about to enter the New Mexico Military Academy in Roswell. He wanted to pursue a military career like his father, and both believed that this school would provide better prep for Annapolis.

Andy was becoming the family athlete, playing soccer like his grandfather. With his father's encouragement, he turned to golf. His father gave him a set of Jackie Burke clubs that had belonged to Jack Swigert, who had made them a gift to Abbey. "Let's just say it was not the best set," Andy said later. He became a skilled golfer in spite of the equipment.

Abbey still traveled to Seattle whenever possible, to see his parents and his sister, Phyllis, who was battling breast cancer. Since their three brothers were so much older, Phyllis had been Abbey's closest sibling, the one he spent summers with. She had married a navy man named Ray Williams and spent several years going from base to base with him, until he settled down in Seattle in civilian life, working for a grocery firm, then in administration at Boeing.

Phyllis took a job at a bank, "eventually almost running the place," Abbey said. But in the early 1970s she was exposed to a toxic hair dye at a salon. The stylists were supposed to test it first, since some women were known to react badly. But they didn't do the test—and Phyllis had a terrible reaction, winding up in the hospital. She later developed breast cancer.

After a valiant battle, she died in Seattle on December 23, 1981.

36 | Operational Flight

IN 1981 THE SHUTTLE had accomplished two flights and proven that it would work as designed. Four missions were planned for 1982; in early 1983 a second orbiter, *Challenger*, would be ready . . . and the pace of launches would ramp up.

Lousma-Fullerton and Mattingly-Hartsfield were assigned to STS-3 and STS-4.

ASTP veteran Vance Brand and former MOL astronaut Bob Overmyer were named CDR and PLT for STS-5, the first "operational" shuttle mission, scheduled for launch in November 1982 and intended to deploy two commercial communications satellites. Apollo-era scientist astronauts Joe Allen and William Lenoir were added as mission specialists.

Skylab veteran Paul "P.J." Weitz was given command of STS-6, with former MOL astronaut Bo Bobko as PLT. Another MOL pilot, Don Peterson, was assigned as a mission specialist, as was Apollo scientist-astronaut Story Musgrave. STS-6 would launch the first Tracking and Data Relay Satellite, part of a planned network that would allow shuttle crews to remain in almost constant contact with mission control.

Jack Lousma and Gordon Fullerton had been assigned to STS-3 in the spring of 1981, though there had been no public announcement. (They had been training together since the fall of 1979.)

The training went well. *Columbia* was processed for a new flight in a record time of seventy days.

STS-3 was designed to be a major step in the evolution of the shuttle system, a weeklong mission that would expand the test program, concentrating on the RMS.

The development flight instrumentation package was mounted in the payload bay as was a set of experiments for NASA's Office of Space Science. For the first time the orbiter also carried small "Getaway Special" packages, and the McDonnell Douglas continuous flow electrophoresis system.

The crew would be in orbit for seven days and would perform 270 different maneuvers.

One complication arose: severe rains in southern California flooded the Edwards area, inundating the lake bed used for shuttle landings in the test program. A week before launch, Kraft and Abbey went to see Lousma and Fullerton in quarantine to tell them. Future missions would land at the runway at the Cape, but no one believed that the system was yet mature enough for that.

The options were to postpone STS-3 until Edwards was ready or land the orbiter at the Northrup Strip site at White Sands, New Mexico. This was where all shuttle commanders flew dozens of practice landings in the STA—Lousma and Fullerton knew that approach better than they knew Edwards, so both agreed to make the switch.

Forty train car loads of support equipment were shipped to New Mexico for the event.

Columbia launched on March 22, just seventy days after the orbiter arrived at the processing facility at the Cape—a record turnaround, though one that would still have to be improved if the shuttle ever hoped to meet NASA's increasingly ambitious schedule.

Lousma and Fullerton settled into a happy routine of operating a suite of scientific experiments in *Columbia*'s payload bay, testing the RMS, and operating the first collection of science experiments inside the flight deck.

One of STS-3's development test objectives was to leave the autopilot engaged to a low altitude, a decision that, given the winds and last-minute adjustment of *Columbia*'s approach to the runway, did not give Lousma enough time to retake control of the vehicle.

The main gear touched down and Lousma felt the nose sinking too quickly, so he used the hand controller to pitch up. Seeing no response, he repeated the input.

Columbia's nose rose again in a "kind of a wheelie." Advised by Fullerton to let the vehicle fly itself, Lousma did just that. *Columbia* quickly settled down, slowly, on its front wheel.

Fullerton would blame the software, noting that while "The gains between the stick and the elevons" were "good for flying up in the air," they were "not good when the main wheels were on the ground, and he thought he had ballooned."

He dismissed it as a minor problem, easily corrected. To Abbey, it was another case where "Fullerton saved the day again."

In early 1982, with HQ resisting his desires to recruit yet more astronauts, Abbey had had to produce a study justifying his manpower needs for the next five or six years, based on the shuttle manifest, which called for up to a dozen flights by 1985.

The study provided some insight into his thinking at the time, since it projected the attrition of most veteran astronauts, including Young, after one flight. The ex-MOL astronauts would fly as PLTs once, then as commanders. The last two might get multiple commands.

Even the first TFNGs to win piloting spots would rotate to a single command before leaving. Those who followed them, however, might go on to a second pilot assignment and then four or five commands through 1988.

The first handful of TFNG mission specialists would fly two missions in quick succession followed by two more before attrition. The next ten would fly as many as eight missions, sometimes two a year. The last in line would fly two or three science missions during the same span.

The document showed that the shuttle would require on the order of thirty-five new astronauts beginning in 1984, with new arrivals eligible for flights as early as 1986.

Just as Abbey was submitting his proposal, Jim Beggs finally struck at Chris Kraft.

Kraft had been serving as JSC director as a "retired annuitant," a status that allowed him to collect his civil service pension while being paid a salary—much like a retired military officer working for a civilian company.

Max Faget and Deke Slayton had taken Kraft's advice and opted for the same deal.

The catch was, annuitants had to be renewed annually. And in March 1982, Beggs announced that they would not be. Meaning that Faget and Slayton were out. And in July, when his deal expired, so was Kraft.

Kraft's firing wasn't all due to embarrassing his boss on STS-2. He and Beggs were in conflict over future shuttle operations. Beggs and Mark were eager to commercialize shuttle processing at the Cape, putting the contract out for bid even if it meant awarding it to a company other than Rockwell. According to Neil Hutchinson, "Chris and a lot of other people, me included, were very uncomfortable with turning over this very sophisticated, one-of-a-kind, hard-to-understand, complicated piece of hardware to a contractor, to turn it around between flights, who had not built it. I probably am not privy to the kinds of decision processes that went on, but Chris lost his job."

With STS-3 complete, only one orbital flight test remained, STS-4 in late June 1982. Abbey had promised that flight to Mattingly and Hartsfield, and they were officially assigned.

Crews of four for the first two "operational" flights, STS-5 in November and STS-6 in early 1983, were in training. All eight astronauts were veterans of the Apollo and Skylab eras.

If those missions went as planned, the launch rate would rise: STS-7 in May 1983 (deploying a pair of commercial satellites), STS-8 two months later (deploying the second Tracking and Data Relay Satellite), STS-9 with Spacelab 1, and a DOD mission, STS-10, before the end of the year.

There was still the question of how and when to fly the other holdovers from Apollo, former scientist-astronauts Kerwin, Lind, England, Henize, and Thornton, as well as pilot Bruce McCandless. McCandless was a possible candidate for PLT on STS-7, but with his work on the Manned Maneuvering Unit was also an intriguing candidate for that test flight, likely on STS-11.

As for the others, Spacelab missions beckoned. For one thing, these eight- to ten-day science flights fit well with the holdovers' skill sets.

Abbey was also reluctant to assign too many TFNGs to Spacelab, since the lengthy training program—eighteen months to two years—would be a diversion from vital early operational experience. Assigning the Apollo holdovers to Spacelab 2 and Spacelab 3 would solve both problems.

Spacelab 3 was dedicated to life sciences and space medicine, and seemed perfect for veteran Joe Kerwin. But when Abbey offered the assignment, Kerwin was unwilling to commit. Although Abbey never heard it directly, word was that Kerwin wanted to fly on a more challenging shuttle mission, possibly the Solar Max satellite retrieval scheduled for STS-13.

Kerwin left the meeting with the issue unresolved. Within a month he accepted a post of NASA senior science representative to Australia, an assignment that took him away from JSC for two years, and out of the flight assignment pool essentially forever.

Assigning Bill Thornton to Spacelab 3 was not a problem. The gruff, determined physician had made himself JSC's resident expert on space adaptation syndrome. After fifteen years in Houston, he wanted a flight.

Spacelab 2 was trickier. This mission was dedicated to astronomy, so would be a likely place for astronomer Karl Henize. In the spring of 1982 Henize was already fifty-six years old—clearly not a candidate for a long career in the shuttle program. But he was capable and well-liked, and a logical choice for Spacelab 2.

Another candidate was Tony England, the youngest of the Apollo holdovers—in fact, he was four years younger than TFNG John Fabian. England had served as mission scientist for Apollo 16, then left the agency to spend the next seven years with the US Geological Survey.

Then there was Don Lind. Already in his early fifties, Lind had been an astronaut since 1966, selected with Vance Brand, Paul Weitz, and Joe Engle. Even though he was a PhD physicist, Lind had gained enough flying experience with the navy to qualify as a NASA pilot-astronaut and had been classed that way for a decade, serving as backup to Jack Lousma and Bill Pogue on the second and third Skylab missions.

The first time Abbey had a conversation with Lind, sometime in 1970, the astronaut had openly complained about Deke Slayton and Al Shepard, who had not only transferred him from Apollo lunar landing to Skylab but had the temerity to assign geologist Jack Schmitt to Apollo 17 instead of Lind. Schmitt was then among Abbey's closest friends in Houston, and as a geologist was the consensus best choice to be the first scientist on the Moon.

In the big post-Skylab reorganization of the astronaut office, Lind was reclassified as a scientist, not a pilot, and kept up his complaints on any and all subjects. (He had gone to the press back in 1969, to protest his transfer to Skylab.) Abbey heard it, or as much as he needed, and took consolation in the fact that Lind lumped him in with Slayton and Shepard. Good company.

He was willing to give Lind a chance to fly, but he had to grit his teeth. Where and when would depend on a shuttle commander willing to accept Lind.

It was time to start flying the new generation.

The decisions would be subject to Kraft's approval. Abbey would talk them over with Young and with department heads in engineering, mission control and science, and especially in aircraft ops.

But these assignments, like those for the first six missions, were all Abbey's.

First up were commanders. Crippen had distinguished himself on STS-1 and so had Truly on STS-2. They had also learned from their more experienced mentors, Young and Engle, and were ready to pass in their lessons. They would lead crews on STS-7 and STS-8, with Young taking the first Spacelab, STS-9.

As pilots, Abbey selected Rick Hauck for 7, Dan Brandenstein for 8, and Brewster Shaw for 9. All three were capable and struck Abbey as ready to serve as commanders on their second flights. Four mission specialists were needed for 7 and 8. (The two scientific mission specialists for STS-9 had been assigned back in 1978.) Abbey's first goal was to assign MS who were established leaders, and who could fly again soon on more complex downstream missions. His second goal, however, was to fly the first American woman and the first African American. His personal choices were Ride and Bluford, and he presented them to Kraft, who immediately wanted to discuss the choice of Ride.

"What about Anna Fisher?" Kraft suggested.

So, as often happened with Kraft, Abbey had to justify his selection. He worked with Pete Clements to create a matrix evaluation of the six women and two African American mission specialists (apparently TFNG Fred Gregory was not considered a candidate for PLT on either mission) as well as Abbey's preliminary choices for white male mission specialists, noting their general background (orbiter systems, payload, control center and support crew work), STS-7/8 mission-specific experience (RMS, proximity operations, general purpose computers and EVA), motion sickness tendency and medical qualifications, and general comments.

In this document Anna Fisher, Judy Resnik, and Sally Ride were the three leading candidates. Resnik and Ride were judged to be the most qualified, with Ride having a slight edge over Resnik on orbital systems. The other women all had squares to fill, though Fisher was cited (as Resnik and Ride were) for "outstanding public presence." The clincher for choosing Ride, for Abbey, was her skill at operating the RMS.

Bluford and McNair were rated equally, with McNair cited for his outstanding public presence. Bluford had an overwhelming edge in operational experience, however—he was a pilot and a combat veteran. He was also, like Ride, skilled on the RMS.

Ride would fly on STS-7; Bluford on STS-8.

Clements's study also rated Abbey's STS-7 choice, John Fabian, as a "leader, well-organized" with experience in prox ops and RMS. STS-8's Dale Gardner ("extremely well-qualified on STS software") was also evaluated.

During the first week in April, Abbey called Hauck, Fabian, and Ride into his office individually. When Ride gleefully accepted the assignment, Abbey took her to meet with Kraft so the JSC director could give her some idea of the media scrutiny she would face.

Truly's crew learned the good news as a group. Guy Bluford recalled being taken aside by John Young after a meeting and told to report to Abbey's office at 11:00 AM. On the way over, he ran into Brandenstein and Gardner, who were heading there too.

Truly was with Abbey, who praised their work on support crews and said, "We need a crew to fly STS-8. Are you interested?" After the three TFNGs said yes, Truly smiled and said, "Can I fly with you guys?"

Brewster Shaw was also called into Abbey's that week and told he was going to fly as John Young's pilot on STS-9, the first Spacelab.

With no warning to the others in the office—though some suspected that new assignments were going to be made—Abbey appeared at the pilot's meeting on Monday, April 19, and announced, "We've made some crew assignments." Then he simply read the names for STS-7, STS-8, and STS-9. "Hopefully we'll get some more people assigned soon."

He left, and missed the mixture of congratulations and disappointment. Most were genuinely pleased that their comrades were flying, hoping that it meant their own turn was coming soon.

But, as Mullane said, "With Abbey's words, TFNG comraderie vaporized."

Nevertheless, that night there was a party at the Outpost to celebrate the new assignments. After a beer or two, Hoot Gibson—who had expected Hauck to be the first TFNG pilot to be assigned, because "he was that sharp and that good"—grabbed Hawley and dragged him over to Abbey.

"Stevie and I wanted to tell you that you screwed up today," Gibson remembers saying.

"How so?"

"You didn't assign us." Gibson could tell that Hawley "wanted to be somewhere else."

Abbey smiled and told Gibson to come see him the next day. Gibson hoped it might be word of an assignment to STS-10, the next mission in line. But it was to be the new deputy for aircraft operations.

There was real grumbling, of course, and it reached Abbey. Not that he was upset or surprised. "They were competitive people. Obviously, everyone wanted to be first."

One orbital flight test remained to be flown—STS-4 in late June 1982. It was to be the first Department of Defense mission. As early as 1979 the air force had booked the June 1982 shuttle mission for an undisclosed payload. At the time, June 1982 was supposed to be STS-19. As the first shuttle mission slipped to the right, it became STS-7. And finally STS-4.

The original payload was to have been a pallet-mounted imaging reconnaissance system called ZEUS, but by early 1981 that had been scrapped and replaced with a test bed for an infrared detector. Mattingly would later call it a "rinky-dink collection of junk," but it served as a worthwhile test run for NASA in its dealings with the air force.

For example, what information could be released about the payload, and mission operations?

Would air-to-ground communications be encrypted? NASA wanted them in the open. Air Force Space Command compromised by ordering the astronauts and capcoms and paycoms to speak in code. One of the paycoms, Jeff DeTroye, recalled a debate over how the Satellite Control Facility was to identify itself. Air-to-ground protocol had the orbiter identified as "Columbia" and mission control in JSC as "Houston," but apparently one general disliked the idea of hearing "Sunnyvale." "What do we use?" DeTroye wondered. "Saratoga?"

The codes resulted in some comic moments during the flight. Hartsfield would remember that near the end of the mission, he and Mattingly locked up all the classified flight materials. Thirty minutes later, Sunnyvale requested the astronauts to "do Tab November" on their checklist. Neither astronaut could recall what this meant, so they unlocked the drawer, removed the materials and looked for Tab November on the checklist. It ordered them to stow the classified materials.

STS-4 ended with a spectacular landing on the concrete runway at Edwards on the Fourth of July, with an estimated half a million spectators present, including President Ronald Reagan and his wife, Nancy.

The next new orbiter, *Challenger*, was parked on the runway, too.

Abbey learned of one problem, however. During final approach, Mattingly suffered a surprising bout of vertigo. The worst of it passed, but Mattingly gave up on his plan to let Hartsfield take the controls, forcing himself to concentrate and listen to his pilot's callouts. Even so, "it was slower than normal," and *Columbia* landed further down the runway than Mattingly wanted.

Given Lousma's similar issue on the STS-3 landing, Abbey and Young were concerned, ultimately deciding to change the flight plan to give the commander control of the orbiter much earlier.

Mattingly had always wanted to return to the navy after his NASA tour and had thought STS-4 would be his farewell to flying in space. But, confident that Mattingly's issues weren't unique or unsolvable, Abbey asked him if he wanted to fly the next DOD mission, STS-10. With a new flight on the table, and a projected launch date less than a year away, Mattingly accepted it and postponed his departure.

Abbey assigned three TFNGs as Mattingly's crewmates: air force test pilot Loren Shriver, marine pilot and engineer Jim Buchli, and his good friend air force flight test engineer El Onizuka. There would also be an air force manned spaceflight engineer at some point.

With the end of the orbital flight test program, and the increased flight rate, Abbey and Young realigned and reorganized the astronaut office. There were seventy-five active astronauts, two dozen of them assigned to actual missions, with the other fifty in a variety of technical assignments.

There would be a mission development branch, led by Mattingly, that would concentrate on upper stages, EVA servicing and maintenance, life sciences, materials science, small payloads, and mission integration. And also a larger mission support branch managing capcoms and Cape Crusaders (astronauts assigned to Cape Canaveral), SAIL and software, flight data files and training.

There was a group of ongoing "firefighter" tasks that included orbiter displays, landing and rollout matters, cockpit upgrades, and Shuttle Training Aircraft. Astronauts responsible for these matters reported directly to Young and Crippen.

Finally, there were astronauts who had detached duties such as deputy director for aircraft ops at Ellington, head of safety, lead for astronaut appearances, technical assistant in Abbey's office, and representative to the shuttle program integration office.

STS-5, on *Columbia*, would be the first "operational" shuttle flight, a designation intended to show that the system had completed flight tests and was now suitable for regularly scheduled "safe" flights. (The "operational" term was usually applied to military or commercial aircraft after hundreds of test flights, not four.)

Abbey loathed the term. In truth, all it meant was that NASA was now willing to launch crew members who had zero chance of escaping a disaster on ascent. Not that the flight deck ejection seats provided any real safety—the pilots of the flight tests could not have used them if a problem occurred with the solid rocket boosters. The seats were still installed, but their explosives were removed, and they were "pinned," unable to launch. The crew debated the matter in one of its first meetings, with Allen and Lenoir both telling Brand and Overmyer to keep the seats active. "If it's nobody gets out versus two get out—"

Brand was adamant. If there was a problem, nobody would get out. "It's not selfless," he told the crew. He didn't believe he could live with himself if he and Overmyer ejected, leaving Allen and Lenoir to die.

The crew would no longer wear the orange high-altitude pressure suits, but rather blue flight uniforms—trousers, polo-style shirts, and jackets.

For protection during on-pad aborts, and against the sheer noise of liftoff, they would wear helmets.

In the crew, Allen and Lenoir were both noted for their humor. Looking at his crew, Allen joked that he was "the impedance matching device between Bill Lenoir, an extremely smart, well-disciplined, very tightly wound individual, and Vance Brand and Bob Overmyer." These two pilots, both with marine corps backgrounds, according to Allen "had a much higher tolerance for people being not quite so intense."

(According to Allen, "impedance matching is an engineering term for getting very unlike electrical circuits to communicate.")

When he uttered this to Abbey, "he did not find it as amusing" as Allen thought it was.

At a press conference, reporters noted that Brand would be commanding, Overmyer would be piloting, and Lenoir—seated behind them on the flight deck—would be the flight engineer. But what would Allen be doing alone on the middeck? "I think I'm just in charge of religious activities," he said. That is, praying.

Lenoir's wit was a bit edgier: during the walkout, he surprised Abbey by producing a crude placard—a plain white sign with the words NASA OFFICIAL GEORGE ABBEY and a pointer in Abbey's general direction. Lenoir was responding to a recent newspaper article that quoted Abbey, but not by name. Abbey wasn't amused. He didn't care about being singled out but felt that Lenoir wasn't demonstrating the right awareness of his situation: shuttle launches were dangerous business.

There was another far more serious element to the crew walkout—SWAT teams had been deployed in response to a security threat. There was even an armed officer in the Astrovan.

Allen turned to Abbey and joked that the flight ops director didn't need such high-pressure tactics. "We *are* going through with this launch."

Columbia launched at 7:19 AM on November 11, 1982, the first time a shuttle had lifted off on its first scheduled attempt.

STS-5 was a qualified success. The orbiter operated well and the crew returned safely. The crew demonstrated the first deployments from the orbiter, launching the SBS-3 and Anik C3 communications satellites atop Payload Assist Modules.

Another major goal was not met. Allen and Lenoir were to have performed the first EVA or space walk from an orbiter, but the first scheduled attempt had to be postponed because Lenoir was ill—suffering what NASA doctors later labeled space adaptation syndrome.

When Lenoir was judged ready to go, Allen's space suit developed a problem. A ventilation fan started slowly, made ugly noises "like a motorboat," and then shut down. An oxygen regulator on Lenoir's suit failed to generate sufficient pressure. And some of the helmet-mounted lights on both suits were also bad.

Commander Brand canceled the EVA. Nevertheless, *Columbia* and the crew of STS-5 landed safely at Edwards on November 17, 1982.

While engineers addressed the space suit issue, which turned out to be due to carelessness by the contractor, Hamilton-Standard, Abbey had to deal with space adaptation syndrome.

Astronauts on Apollo and Skylab had shown signs of space sickness, and now so had half of those who had flown on the shuttle. Program managers felt that the subject deserved immediate and direct study, and Abbey had a solution: add astronaut-physicians to upcoming crews. There was no time to slide a fifth crew member into STS-6, scheduled for January 1983, but STS-7 wouldn't fly until April and STS-8 in July.

This wasn't just an attempt to investigate a potential medical issue. Abbey had felt that the workload on a shuttle mission required three mission specialists, not two—and certainly not the single MS originally envisioned back in the 1970s. Increasing crews to five for STS-7 and STS-8 would give him a chance to test this theory.

TFNG Norm Thagard and Apollo holdover Bill Thornton were already pointed at the Spacelab 3 mission in late 1984. Abbey was able to give both the good news that they would be flying earlier—Thagard on STS-7 and Thornton on STS-8. Thornton was considered one of JSC's specialists in space adaptation syndrome.

Into his late eighties and early nineties Sam Abbey remained active and vigorous in an old-school manner. Even in the coldest Seattle weather, he insisted on leaving some windows open, for fresh air. In December 1982, however, the habit—and age—caught up with him. He contracted pneumonia and was hospitalized. Shortly after Christmas, Brenta telephoned George in Houston and told him he needed to come home.

Abbey flew to Seattle and went to the hospital, where he was only able to watch as this fit, fierce, busy man grew weaker and weaker. He was in the room when his father died on New Year's Day. Sam Abbey was ninety-two.

Chris Kraft announced his retirement from the JSC directorship in August 1982. The official statement declared that he had stayed on through completion of the orbital flight test program—there was no mention of his conflicts with Beggs.

The announcement was a formality—Abbey had known that Kraft was leaving for months and had no clear idea of who would succeed him. Charlesworth, the deputy director, was not a likely candidate.

Then Abbey heard that Gerry Griffin was Beggs's choice.

Griffin was a former flight director, having proved himself on the tricky lightning-struck Apollo 12 launch and as one of the heroic team on Apollo 13. He had been lead flight director for two lunar landing missions, Apollo 15 and Apollo 17. He had gone on to a tour as legislative liaison, then deputy associate administrator for space ops at NASA HQ from 1973 to 1976, before transferring to NASA Dryden as deputy director. He had worked there under former astronaut Dave Scott during the shuttle ALT program.

Later tours took him to the Kennedy Space Center, back to HQ (where he was part of the Reagan administration transition team, which resulted in the appointment of Beggs), before leaving the agency in 1981 to work again for Scott in a private space firm.

And his twin brother Richard, an air force colonel, was heading the service's detachment at JSC.

Griffin was a logical choice. But one of his first decisions was to reinstate the old separation of mission operations (flight control) from flight crew, making Gene Kranz head of that division and Abbey director of flight crew operations.

"The moment I learned about the decision, I knew that Gerry would reorient the divisions. He had memories of the flight controller battles with Deke during Apollo. Kraft knew all that, too, and had shared Gerry's position—until he spent a couple of years in center management. By the time he put me in the job, he ordered me to bring flight crew and mission control together in one organization and make them work together.

"But you had Gerry coming in, Charlesworth was from mission control, and Gene agitating for a separate operation."

Kranz made immediate use of his new authority—responding to the increased rate of shuttle missions by selecting eight new flight directors in March, among them Milt Heflin, Chuck Shaw, and Cleon Lacefield, whom Abbey would work with in a variety of modes over the next two decades.

Even though the program had only completed one operational flight, Abbey saw the pace of launches ramping up, and needed to put more crews into the training flow—his general rule was to assign them a year from scheduled launch.

STS-11 in January 1984 was scheduled to deploy a pair of comsats and also to make the first tests of the free-flying Manned Maneuvering Unit (MMU), the so-called Buck Rogers backpack. STS-12, in March, would deploy satellites on the first flight of the third orbiter, *Discovery*.

During the first week of February 1983, Abbey called Vance Brand, veteran of Apollo-Soyuz and STS-5, to discuss STS-11 and to finalize the crew.

First up was the pilot. TFNG Hoot Gibson got called to Abbey's office, where Abbey and Brand waited. Gibson had been exiled—he thought—to Ellington as deputy for aircraft ops. He wanted a technical job that put him closer to shuttle ops, so when Abbey asked if he had something in mind to follow his current assignment, Gibson said, "I'd like to be capcom."

Amused, Abbey said, "Do you have any interest in any of the upcoming missions?"

Still not understanding his situation, Gibson said, "I guess."

"Well, if you're not against it," Abbey said, "we were thinking that you might like to go with Mr. Brand here on STS-11."

Abbey then turned to the mission specialists, assigning TFNG Ron McNair as the RMS operator, and army test pilot Bob Stewart to test the MMU.

As for the lead MMU astronaut, Abbey was delighted to offer it to veteran navy pilot Bruce McCandless.

Selected in the April 1966 group with Joe Engle and Don Lind, McCandless had gone on to serve in support of Apollo (he was the capcom for Armstrong and Aldrin's lunar EVA) and in a backup assignment for the first Skylab mission. For Skylab he also worked as coinvestigator of an astronaut maneuvering unit that was flight-tested inside the station. He had then worked on shuttle development, on upper stages, the Hubble Space Telescope, and EVA procedures.

With mounting interest and concern, he watched Crippen, Truly, Fullerton, and other former MOL pilots—all of whom had come to NASA after him—receive flight assignments and fly missions. He told himself they were test pilots, even as he continued to fly T-38s and the Shuttle Training Aircraft.

Abbey knew this, of course. As with Don Peterson, he had not really considered McCandless a great candidate to be a shuttle commander so had not assigned him as a pilot. But given McCandless's years of involvement with the MMU and his knowledge of the device, chose to give him the first shot at test-flying it.

One morning in February 1983, Abbey called McCandless up to the eighth floor to discuss the mission. "You can fly the MMU as a mission specialist," Abbey told him. "Or you can wait for an opening for pilot."

McCandless didn't hesitate—the MMU flight would be a notable event, the free and untethered flight of a so-called "Buck Rogers" backpack. He would be happy to take that assignment.

Hartsfield's STS-12 crew was next, with TFNG Mike Coats as the pilot, and mission specialists Steven Hawley, Judith Resnik, and Mike Mullane.

Two weeks later Abbey was assigning more crews—for STS-13 and for Spacelab 3, which had slipped ahead of Spacelab 2 in the schedule.

STS-13 was an ambitious flight, featuring the EVA retrieval and repair of the errant Solar Maximum Mission satellite, which had been launched February 14, 1980, and suffered a critical failure soon after.

Abbey chose to give this assignment to Crippen, even though Crip had yet to fly STS-7. The veteran of STS-1 and distinguished member of the Ace Moving Company was proving to be a true leader; he also possessed an Abbey-like memory for technical and personal matters.

The rest of the crew would typify Deke Slayton's adage that while "all astronauts are created equal, some are more equal than others."

The mission would see a tricky rendezvous with an orbiting satellite, so Abbey and Crippen chose Dick Scobee, a former air force test pilot who had flown the X-24 lifting body. He was highly respected and well liked.

For RMS operator, Abbey and Crip chose Terry Hart. For the EVA astronauts, the choices were George "Pinky" Nelson, the former Minnesota baseball star turned astronomer, and James "Ox" van Hoften, the F-4 fighter pilot turned civil engineer.

At six feet four, van Hoften was at the outer limit of astronaut height specs. It put him in the ninety-ninth percentile, and NASA simply didn't have a space suit for him. (The same thing applied to the smaller astronauts at the other end of the scale.) But Abbey and Crippen wanted van Hoften to be one of the EVA astronauts on 13, so Abbey simply ordered the crew equipment people to make two suits for that ninety-ninth percentile.

By now twenty-two of the thirty-five TFNGS had been assigned.

Looking at the training flow, Abbey realized that MS like Fabian would be flying a second time before their pilot colleagues flew once—which was bad for morale. With Griffin's concurrence, he decided to assign two pilots as mission specialists in order to give them flights. Dave Griggs would fly on a Bobko crew, and Steve Nagel on one led by Brandenstein.

Critics would later complain that again Abbey was somehow punishing air force astronauts, which he found baffling: Griggs was a civil servant, but a former navy test pilot and a naval reservist. Abbey was giving Griggs and Nagel an extra early flight opportunity.

No matter what Abbey did, or why, he always made someone unhappy. And each unhappy astronaut made his or her displeasure known, according to Griffin. He would say, years later, that he was surprised at how much time he spent on this single issue. "Every time I'd go in," Griffin said, he would ask his assistant, "'What have we got on the calendar today?' And if she'd tell me that 'Astronaut So-and-so' (whoever that is) 'wants to see if he can set up a time to come see you,' I knew what it was about before he ever came in. And I always had my pat answer, which was: 'I don't make crew assignments.' But I'd have to listen and cajole and commiserate with them."

STS-6 was the first flight of OV-101 (Orbital Vehicle 101), *Challenger*. Its crew of four was commanded by Skylab veteran Paul "P.J." Weitz, with Bo Bobko as PLT and Don Peterson and Story Musgrave as mission specialists. The primary goal of the mission was deployment of the first Tracking and Data Relay Satellite using the inertial upper stage (IUS) developed by the US Air Force. (*Challenger* was also the first orbiter to use new, lightweight solid rocket boosters and a lighter external tank.)

Originally scheduled for launch on January 20, 1983, STS-6 was delayed three months by a number of issues, including a hydrogen leak in its main engines during a readiness firing on the pad, fuel line cracks in two main engines (requiring replacement) and contamination of the TDRS on the pad during a severe storm at the Cape on February 28.

It finally lifted off on April 4, 1983, and within hours Peterson and Musgrave had deployed TDRS, which was mounted in the first Boeing inertial upper stage to be flown on the orbiter.

The TDRS/IUS required two burns in order to reach its target GEO. The first one went fine. But the second burn failed, the payload tumbling out of control until stabilized.

Flight controllers at NASA Goddard separated TDRS from the stage and eventually developed a plan to slowly nudge the satellite to its target orbit by burning its smaller steering rockets with a single pound of thrust each. It took months.

On flight day two, Musgrave and Peterson completed the first shuttle EVA. But the stranding of TDRS was a reminder that satellite deployments were far from routine. And it also had severe effects on the shuttle schedule. STS-7 would not use an IUS, but NASA had to remove TDRS-B from STS-8, and to delay the secret STS-10.

As for the crew, Weitz had expressed a desire to give up the training grind and move into management. He became deputy chief of the astronaut office.

Bobko was in line for promotion to commander on a downstream mission. Abbey had no immediate plans for Peterson or Musgrave.

In April 1983, just after STS-6, Griffin announced his new organization, giving Charlesworth the added title of "director of space operations." Abbey, Kranz (head of the newly independent mission operations division), and Jerry Bostick, head of mission support, would report to him.

The new JSC director also ordered a new system for naming shuttle missions.

After the IUS failure, STS-10 had been postponed for at least a year and was likely to be launched after STS-14. STS-12 was slipping behind STS-13. Griffin and Lunney in the program office were facing a flight order that went 11, 13, 12, maybe 14, 10. Missions beyond that were in total schedule chaos.

Given the mountain of paperwork—the joke was that a shuttle couldn't be launched until the weight of the documents equaled the weight of the vehicle—designations needed to be locked a year or even two in advance. It was impractical to simply renumber the missions as they were about to be launched in order to preserve an arbitrary sequence.

So Griffin's team came up with a new system described officially like this: "Mission designations consist of a numerical prefix indicating the year in which the launch will occur, a numerical designator for the launch ('1' for a KSC launch, '2' for a Vandenberg launch), and letter suffix which reflects the originally scheduled order of launch."

A vital word—"site"—was missing from the notice.

Helpfully, the statement continued, "Mission 41-D, for example, is a 1984 launch—'4'; to occur at KSC—'1' and was originally manifested as the fourth mission of that fiscal year—'D.' If the launch moves forward in the sequence, the mission designator will not change."

This was all well and good, but there was another motive. According to Pete Clements, "Gerry did not want an STS-13." As a flight director, Griffin had been involved with Apollo 13, launched at 1:13 PM (EST), and which suffered its mission-altering accident on April 13, 1970.

The new system was announced on September 9, 1983, to universal bafflement, and Abbey's exasperation. He didn't believe that anyone cared about the sequence of numbers.

37 | Frequent Flying

LIFE WAS STILL COMPLICATED in the Abbey household. Many days Abbey would run home to put something in the crockpot for dinner, then return to the center. At times he forgot to give a child lunch money, so his secretary Mary Lopez would deliver it.

Abbey was just too busy. The shuttle launch schedule never truly became predictable or operational the way military or commercial airlines would be, so Abbey was forced to scramble every few days. He had multiple meetings with his direct reports: flight control, aircraft operations, crew equipment, and the astronaut office. He attended every flight readiness review.

He would then fly to the Cape the same day as the crew, overseeing their final activities, whether social (a preflight family and friends night at the beach house) or operational (briefings and sims).

He would have breakfast with them on launch day, then accompany them as far as the launch control center—and back, if the day's launch got scrubbed.

After each launch, he flew back to Houston the same day, listening to the air-to-ground communications between the crew and mission control in his office. On landing days, Abbey made sure to be at the Cape or Edwards to greet each crew, then accompany the astronauts back to Houston. He was also involved in everything involving the astronaut team, from speaking engagements to the design of mission patches.

For STS-7, the first flight of his new generation of astronauts, Abbey flew to the Cape the same day as Crippen's crew, June 15, 1983, hitching a ride in the Shuttle Training Aircraft with Young. The weather promised to be perfect, a good thing given the size of the crowd for Ride's historic launch.

Prelaunch prep of *Challenger* went smoothly. As for the crew, as usual Abbey paid special attention to the first-time fliers—four in this case—as they went through final checks and exams in the crew quarters, and at the private beach house not far from pad 39A.

It was at the beach house the day before launch that Abbey sensed unusual nervousness in Ride. "She was pacing a lot." So he turned to her list of guests and asked college friend Molly Tyson to come to the house and talk with Sally.

Early on the morning of June 18—launch was scheduled for 7:33 AM Cape time—Abbey went through his familiar ritual of waking the crew members and accompanying them through breakfast and final weather briefings. He rode with them to as far as the launch control center, then offered good luck before they pressed on to the pad.

Upstairs, in the *Challenger* cockpit, Anna Fisher was the astronaut support person doing final switch checks, one of a crew of seven techs who would strap Crippen and his crew into the vehicle, then depart to a safe vantage point.

Right on time, *Challenger*'s main engines ignited, followed five seconds later by the twin solids, and the shuttle took off, rising in the Florida dawn.

Abbey watched it go, hearing Crippen's callouts, picturing Ride in the jumpseat behind Crippen and Hauck as mission specialist 2, the flight engineer. Once on orbit, the California native likened the event to "an E-ticket" from Disneyland.

Abbey would watch from mission control as Ride and her other first-time crewmates accomplished two commercial satellite deployments and the deployment and retrieval of a pallet containing several scientific instruments.

The only irregularity in the mission was its diversion from a Cape landing to Edwards on the final day, June 24.

Just like that, America had finally sent a woman into space.

It was obvious to anyone who attended Ride's launch that STS-7's prime astronaut support crewmember, Anna Fisher, was sporting a baby bump. Astronaut pregnancy had been one of those looming challenges facing Abbey and JSC management. How would they deal with it?

Fisher and husband, Bill, had kept the news to themselves for several weeks in the spring of 1983, but Anna began to show, and to realize that flying in high-altitude T-38s was potentially harmful. (Her technical assignment was flight support at the Cape, so she was commuting weekly.) By that time, word had gotten around to Abbey. After confirming Anna's condition, he took her off T-38 status but allowed her to keep working at the Cape.

Early in July, with Anna in her final month, Abbey called both Fishers to the eighth floor. He wanted to assign Anna to a flight scheduled for launch in the fall of 1984, but wanted to be sure that Anna felt ready. Anna wanted the flight, and was named a mission specialist in a crew commanded by Rick Hauck, the first TFNG to lead his own mission. Dave Walker was announced as pilot, with Dale Gardner and Joe Allen as the other mission specialists.

On Thursday, July 28, Anna spent the entire day in a cargo integration review for the mission, though pains suggested that she was indeed going into labor. Bill was scheduled for a stint at Clear Lake Hospital, where he worked in the emergency room. He had already begged off several nights in a row, expecting Anna to go into labor. Now she did, and he was on duty elsewhere.

After what she described as a "long labor," Anna gave birth to a daughter, Kristin, at 9:30 Friday morning, July 29, 1983. Knowing the risks of a prolonged hospital stay, she had made plans to be discharged as soon as possible. But while she and Bill were still asleep in the recovery room, fellow TFNG and new crewmate Dave Walker arrived with a gift basket that included stuffed bears and a note saying, "Bears for the bairn and the bearer who bore her." The gesture made Anna reconsider her plan to be discharged the same day. "I probably ought to get one night of good rest before I go home," which forced the hospital to scramble for an extra room.

By the time she got home she was feeling great, with a new baby and a flight assignment, and to everyone's surprise showed up at the Monday pilots' meeting as if nothing had happened—though she was carrying her doughnut seating pad.

During the next several weeks Fisher began to work as a capcom, something Hauck initially resisted. But she felt that the assignment made her a better crew member, given that she still couldn't train full-time, and Hauck eventually agreed. Fisher worked on STS-9, the first Spacelab mission, before turning to full-time crew training.

On August 27, Abbey was back in Florida with the crew of STS-8—commander Dick Truly, pilot Dan Brandenstein, and mission specialists Dale Gardner, Bill Thornton, and Guy Bluford. A thirty-nine-year-old air force major, Bluford would be the first African American in space.

The mission had lost its primary payload, the second TDRS satellite, but was still scheduled to launch an Indian communications satellite, and to perform rigorous tests of the remote manipulator arm with a dummy payload.

Orbital mechanics dictated a launch and landing at night—two more firsts for the program.

Abbey gathered the crew at 10:00 PM for "breakfast," then took them through the prelaunch prep. He had been less concerned about Bluford's state of mind than about Ride's—the air force pilot and engineer had flown 155 combat missions in Vietnam.

Following a flawless launch, the crew settled down to its tasks, deploying INSAT (Indian National Satellite) on flight day one. RMS work by Bluford and Gardner followed, while Thornton realized his professional dream of performing research into space adaptation syndrome in flight.

Even though the mission had lost its TDRS payload, Truly's team was able to help with the checkout of TDRS-1 in advance of November's STS-9 mission.

Truly and Brandenstein conducted the first nighttime landing in the shuttle program, gliding out of a darkness lit only by a few runway lights, and touching down at Edwards.

The next shuttle mission, STS-9 carrying Europe's Spacelab, would see the first flight of noncareer astronauts—payload specialists Byron Lichtenberg from MIT and Ulf Merbold from ESA. They would join the NASA flight crew of John Young, pilot Brewster Shaw, and mission specialists Owen Garriott (a veteran of Skylab) and Robert Parker.

Lichtenberg and Merbold would be the first in a long line of payload specialists. The scientific working groups for Spacelab 2 (astronomy) and Spacelab 3 (life sciences) had identified four candidates each, with final selections to come in early 1984, in advance of flights a year later.

The air force had its cadre of 13 MSES, half of them already tentatively assigned as prime or backup crew members for three DOD flights.

Owners of commercial satellites scheduled for shuttle deployment, companies like Hughes Aircraft and RCA—and foreign nations, like the United Kingdom, India, Indonesia, and Saudi Arabia—were entitled to send payload specialists into space. NASA had also agreed to another type of payload specialist for the shuttle—fliers from the commercial world—establishing a policy in 1980 to that effect.

A division of McDonnell Douglas Astronautics—the same company that had built the Mercury and Gemini spacecraft for NASA—had focused on a program with NASA Marshall to develop pharmaceuticals in microgravity using a process called continuous flow electrophoresis (CFES).

CFES was originally designed to be flown on Spacelab 3, but when that mission slipped by several years, McDonnell Douglas Aerospace Corporation's (McDAC's) Jim Rose—a former NASA official—and Glynn Lunney began to consider flying the experiment on the orbiter's middeck.

An early version of the CFES was installed on *Columbia* for STS-4, and operated by pilot Hank Hartsfield, where the equipment worked well enough that a second flight was ordered.

Rose and McDAC's engineer, Charles Walker, met with Lunney to discuss the follow-on flights, which Lunney approved. According to Walker, Rose then said that while astronauts like Hank Hartsfield were capable pilots and engineers, they were not and never would be experts in CFES. The device would be most productive if operated on orbit by a dedicated crew member, a commercial payload specialist.

Lunney chewed on his cigar for a moment, then allowed that NASA was prepared to consider a commercial payload specialist, subject to approval from HQ. "Who did you have in mind?"

Rose turned to Charles Walker. "You're looking at him."

Walker had actually applied for the TFNG group in 1977 but had not made it to the interview round.

In September 1982 a formal application was filed by McDAC, where it went to HQ first, and then to JSC management, including Abbey. Walker continued to work on the next CFES flight, which would be STS-6 in April 1983—and waited.

He actually learned of his approval from *Aviation Week* reporter Craig Covault, who had received a leak from NASA HQ.

Abbey met with Walker and, looking at the 1984 missions, judged STS-12 (later 41-D) and Hartsfield's crew to be the best fit. After all, Hartsfield had already operated CFES.

Some astronauts welcomed the new arrivals. Not all. One of the air force MSEs had an encounter with Sally Ride where she "chewed my ass about useless payload specialists who were taking jobs from astronauts."

As for STS-9, it was launched in spite of the fact that NASA only had one TDRS satellite available to downlink scientific data—fortunately, the errant comsat had reached GEO and proved useful.

The crew on *Columbia* worked not only with mission control in Houston, but with the Payload Operations Center at NASA Marshall in Alabama. There were seventy-two experiments in a variety of fields, from plasma physics to materials sciences. The ten-day mission proved to be a good pathfinder for later more specialized Spacelab flights.

Young and Shaw guided the orbiter to the runway at Edwards on December 8, 1983. The problem was, *Columbia* was on fire. Two of its three auxiliary power units had started to burn owing to a hydrazine leak. The astronauts exited the vehicle after ground crews put out the flames.

Abbey looked back over the year 1983 and saw many successes, but also nagging failures such as the IUS on STS-6 and the potentially catastrophic hydrazine leak on STS-9 as well as software and hardware issues, and reminded himself that the shuttle was still an experimental vehicle.

And yet, here were Beggs and Mark, and allies at JSC, Marshall, and other centers, pressuring shuttle teams and associated contractors to make shuttle operations more routine and hence cheaper. Their motivation was to free up money for newer programs beginning with a space station.

Jerry Bostick, a senior JSC engineer, recalled that Mark quizzed him about the need for software upgrades. And at one point, Gerry Griffin seemed to buy into the idea that flight readiness reviews were another costly, time-consuming effort. Abbey, Young, and others in the program found this hard to believe. They knew that the shuttle was never going to be an airliner. It was too complex, too fragile.

Eliminating the flight readiness reviews was simply an appalling idea. No one with any experience or sense would even consider the idea, and it died swiftly.

But Beggs did win one major battle, on shuttle processing.

Rockwell International, under its space division headed by Rocco Petrone, had not only built the shuttle orbiters but also processed and serviced them at the Cape. In 1982, however, Beggs wanted companies to compete for this business, to streamline it, get out from under civil service regulations, saving the taxpayer money.

Rockwell's John Tribe had worked in Apollo, then shuttle launch processing from the very beginning. "We had what we called the great partnership. It was Rockwell, Martin, Thiokol, the three element manufacturers, and we got together and said, 'We'll produce an operational team that will conduct all the shuttle operations for NASA.' In the meantime, Lockheed came in as another bidder with no background at all on the vehicle."

Lockheed was a huge aerospace contractor with vast experience in National Reconnaissance Office satellites, but it had not built, operated, or serviced anything like the shuttle. Lockheed won the contract, effective October 1, 1983. It was quite lucrative, expected to run fifteen years (with renewals every three) and earn as much as $6 billion. It was a business decision that in Abbey's mind had no benefit to the shuttle program.

Tribe called it "one of the biggest mistakes NASA made."

38 | 1984

ON JANUARY 25, 1984, one week before the scheduled launch of 41-B, President Reagan delivered his annual State of the Union address in which he announced that he was "directing NASA to develop a permanently manned space station and to do it within a decade."

Abbey knew the announcement was pending. While he saw the need for a future space station, he didn't trust any of the proposed configurations emerging from the JSC team. He preferred Max Faget's concept of a modular free-flying unit that could serve as the first of a series of elements that would eventually become an orbital outpost.

He also didn't trust the proposed cost of Reagan's new program—$8 billion over nine years.

And he was too busy with the shuttle, with a dozen flights scheduled for 1984, most of them satellite deployments. A new orbiter, *Discovery*, would be introduced. The Manned Maneuvering Unit would be tested, and if that worked, used to retrieve an errant scientific satellite.

Two DOD missions were on hold because of the ongoing troubleshooting with the inertial upper stage, among other reasons. Even without them, it looked to be a busy year with the shuttle expanding its capabilities.

First up was the former STS-11, now known as 41-B, with a pair of satellite deployments and the MMU test. The orbiter *Challenger* launched successfully on February 3, 1984, with a crew of Vance Brand, Hoot Gibson, Ron McNair, Robert Stewart, and Bruce McCandless.

The first satellite, Westar VI, was propelled out of *Challenger*'s payload bay by the spring assist system. Half an hour later, its payload assist stage fired . . . and then shut down twenty seconds into a burn designed to last much longer, leaving the satellite stranded in low Earth orbit.

While flight controllers assessed that problem, the crew attempted to deploy the same pallet satellite that had flown on STS-7. But a joint failure on the remote manipulator arm foiled that plan.

On flight day three, with permission from the satellite owners, and assuming that the payload assist module (PAM) failure was one of a kind, the crew deployed Palapa, the second comsat.

It failed, too.

Fortunately, to break the streak of bad luck, the next day McCandless and Stewart commenced the second EVA in the shuttle program. McCandless attached himself to the MMU backpack and, after careful tests inside the orbiter's payload bay, flew it away from *Challenger* to a distance of three hundred feet.

Photos of McCandless as the first "human satellite" became one of the most iconic images of the shuttle era. Additional tests by Stewart the next day proved that the MMU was ready to go to work. And for George Abbey, the child who had thrilled to Buck Rogers, it was a fantasy come to life.

Just into the new year of 1984, Abbey moved forward with a new astronaut selection, his third. The first group began its interviews on Monday, February 9. There would be five more groups every week through mid-March.

Navy lieutenant James Wetherbee, a test pilot based in Lemoore, California, at the China Lake test site, was the last to be interviewed that first week. He remembered it as intimidating. "You walk in and there's John Young, who always said—I learned—that he had 'worked on the Moon' not 'walked on the Moon.'

"What I remember is that George, who was the chair of the panel, spent the first part of the interview not asking me technical questions or about my test pilot work but about my change in travel plans. Did I have new arrangements to get back to SFO? Was I getting help?"

Wetherbee's interview had been delayed so long that he had missed all flights back to the Bay Area in California. "I looked at my watch and realized that twenty minutes of my hour had already gone by, and then it struck me: George already knew my technical background. (A former Annapolis classmate was then chief test pilot for the navy.) What he was doing was putting me at ease. It showed me that he was interested in taking care of his people."

Ever since arriving at NASA in the 1960s, John Young—who thought of himself as more of an engineer than a pilot—had written frequent lengthy memos about spacecraft development and other issues.

As center director, Robert Gilruth had welcomed them, telling Abbey that Young's memos were the only material he wanted to receive directly. Kraft had appreciated them too.

But not Gerry Griffin, who had several departments complaining about them. One morning in May he summoned Abbey to the ninth floor. "Make him stop," he said. Young's memos circulated beyond JSC to HQ, and who knew where else?

Abbey felt that Griffin was wrong. He also knew that even if he ordered Young to stop, there was no guarantee that the chief astronaut would obey. He mumbled a vague statement about understanding the order and left.

That afternoon, Abbey and Young drove to MD Anderson Cancer Center to visit George Low. Low had left NASA in 1977 to return to his alma mater, Rensselaer Polytechnic in Troy, New York, where he had a spectacular career as the school's fourteenth president. Health problems—a diagnosis of melanoma—had been the reason for Low's departure from NASA, and now, in 1984, his prognosis was grimmer. But on this day in early May, Low was in a good mood, talking of new hobbies he would pursue when he returned to Upstate New York. "I'm going to take up golf, since I get the impression a lot of important decisions are made on the golf course."

Low saw what he took to be disapproval on Young's face. "Oh, John, I really didn't mean it." He thought Young wanted him to concentrate on engineering.

Young protested. "No, go ahead and play golf!"

At that moment, Gerry Griffin and his wife arrived to see Low. After the usual exchange of greetings, Low suddenly turned to Young. "There's one thing I've always meant to tell you, and that's how much your memos meant to me on Apollo. I never thanked you."

Abbey tried not to look at Griffin. Griffin never protested Young's memos again. And Abbey, not that he'd ever intended to, never raised the matter with Young.

One small gesture Abbey made was telling George that his son, G. David Low, was going to be selected as member of the 1984 group of astronaut candidates. Low returned to New York, publicly convinced that he would still beat cancer. He was miffed at Rensselaer Polytechnic Institute's administration for prematurely naming a center after him.

He was awarded a Presidential Medal of Freedom on July 16, 1984, the fifteenth anniversary of the launch of Apollo 11. He died at home in his sleep two days later.

Mission 41-C launched in April, with a crew led by Crippen on his third flight and second command. TFNG Dick Scobee was the pilot, with three rookie mission specialists: Terry Hart, James van Hoften, and George "Pinky" Nelson.

One goal was the deployment of the Long Duration Exposure Facility, a scientific package designed to be retrieved on a later mission in six months or so. Also on the agenda—the capture and repair of the Solar Maximum satellite, which had been launched in 1980 and suffered a failure later that year. It was now uselessly orbiting the earth, but the team at Goddard had come up with a plan to have an astronaut replace several fuses that would restore its systems.

After Crippen and Scobee piloted the rendezvous, Nelson flew to the satellite on an MMU. He was to use a unique device called a trunnion pin acquisition device (TPAD) to latch onto Solar Max and steady the satellite so Hart could grapple it and move it to the orbiter's payload bay. But when Nelson triggered the TPAD it failed to latch. He tried again, with the same result, and now Solar Max started to wobble. Finally, he just grabbed a solar panel by hand, and that made things worse.

Hart tried to capture Solar Max with the robot arm, but the satellite was too wobbly. Knowing that the orbiter had finite maneuvering fuel, Jay Greene and his team of flight controllers called off the retrieval and had Nelson fly back to *Challenger*. Overnight Goddard managed to stabilize the satellite, and the next day Hart grappled it with the arm and brought it to the payload bay. Nelson and van Hoften did the repairs. And then Solar Max was released and worked fine for five or six years.

Early on, during Nelson's struggles with the TPAD, Abbey had to defend the crew against critics on Greene's team. But later it was discovered that the TPAD design was faulty: it would not permit a successful latching. Ultimately the mission showed how flexible the shuttle was with a crew using the RMS and EVA.

On Monday, May 21, 1984, Abbey began making his calls to the new ASCANs. He had a list of eighteen, but one potential ASCAN, a female doctor, turned down the offer—which left seventeen, seven pilots and ten mission specialists. There were three women and one Hispanic male.

Jim Wetherbee was one of the pilots; so was John Casper, an air force test pilot originally on Abbey's shortlist for 1978, and held out of the 1980 selection by a Pentagon assignment. The other pilots were marine Ken Cameron, the air force's Blaine Hammond and Sid Gutierrez, navy test pilot Frank Culbertson, and Mike McCulley. McCulley had the distinction of being the first submariner to be selected as an astronaut—he had earned his "dolphin" badge as a sailor before transitioning to flying.

Among the mission specialists were George Low's son David, of course, as well as the army's Jim Adamson, a pilot and engineer currently assigned to JSC, and air force pilots Mark Brown and Mark Lee. Charles "Lacy" Veach was a former air force pilot who had flown low-altitude "Commando Sabre" missions in Vietnam and had been shot down and rescued. Following a tour with the Thunderbirds' demonstration team, he had left the service and joined JSC's aircraft ops as an instructor pilot on the Shuttle Training Aircraft.

The three women in the new group were all mission specialist candidates: physicist Kathryn Thornton from the Army Foreign Science and Technology Center, NASA physician Ellen Baker, and Marsha Ivins from JSC's aircraft operations.

There was a navy SEAL, William Shepherd, an ocean engineer and special operative who, according to legend, amused the selection panel by answering the question "What is it you do better than anything else?" with this statement: "Kill people with a knife."

And a naval flight surgeon, a test pilot, and a former pro soccer player—all in one individual named Manley Lanier Carter Jr., better known as "Sonny." Carter had paid his way through Emory Medical School in Atlanta by playing professional soccer. He had then gone on active duty with the US Navy, completing the flight surgeon course, then earning a coveted slot in aviator training. During an operational tour with the Marine Corps' Attack Squadron 115 aboard the carrier USS *Forrestal* in the Med, he was in the air in an F-4 on August 19, 1981, when two of the *Forrestal*'s aircraft engaged and shot down a pair of Libyan aircraft. He had then gone on to the famed Top Gun school and graduated from the Naval Test Pilot School. His service with a Marine attack squadron led to his "adoption" by his comrades in that branch of the service. Carter was, Abbey had heard, "the finest pilot in the Marine Corps."

The new ASCANs were asked to report at the end of July.

One of the unsuccessful applicants, marine test pilot Robert Cabana, was told by Paul Weitz, "We're going to select more astronauts next year. You should apply again."

To Abbey, that was the bonus of having annual selections: you could allow promising candidates to acquire a bit more experience without losing them forever.

There was a downside to expanding the team, however. Abbey had been under pressure to contain the costs of the T-38 fleet. Pilots and mission specialists were logging fifteen hours a month, a notable amount that kept a team of instructor pilots busy, as well as the contractor maintenance crew.

Abbey had also allowed NASA flight surgeons to log several hours a month in the T-38. He had already connived to add a couple of T-38s to the fleet of thirty aircraft, calling back planes that had been on loan to other centers.

With increased potential users, a finite number of aircraft and pilots, and no more money, there was no alternative to reducing the number of hours mission specialists would fly. The fifteen per month became fifteen per quarter, still above the minimums the air force required of pilots (four hours a month), but not what Abbey wanted.

When members of this new group made their first trip, to Pensacola Naval Air Station for aircraft survival training, one of the 1984s, marine Ken Cameron, began referring to the group as "maggots," a marine term for recruits. The name stuck, and eventually all ASCAN classes would earn a derogatory nickname—"Hairballs," "Sardines," "Penguins"—chosen by the preceding group.

Hank Hartsfield's crew, and payload specialist Charlie Walker, were strapped into the cockpit of *Discovery* on June 26, 1984, prepared to launch on their six-day Mission 41-D to deploy three commercial satellites.

Abbey listened as the countdown reached T minus six seconds, the orbiter's three main engines ignited one after another, and almost immediately shut down in a cloud of steam that was followed by a tense exchange between the crew and mission control. "NTD we have an RSLS." That is, to a NASA test director, the vehicle had experienced a redundant set launch sequencer abort, its computers having detected a problem with the main engines, shutting them down before ignition of the solid rocket boosters.

From the cockpit, pilot Michael Coats radioed that he had an indication that one engine had not shut down. It had, but no one was quite sure what the problem was. Then a fire broke out at the base of the launch structure, triggering fire suppression sprinklers.

Hartsfield ordered his crew to unstrap and prepare for emergency egress, something that wouldn't be possible until *Discovery* stopped rocking. (The entire vehicle twanged several feet when the main engines started up. The solids normally ignited when it rocked back to vertical.) Judy Resnik, looking out the hatch window on the orbiter's middeck, could see the white room access sliding back and forth.

According to his crewmates on the flight deck, mission specialist Steve Hawley joked, "I thought we'd be higher at MECO [main-engine cutoff]."

The crew stayed put until the rocking diminished. Crossing to the access arm and relative safety, they were soaked by the sprinklers. Cold and wet, they were picked up by Abbey at the base of the structure and driven back to their quarters, their mission on hold indefinitely. And here was yet another reminder that the shuttle was nothing like an airliner.

Shortly before the 41-C launch, NASA's associate administrator for manned spaceflight, Maj. Gen. James Abrahamson, was called to the Pentagon, where Secretary of Defense Caspar Weinberger offered him the directorship of the new Strategic Defense Initiative.

Abrahamson was happy at NASA, but he believed he could contribute to SDI, so he accepted the job . . . but wanted to stay with the agency through 41-C.

Beggs and Mark brought in Jesse Moore from the HQ staff, where he had worked in Space Science and the Spacelab program. "Jesse was a nice guy," Abbey would say. "But he had no business being put in charge of human spaceflight."

Not long afterward, Hans Mark also chose to leave NASA, announcing in August that he was accepting the chancellorship of the University of Texas at Austin. Like many civil servants, Mark had found himself financially strapped by the burden of sending children to private schools. A lucrative job in the private sector was welcome.

One of Mark's goals had been to make the shuttle operate like an airline—and to take VIP passengers.

In late April 1981, shortly after STS-1, acting NASA administrator Alan Lovelace had appeared before the Senate Subcommittee on Science and Applications, whose chair was Sen. Jake Garn of Utah. According to staffer Jeff Bingham, at one point in the proceedings, Garn leaned forward to his microphone and said to Lovelace that he had a "more serious question" than the budgetary matters under discussion. "I want you to think more carefully than maybe the others in answering."

Having no idea what Garn wanted, Lovelace shifted in his chair. Then Garn said, "When do *I* go?"

The idea of flying dignitary or other passengers had been baked into the shuttle concept from the beginning. Former astronaut John Glenn, now a senator from Ohio, had spoken openly about his plans to return to space. CBS broadcaster Walter Cronkite was eager to become the first journalist to fly.

And now here was Garn—though he claimed he was kidding. A short time later, newly appointed administrator Beggs and deputy Mark paid a courtesy call on Garn, and the senator jokingly repeated the request. To Garn and Bingham's surprise, Mark said, "But of course you should fly. The whole purpose of the space shuttle is to have routine access to space. What better way to demonstrate that than to fly a member of Congress?"

According to Bingham, "that was all Jake needed to hear."

Before long, word of the proposed political payload specialist reached Abbey in Houston. He was not enthused to have the chairman of the senate committee that controlled NASA's budget scheduled for a seat on the orbiter.

He had nothing personal against Garn. The senator was a naval reservist with thousands of hours of flying time and could surely work with an astronaut crew aboard the orbiter. It was the principle. Soon a list of six congressmen was circulating, and included Congressmen Don Fuqua of Texas, Bill Nelson of Florida, and Edward Boland of Massachusetts.

Garn was formally invited to fly on the shuttle by Jack Murphy, head of NASA's legislative office, on November 8, 1984—one day after the election in which President Reagan trounced Walter Mondale.

Garn and Bingham flew to JSC for a December 17 meeting with Gerry Griffin. Jay Honeycutt represented Abbey, and soon became known as "short straw," the JSC official who had the misfortune to be responsible for the senator's activities. Abbey assigned Mike Smith to serve as Garn's training guide, what Garn would call his "mother hen."

Abbey also reached out to Bingham. "He made a decision, I think, to adopt me. I received invitations to all kinds of astronaut-related events, and yes, beer calls. It was as if he wanted to help me see the world, so to speak. I realized even then that it was strategic, too—he could hear from me what Jake was thinking and doing, and send messages to Jake through me."

Garn and Bingham moved into building 4. In order to keep up with his senatorial duties, Garn obtained a portable phone, but the signal was so weak that Bingham and he had to find just the right spot on the platform leading up to the shuttle simulator in building 9 to complete a call.

The senator became a student, making an effort to fit in with his new crewmates—commander Bo Bobko, pilot Don Williams, and mission specialists Dave Griggs, Rhea Seddon, and Jeff Hoffman. According to Bingham, at his very first crew meeting, Garn told Bobko and the others, "Call me Jake. I am just a payload specialist. If you, the commander, say 'jump' I will say 'how high.'

"At which point Bo said, 'Yes, sir.'"

Challenger lifted off for mission 41-G on October 3, 1984, with a crew of seven—Crippen as commander for the fourth time, pilot Jon McBride, Sally Ride on her second flight, Kathy Sullivan and Dave Leestma on their first. There were two payload specialists, oceanographer Paul Scully-Power and Canadian astronaut Marc Garneau.

Intended to deploy and retrieve a pallet containing Earth sciences instruments, 41-G would also test satellite refueling (an NRO program, though not stated) with an EVA by Sullivan, who was expected to become the first woman to walk in space, and Leestma.

The lead-up to the mission—and the flight itself—was chronicled by *New Yorker* staff writer Henry S. F. Cooper (a descendant of James Fenimore Cooper and author of a well-received book on Apollo 13), who had approached Abbey. Mindful of history, Abbey had encouraged a reluctant Crippen to give the writer access.*

In addition to the astronauts, much of the narrative focused on the team from the training division under the supervision of their lead, Myron Fullmer. Cooper also provided the first published portrait of Abbey in his position, describing him as "a stocky man with crew-cut, almost black hair" who had

* The narrative never appeared in the *New Yorker*, probably because of the *Challenger* disaster. It was published in book form in 1987 under the title *Before Lift-Off: The Making of a Space Shuttle Crew.*

"strong ideas about what astronauts should be like, how they should behave, and what they should wear."

In comparing him to John Young, Cooper wrote, "Young is generally considered a straightforward fellow without a shred of mystery. Abbey, on the other hand, may also be straightforward, but he has also cultivated an air of considerable mysteriousness." Cooper alluded to Abbey's ability to remain "out of the limelight," citing Bill Lenoir's NASA OFFICIAL GEORGE ABBEY gag on the walkout to STS-5.

Abbey did give Cooper some insight into the crew selection for 41-G, especially the fast recycling of Crippen, who was still deep in training for the Solar Max retrieval mission (41-C) scheduled for April 1984. "As missions were becoming more frequent, and as the time between them was diminishing, it seemed increasingly likely that a commander coming off one flight would join its crew late in its training." That is exactly what happened with 41-G: McBride, Ride, Sullivan, and Leestma, who were assigned to the mission in November 1983 rarely saw Crippen until May 1984—still a good five months before launch.

Before Sullivan could claim the honor of first woman to make a space walk, however, the Soviet Union launched Soyuz T-12 with a crew that included Svetlana Savitskaya. The vehicle docked with Salyut 7 and its resident crew of Kizim, Solovyov, and Atkov.

On July 24, Savitskaya and Dzhanibekov ventured outside the station to perform a three-and-a-half-hour EVA in which the cosmonauts tested space welding techniques . . . and, not so incidentally, beat Sullivan to become the world's first female space walker.

In October 1984 Abbey brought Steve Hawley to building 1 as his technical assistant—his new bubba. Bubbas weren't allowed into every decision. "There were many times when George closed the door," Hawley recalled. In fact, the job was dealing with action items—responding to memos, letters, or phone calls. "If there was good news, George gave it," Hawley said. "If it was a no, it was your job."

One of the action items that landed on Hawley's desk was a complicated form calculating flight crew operations' travel budget for the upcoming fiscal year. With no idea how to plug figures into the many boxes, Hawley brought the form to Abbey. "He looked it over, reached for a Post-it, wrote a dollar figure on that, and stuck it on the form."

"What about these boxes?" Hawley asked.

"Do whatever you want," Abbey told him, "but that's the number."

Hawley, who would later become director of flight crew operations, found this a valuable lesson. "There are tasks that demand a lot of attention, and there are those that aren't worth five seconds. This was one of the latter. I realized that NASA was going to give us whatever we needed."

Shortly after this, Abbey expanded his eighth floor team, assigning shuttle commander Robert Crippen as deputy director of flight crew operations—a new position. "Charlesworth never could get hold of him when they wanted him." Crippen said. "So one night George took me out for a drink, and he says, 'Crip, they tell me I got to have a deputy, so you're it.'"

Abbey mildly disputes the reasoning: "Kranz had been my deputy from 1976 to 1983," which was when flight operations was again split into into flight crew (Abbey) and mission operations (Kranz). "Within a year, with increased flight rates and a larger astronaut corps, it was clear I needed more help. It was more my idea than Charlesworth's." He also believed it was a good way to give senior astronauts experience in management.

Whatever the motivation, Crippen moved in. "I was there primarily to carry the fire when the ninth floor called and George wasn't there, or even a lot of times when he was there."

Crippen would find himself dealing with budgets and other administrative work, but he was still an active astronaut with a new assignment: commander of 62-A, the first shuttle mission scheduled to be launched into polar orbit from Vandenberg Air Force Base in California.

Abbey knew that Crippen was burned out by eight years of life in simulators but also knew that he was eager to launch from Vandenberg—that was what he had trained to do in the MOL program before coming to NASA.

(Abbey also knew that while 62-A was on the manifest for spring 1986, it wasn't likely to fly until late that year or even in early 1987 . . . Crippen would have plenty of time to recharge.)

Guy Gardner, one of the 1980 group, was assigned as PLT, with Mike Mullane, Dale Gardner, and Jerry Ross as mission specialists. The air force had a pair of payload specialists already working on the mission—Brett Watterson and Randy Odle. That is, until air force secretary Pete Aldridge decided to accept a Beggs invitation. Odle, whose job was to supervise the deployment of TEAL RUBY, was demoted to backup PS—and tried to resign his commission.

Maj. Gen. Ralph Jacobson at Special Projects raised a mild objection to the notion of a navy captain as commander of this air force mission. According to Watterson's memory, Abbey simply said, "Do you want this crew to be all air force, or the best?"

Watterson would say later, "We were using new filament wound SRBs, running the SSMEs to 109 percent, we had gaps in our ability to abort. Also, the pad itself was a problem—the air force ran nitrogen lines through the blockhouse, and there was some chance of hydrogen envelopment at SSME ignition—or sonic overpressure.

"We could have died."

Crippen had a different view. Yes, 62-A presented challenges. "But that's why you test."

After several changes of mission and payload, Rick Hauck's crew finally landed on mission 51-A, targeted for launch in early November 1984. Its primary goal was to deploy two new satellites.

It was also given the job of retrieving one or both of the comsats left stranded in useless low orbits in February.

Just before launch, a new NASA associate administrator for public affairs met with Hauck and his crew—an unusual event in any case, and particularly fraught this time. According to mission specialist Joe Allen, "An unnamed NASA spokesperson had quoted the crew as saying, 'The likelihood of capturing both satellites was very high.' We'd read the article and amongst ourselves, were curious, because none of us had said it."

With Abbey looking on in a preflight meeting, Hauck bluntly told the new associate administrator that the crew had nothing to do with the quote, didn't appreciate it, and felt that they would be lucky to retrieve *one* of the satellites. "If we can get both, it will be a fucking miracle." Then he announced that the meeting was over, and led his crew out.

They launched on November 8, and successfully deployed satellites on flight days one and three, then tackled the potential miracles. Dale Gardner's space walk using the MMU succeeded in grabbing Westar.

Allen used the MMU to capture Palapa a day later. He and Gardner posed in the orbiter's payload bay with a sign offering satellites for sale, crediting the "Ace Repo Company," obviously a subsidiary of Abbey's 1970s drinking group.

The crew of 51-A had pulled off their miracle.

With the successful conclusion of 51-A, Abbey had a challenging new assignment for commander Rick Hauck and pilot Dave Walker. He wanted them to lead the crews for missions 61-F and 61-G, respectively. Both were scheduled for launch in May 1986.

Both would employ a new upper stage for the shuttle—a General Dynamics Centaur that had been flying since 1962. Centaur used a combination of liquid oxygen and liquid hydrogen, which made it more powerful than the PAMs and even the IUS. NASA needed the Centaur in order to boost the heavy Ulysses and Galileo space probes. Because of planetary alignments, both had to be launched within five days of each other, May 15–20, 1986.

Abbey, Young, and most of the astronaut corps were strongly opposed to flying the Centaur upper stage in the orbiter. The vehicle was relatively fragile, and the bulkhead between its liquid oxygen and hydrogen tanks was so thin that a change in pressure in either might cause the bulkhead to fail. Bad enough, but the real problem was that the instrumentation showing pressures in the tanks, and shutting them down if the figures got too high, did not provide direct measurement.

Another major concern was fueling the stage before launch. Liquid oxygen and hydrogen for the Centaur had to be fed through the same lines that fueled the orbiter. JSC safety deputy Gary Johnson feared that a sudden stop in feeding would send a "hammer-type pressure," a dangerous pulse through the lines, rupturing them and likely destroying the orbiter on the pad.

This was all in addition to frightening abort scenarios, the weight constraints, and the requirement to run the SSMEs at 109 percent rated power.

The program was run by NASA's Lewis Research Center in Cleveland, which presented new problems. Gary Johnson found himself fighting both Lewis and General Dynamics. Eventually it was open warfare between GD, Lewis, Marshall, and JPL with Houston.

Young gave Centaur the name "Death Star," and it stuck. Nevertheless, NASA chose to fly the Centaur, and Abbey had to assign astronauts to the two missions.

39 The Breaking Point

T. K. MATTINGLY HAD delayed his return to the navy by a year in order to command STS-10, originally scheduled for launch in December 1983.

With the delays to the inertial upper stage, its classified payload slid far to the right on the schedule, first to STS-15, then STS-16, then, acquiring a new name (mission 51-C), it finally reached the launchpad in January 1985.

During the long wait, Mattingly had served as the astronaut office coordinator for military and national security missions. One of his accomplishments was forcing NRO to disclose the nitty-gritty details of the 51-C payload, which was more highly classified than the collection of experiments he had flown on STS-4.

The 51-C payload has long been rumored to be a new generation signal intelligence satellite code-named ORION. NRO officials initially wanted Mattingly and the other NASA astronauts to have limited information about what they were carrying. Air force payload specialist Gary Payton, who represented the ORION program office (and had actually served on the 1979 source selection board for the vehicle) would be the only one with detailed knowledge.

But Mattingly cited the case of the USS *Pueblo*, an intelligence-gathering ship that had been attacked and captured off the coast of North Korea in 1968. The commander of the vessel, navy officer Lloyd Bucher, had no knowledge of what surveillance equipment he carried, and failed to take the proper measures to destroy it. Mattingly didn't want to make an operational decision in ignorance.

Mattingly's victory was another first for the program, and a good precedent for Abbey and shuttle commanders on upcoming DOD/NRO missions.

There were other challenges for 51-C. In the last months of training, pilot Loren Shriver developed a medical condition that threatened to ground him. While Shriver maintained his training, Abbey assigned Mike Smith as a potential backup. (Mattingly had filled a similar "shadow backup" role on the Apollo 11 crew.)

Shriver's problem cleared up, however, and he launched with Mattingly, Buchli, Onizuka, and Payton on a freezing Florida day.

The secret payload was successfully deployed—the air force lifted the veil of secrecy far enough to admit that the IUS had been used.

Three days later *Discovery* landed back at the Cape.

NASA hoped to launch another eight or nine shuttle missions in 1985, almost double the number in 1984. In 1986 the schedule called for as many as twelve.

Headquarters officials like Beggs and Moore, and JSC officials like Lunney, Griffin, and Charlesworth, wanted Abbey to create more standardized crews—commander-pilot teams that would fly two missions a year for two to three years, for example. And even mission specialist teams who would do the same.

Abbey resisted this, as he resisted most suggestions from above. He was able to prevail in arguments because he and Young presented a united front—they were, in Paul Weitz's term, "always in sync."

Abbey also possessed more hard facts about the nature of the job of being a shuttle astronaut—reports from those who had trained and flown, from their instructor pilots and training supervisors, even from office secretaries and support people. It was his belief that the training schedule was too arduous and the flights themselves so dangerous that astronauts shouldn't face such risks that often. He was also convinced that since each mission had unique requirements in payloads or operations that consumed several months of training, there was no way—yet—to standardize Spacelab crews, for example, or those who would fly IUS missions.

Young and he also resisted suggestions that they assign crews two years ahead of scheduled missions. It gave them less flexibility in responding to individual astronaut performance—filling perceived gaps in an astronaut's skill set, for example—and to Abbey's special project in career management.

There was also the matter of training teams and simulators. In early 1985 it was only possible to train six crews at a time. So Abbey only assigned crews as close to launch dates as possible, generally a year out. Crews did generic training for the first few months following assignment, then moved into dedicated mission training at L minus six months.

As for flexibility, constant changes in shuttle schedules made it mandatory. For example, in January, NASA HQ announced that Senator Jake Garn would fly aboard mission 51-E, scheduled for launch in March with Bobko's crew. The primary goal of 51-E was the deployment of the second Tracking and Data Relay Satellite. But then technical problems were discovered in TDRS, so it had to be pulled from the schedule. Mission 51-E's other satellite was moved to the next flight, 51-D, and so was the Bobko/Garn crew. The original 51-D crew commanded by Dan Brandenstein was bumped to the next mission, with similar changes rippling through the schedule.

Looking ahead, Abbey knew that two of the year's late-autumn missions would include dignitary payload specialists. One would be the second political passenger, Congressman Bill Nelson of Florida, and the other would be the first teacher in space, still to be selected. HQ and the program office decided that mission 51-L in December would be the ideal spot for the teacher. Knowing this, Abbey handpicked a crew.

By now he had developed a set of practices—not hard rules—in forming crews. It was part rotation—astronauts coming off one mission largely earned new missions in the same order—and part matchmaking of both astronaut to mission (a satellite retrieval, for example, would require two highly rated EVA crew members) and also astronaut to astronaut (Abbey frequently teamed a mission specialist with a pilot or commander who had worked well on a past assignment). Luck and timing played a role. An astronaut might not be available for a medical or family reason.

The process also allowed for the introduction of fresh faces. For example, Dick Scobee had flown with Crippen on 41-C in April. Abbey wanted to give him a crew of his own, and the next open command was 51-L.

Among the unflown pilot candidates, Mike Smith had just come off the 51-C shadow backup and also the Jake Garn mother-hen job. In addition to superior flying skills, he was also, Abbey knew, patient—the very reason Abbey had given him the Garn assignment.

For mission specialists, Ron McNair was due for another flight, and so was Judy Resnik. Resnik was a superior robot arm operator, and McNair had qualified as an EVA astronaut. Both were also good with the press, a vital skill for a mission that would be the focus of so much attention.

Like Smith, El Onizuka was a patient and generous soul, ideally suited to babysit a teacher in space. He would be the other EVA mission specialist.

So there was a core crew. But there was still a battle to be fought, since HQ and the program office wanted Nelson and the teacher on the same mission, 51-L. Abbey thought it put too much pressure on a flight crew and lessened the experience for the non-astronauts.

He had a solution. In November 1984 he had assigned Hoot Gibson as commander of mission 51-I, scheduled for launch in August 1985. With subsequent cancellations and changes, Gibson's team had moved to 61-C, to be launched in December 1985, just ahead of 51-L.

The other crew members were pilot Charles Bolden and mission specialists Pinky Nelson, Franklin Chang-Diaz, and Steve Hawley. A commercial satellite payload specialist, Gregory Jarvis for Hughes Aircraft Company, was also on board—but there was an open seat. According to Bolden, "Mr. Abbey, in his infinite wisdom, decided that Hoot Gibson and his crew of merry men could better handle the congressman than most other people out there." The congressman joined the crew of 61-C.

Finally, after a dizzying set of changes in mission and payload, Bobko's crew launched aboard *Discovery* on April 12, 1985. The first mission goal, launching a Telesat communications satellite aboard a PAM, went well.

But the second scheduled deployment, of a Hughes Syncom IV vehicle, failed when an arming mechanism failed to engage, leaving the satellite floating uselessly. Here flight controllers and the crew showed what they had learned from 41-C and 51-A, coming up with the notion of sending EVA mission specialists Griggs and Hoffman out to Syncom with an improvised tool that might allow them to enable the satellite's failed trigger arm.

Carrying a "flyswatter," the two astronauts ventured out of *Discovery* on flight day five to install the items on the orbiter's remote manipulator arm—a procedure they had not simulated on the ground. The attachment was a success, but RMS operator Seddon was unable to engage the flyswatter after three tries—all of them fraught because the orbiter was rather close to Syncom if the satellite's kick motor fired. Bobko called off the attempt and backed the orbiter away, leaving Syncom adrift.

As for the rest of the crew, Garn was so violently afflicted with space adaptation syndrome for much of the mission that he became known in the *Doonesbury* comic strip as "Barfin' Jake Garn." Abbey had sympathy for the man—several of his own astronauts had been hit hard, like Bob Overmyer and Terry Hart. And NASA was factoring severe space adaptation syndrome into its flight plans, since it affected about half of all shuttle crew members to some extent. Medications had been developed to minimize the syndrome, but it was still a problem.

Present to greet the crew, Abbey took special care to escort Garn, who was shaky from his ordeal, and the landing, which had been tricky. Touching down at Edwards in a high crosswind, commander Bobko and pilot Williams had to apply extra braking to keep *Discovery* centered on the runway, blowing a tire in the process.

Mission 51-B, the Spacelab 3 team commanded by Bob Overmyer, launched on Monday, April 29, reaching orbit safely.

A problem arose early in the mission. Payload specialist Taylor Wang, a forty-four-year-old Chinese-born physicist from the NASA Jet Propulsion Lab, had developed the drop dynamics module experiment to test "containerless processing" in zero gravity. An electrical short caused the experiment to fail the moment it was turned on, throwing Wang into a fit of depression. For the first half of the mission the payload specialist had his head inside the cabinet containing the experiment while Thagard, Thornton, Lind, and Lodewijk van den Berg, the second payload specialist, dealt with other experiments as well as two squirrel monkeys and twenty-four rats, the first animal subjects ever carried aboard a manned vehicle.

Wang got the drop dynamics module operating and tried to cram seven days of work into the final three, with some success before the device broke down forever.

Challenger and Spacelab returned safely to Earth on May 9, 1985.

As Bobko's crew worked on orbit, on Sunday, April 14, the first group of applicants for the 1985 astronaut candidate selection arrived in Houston. The second and third would arrive on successive Sundays.

This was the second annual recruitment. Abbey and his selection panel, led by Duane Ross, had simply reviewed applications submitted after July 1, 1984, and identified sixty worth considering. (One potential candidate declined, so only fifty-nine were actually interviewed.)

Abbey knew that he needed more astronauts, not only to provide flight crew members in his model, but also to fill the dozens of vital support positions and technical assignments—capcom, Cape Crusader support, SAIL, tracking ongoing issues such as landing, rollout, and brakes.

He was also aware that soon he would be losing members of the 1978 group. Terry Hart had resigned in the summer of 1984. His previous employer, Bell Labs, wanted him back.

Other TFNGs would surely move on after second or third flights, for family or other reasons.

After a month of reviews and discussion, on Monday, June 3, 1985, Abbey was back on the telephone giving good news to thirteen new ASCANs: Seven were pilots, two navy (Mike Baker and Steve Thorne), one from the marines (Robert Cabana). Brian Duffy and Tom Henricks were from the air force and Steve Oswald, a former navy test pilot, from JSC's aircraft operations. Mission specialists included four civilians with degrees in physics: Jay Apt, Linda Godwin, Richard Hieb, and Tammy Jernigan. There were three military flight test engineers: the army's Sam Gemar, Carl Meade from the air force, and the navy's Pierre Thuot.

Apt, Godwin, Hieb, and Oswald were already working at JSC, a fact that prompted some public criticism that Abbey and NASA were unfairly excluding candidates from industry and academia. Abbey's answer was that the civilian mission specialists were all academics who were motivated to seek jobs at NASA, acquiring experience in flight control and operations that made them superior choices.

They were ordered to report at the end of July.

The 1984 ASCANs—the Maggots—officially graduated to full astronaut status on May 23, throwing a celebratory party at Mark Lee's house.

TFNG Dan Brandenstein had been given a flight and crew of his own in early November 1983. John "J.O." Creighton was his pilot, with mission specialists John Fabian, Shannon Lucid, and Steve Nagel. At that time they were to fly mission 51-A in September 1984.

But, as frequently happened in 1984 and 1985, payloads and schedules shifted. In the case of the Brandenstein crew, several times, ultimately becoming 51-G and adding payload specialists Patrick Baudry of France and Sultan bin Salman bin Abdulaziz Al Saud from Saudi Arabia.

In March, the day NASA announced yet another change in 51-G's mission, Brandenstein proved that he was thinking about his crew and their careers. He dragged Nagel over to Abbey's office.

Nagel had originally been assigned to fly as a mission specialist in fall 1984, plenty of time for him to rotate to his second flight as a pilot of 61-A, the German Spacelab, in October 1985. But the new delay meant that the gap between Nagel's flights was now down to six months. Astronaut Dave Griggs, first assigned as an MS for the 51-D crew in March 1985, had been removed from his downstream pilot spot on 51-F when the gap between those missions shrank to a few months. Brandenstein asked for, and received, a commitment from Abbey to keep Nagel in both assignments.

Discovery launched on mission 51-G on June 17 and soon deployed the ARABSAT communications satellite.

The deployment went smoothly, as did the deployments of two other Hughes-built Comsats. And the mission ended safely after seven days.

Next to the pad was 51-F and Spacelab 2, with a crew commanded by Gordon Fullerton. Roy Bridges was his pilot, and mission specialists were 1967-group scientist-astronauts Story Musgrave, Tony England, and fifty-nine-year-old Karl Henize.

There were two payload specialists, Loren Acton and John-David Bartoe.

After one aborted attempt on July 12, *Challenger* lifted off at 4:00 PM on the afternoon of July 29, following a one-hour hold for a software patch. Within minutes, things quickly went south as flight controllers noted, and *Challenger* commander Fullerton radioed, "Houston, we show the center engine thermal."

A temperature sensor in one of the three orbiter main engines had failed three minutes after launch. A second one failed at 5:12, followed by shutdown of one of the engines. Fortunately, at fifty-eight nautical miles altitude, the orbiter was already high enough for "abort to orbit," in which the remaining pair of engines would fire longer.

Challenger reached orbit safely, though lower than planned, forcing radical changes in timelines and research schedules. The seven-day flight was extended to eight, and ended with a flawless landing at Edwards on August 6, 1985.

In late July Arnold Aldrich was named shuttle program manager, finally filling the vacancy created by Lunney's departure. Aldrich had been heading JSC's orbiter projects office, which would now be absorbed into Aldrich's program office with Richard Kohrs and air force colonel Thomas Redmond as new deputy managers.

Even as the shuttle program office stabilized, higher NASA and JSC management began to fragment. Most troubling, in November 1985, Abbey and the JSC community learned that Jim Beggs was being indicted for alleged contract fraud while in his previous job at General Dynamics. He would eventually be exonerated, but he was required to step aside in favor of an acting administrator. The logical candidate, deputy administrator William Graham, had only begun work at NASA on November 25, 1985. Beggs's leave of absence began on December 4. William Graham became acting administrator of the agency less than two weeks after walking in the door.

In that same month—unrelated to events in Washington—Gerry Griffin announced that he would leave the JSC directorship in January to become president of the Houston Chamber of Commerce.

Griffin's departure came at the end of a string of four wildly different but successful shuttle missions.

In late August, the orbiter *Discovery* carried Joe Engle's crew of five on mission 51-I with a primary goal of deploying three different communications satellites.

On flight day five, with three good deployments behind them, the crew moved on to a more interesting goal, retrieval and repair of the Syncom IV satellite left stranded on the March 1985 mission.

Following a successful grapple by RMS operator Mike Lounge, EVA mission specialists James van Hoften, veteran of the Solar Max repair, and Bill Fisher, Anna's husband, repaired the spacecraft. It was redeployed, and later fired its booster to reach GEO.

On October 3, Bo Bobko's "DOD standby" crew of five lifted off aboard *Atlantis*, the fourth orbiter, on Mission 51-J. Conducted under a news blackout like January's flight with the Mattingly crew, the *Atlantis* team deployed an inertial upper stage that carried two Defense Satellite Communications System military communications satellites.

Atlantis landed safety at the Cape on October 7.

On Halloween 1985 Hank Hartsfield led a crew of eight on mission 61-A, carrying Spacelab "Deutschland-1," devoted to experiments developed by West Germany. There were two German payload specialists in the crew, and also ESA's Wubbo Ockels.

D-1 had a successful seven-day flight, with *Challenger* returning to Edwards on November 6, 1985.

Atlantis was back in orbit on November 26, 1985, just fifty days after returning from 51-J. The new mission was 61-B and had a crew of seven led by Brewster Shaw in his first command. There were two payload specialists aboard—Charlie Walker of McDonnell Douglas on his third flight, and Mexican satellite engineer Rodolfo Neri Vela.

The crew deployed a trio of satellites, including Mexico's Morelos. Then EVA mission specialists Jerry Ross and Sherwood Spring, working with RMS operator Mary Cleave, performed a pair of long space walks, testing methods of constructing structures in space—part of the new space station program.

Atlantis returned to Edwards on December 3.

Abbey and Young hadn't even named complete astronaut crews for all of next year's missions, yet people were just lining up for flights. In addition to Spacelab payload specialists (four on two missions) and air force manned spaceflight engineers (four on three missions), there were dignitary payload specialists, including air force secretary Pete Aldridge.

NASA's highly publicized Teacher in Space program had culminated in a White House Rose Garden announcement that Christa McAuliffe of New Hampshire would fly, with Idaho's Barbara Morgan as her backup. McAuliffe and Morgan had moved into apartments near JSC and were undergoing basic training.

On the commercial front, McDonnell Douglas had a replacement for Charles Walker named Robert Wood.

The British hoped to fly a payload specialist with their Skynet Comsat. Indonesia had a PS lined up for its comsat. India wanted one for INSAT.

Having already flown one PS, Canada had two more in the wings for 1987.

There were two Hughes engineers waiting to fly with Leasat payloads—or just when a seat opened up. (The first of them, Greg Jarvis, had just been bumped from the Gibson crew to allow Congressman Nelson to fly. Jarvis was reassigned to 51-L under Dick Scobee.)

NASA Goddard was lobbying to have one of its staffers assigned to the Hubble Space Telescope mission.

And in addition to the teacher in space, HQ was gearing up to select a journalist in space. The flight would be open to print and broadcast journalists, and the preliminary screening would be performed by the Association of Schools of Journalism and Mass Communication, based at the University of South Carolina. Abbey chose to establish a dedicated payload specialist office in building 32, under Don Bourque. Abbey would have loved to fly, but he also knew there was no justification—"I had no operational role to play."

Not that any of these other VIPs did, either.

Abbey was pleased with the successful flights but also concerned about growing complacency, especially as he and Young looked at the manifest for 1986 and saw thirteen missions scheduled. They were concerned with the stress the flight rate imposed on the astronauts, the training team, and the flight controllers.

The missions flown in 1985 demonstrated great success but also troubling technical problems, as well as major issues with the planned flight of the Centaur upper stage. John Fabian reported a lack of quality control at the contractor's facility. The program was moving too fast, in too many directions.

By December 1985, Abbey could see that the system was being stressed.

40 | The Day the Earth Stood Still

THE LAUNCH OF SHUTTLE Mission 61-C was set for December 18, 1985, but no sooner had Abbey and the crew arrived than they learned of a twenty-four-hour postponement to give technicians more time to finish closing out *Columbia*'s aft compartment. Here was another sign that pad crews were falling behind.

The second attempt, on December 19, ended just fourteen seconds before launch, when flight controllers received an indication that the hydraulic power unit on the right-hand solid rocket booster had exceeded allowable turbine speed limits.

With the Christmas holiday now looming, the launch was then moved to January 6, 1986, in spite of considerable pressure from Aldrich's program office to launch *Columbia.* The orbiter was needed in early March on the Spacelab Astro-1 Observatory mission to observe Halley's comet at its closest approach to Earth.

RCA's Robert Cenker was one of two payload specialists joining the 61-C crew of five. The other payload specialist was Congressman Bill Nelson, a Democrat from Florida whose district included KSC.

Unlike Jake Garn, a naval aviator, Nelson had zero experience in aviation or spaceflight. He was a pleasant individual, but other than that had no earthly justification for flying in space beyond his position in the House of Representatives. He did not devote enough time to training to serve as a capable shuttle crew member. He made so few trips to JSC and spent so little time with the crew commanded by Hoot Gibson that Abbey started phoning Jeff Bingham to get him to ask Senator Garn to urge Nelson to get to JSC.

This attitude hadn't endeared him to the crew, especially to Steve Hawley. Pinky Nelson found that he had to calm his crewmate down. To him, Congressman Nelson "was a model payload specialist. . . . He had no experience either in aviation or anything technical: he was a lawyer, so he had a huge learning curve, but that didn't stop him from trying, and I think he knew where his limitations were."

A *Columbia* launch on December 18 or 19 would have allowed just enough time to refurbish the orbiter and install the three Astro ultraviolet telescopes. A January launch now compressed the available time.

The next mission, 51-L, was due to launch from the newly refurbished pad 39B, the first shuttle mission to do so. Having two pads was necessary if NASA hoped to fly a dozen missions in a single year.

Columbia's launch attempt on January 6 turned out to be one of the most dangerous in the shuttle's operational history. The countdown was halted at T minus thirty-one seconds, following the accidental draining of almost four thousand pounds of liquid oxygen from the external tank. The fill and drain valve, it seemed, had not properly closed when commanded to do so. Launch controllers reset the clock to T minus twenty minutes and efforts were made to reinitiate the liquid oxygen tanking, but it was obvious that time was running out and the window would close before the vehicle was ready. Another twenty-four-hour delay was called.

The next attempt, on January 7, was scrubbed due to poor weather at Transoceanic Abort Landing sites in Spain and Senegal.

The fifth launch attempt, January 9, was aborted when a liquid oxygen sensor on pad 39A broke off and lodged itself in the prevalve of one of *Columbia*'s main engines. Ignition of that engine would have resulted in a catastrophic explosion.

For the crew, it was back to quarters as heavy rains moved in, covering most of Florida. The forecast for launch on January 10 showed a 100 percent probability of rain and thunderstorms, conditions that would prevent a shuttle launch.

Nevertheless, an attempt was scheduled. The evening of January 9 found Abbey and Young arguing that it was senseless to plan a launch attempt and send the crew out to the spacecraft with that kind of forecast. Kennedy Space Center director Dick Smith told Abbey that the weather might change and he needed to have the crew, spacecraft, and launch team ready to go. The weatherman said there was not a chance of being able to launch. However, NASA proceeded with preparations. With the severity of the weather, the crew told their families: don't even bother to come out to the Kennedy Space Center, we are not going to launch.

No matter, next morning, Abbey accompanied Gibson's crew to the Astrovan again, departing at the launch control center in one of the worst thunderstorms he had ever experienced in Florida.

In the *Columbia* cockpit, pilot Charles Bolden could hear lighting crackling in his headset. "You know, you're sitting out there on the top of two million pounds of liquid hydrogen and liquid oxygen and two solid rocket boosters." There was a lightning rod on the pad to prevent damage from a strike, but it had already failed once, on STS-8. Which the crew knew. "We started talking about the fact that we really ought not be out here."

There were multiple holds for weather, meaning that the astronauts wound up spending five hours on their backs. Even then, at one point the countdown reached T minus thirty-one seconds before a final weather hold that led to a scrub.

Years later, Abbey was still upset about the 61-C decisions. "You launch when you are ready and when the weather is acceptable. No one remembers the day you launched, but they do remember if you have a problem."

The strain on the shuttle program was certainly obvious to the crew. "We were a little worried about the vehicle and the ground," George Nelson would say later. "The launch control center was not really smooth." He noted that there were "a lot of new people" in mission control, too.

The crew was stressed in other ways. At that time NASA and the astronaut office gave no financial support to crews and family members who attended launches. Hotel bills "cost me a fortune," Pinky Nelson complained.

When Gibson and his "merry men" went to the pad for a record seventh time, to Abbey's amusement, Steve Hawley wore tape over his name tag and a rubber mask with mustache to fool *Columbia*.

The ruse worked. *Columbia* finally lifted off at 6:55 AM on the Lord's day, Sunday, January 12.

The crew successfully deployed the RCA Satcom K1 and operated a series of scientific experiments. RCA's Cenker also tested a prototype infrared camera for the military.

Then more schedule pressure affected *Columbia*'s return to Earth. *Challenger* with mission 51-L was planned for a January 24 launch, and on March 6, another *Columbia* launch was to take place.

Aldrich and the program office had shortened 61-C's planned duration from five to four days in order to allow 51-L and *Challenger* to launch on January 24, and to get *Columbia* turned around for its Spacelab mission on March 6.

But on January 16, the Cape was blanketed in a low cloud cover, unacceptable for a shuttle landing. So the crew, all equipment stowed, had to remain in orbit an extra day. On January 17, the Cape weather was just as unacceptable as it had been the day before. Another postponement. Weather on the following day was bad, too. But now the 61-C crew was running out of food and clean clothes. *Columbia* had to come down, and there was no other option than to land at Edwards Air Force Base in California, in darkness.

Fortunately, Gibson and Bolden had trained for landings at the Cape or Edwards, day or night. And they put *Columbia* down on runway 33 at 5:59 AM (PST), predawn.

Congressman Nelson was disappointed by the decision to go to Edwards. It was probably for the best. According to Pinky Nelson, "Bill had a really hard time for a few hours after we landed, but, boy, he was a trooper. He was suffering, but he, you know good politician, put on a good face, and we had to do our little thing out at Edwards and all that and get back on the plane. He really sucked it up and hung in there, even though he was barely standing."

At Kennedy Space Center during a press conference after the STS-61C landing, the news media was told that the Edwards landing would not have much impact on shuttle operations for 1986. Abbey knew it was untrue.

On Thursday, January 23, *Columbia* returned to the Cape, completing its ferry ride from California. It was to be processed for launch again on March 6.

That same day, the crew of 51-L left Ellington, heading for the Cape in a trio of T-38s: Scobee with Onizuka, Smith with Judy Resnik, John Young with Ron McNair.

Driven to Ellington, as usual, by Suzanne, Abbey boarded the JSC Gulfstream with teacher in space Christa McAuliffe and payload specialist Greg Jarvis.

He had great confidence in Scobee's crew. McAuliffe and Jarvis, under the wings of mother hens Smith and Onizuka, supported by their families, were ready for their adventure. The first scheduled launch attempt, for Sunday, January 26, was scrubbed the night before because of concerns about weather. (This turned out to be pessimistic: weather during the January 26 window was fine, but it was too late to launch on that date.) On Monday the twenty-seventh, Abbey escorted Scobee's crew to the pad for the first time shortly before 8:00 AM. The seven astronauts entered *Challenger* only to be frustrated when the final latching of the orbiter hatch could not be confirmed by instruments. Inspection by the crew showed that the latches were closed. But then one of three bolts holding the external hatch handle could not be extracted. It appeared to have stripped threads. The pad crew decided to drill out the bolt but had to wait for a drill to be delivered. When it arrived, the drill's battery lacked sufficient charge. Replacement batteries were sent for but did not reach the closeout team for some time, and didn't do the job. An A.C. drill was used, but it wasn't successful, either. As a last resort, the team was told to remove the bolt with a hacksaw. The bolt and hatch tool were finally removed, but at a cost of two and a half hours. By now winds had kicked up and were now in violation of crosswind limits for a potential RTLS abort. The launch was scrubbed. The crew left *Challenger* to return to quarters.

Abbey and the crew had no idea of the battles raging via telephone and fax that night, between NASA Marshall and the Thiokol team in Utah. Thiokol was concerned about the flexibility of O-rings in the solid rocket boosters in such cold weather. That night, Arnold Aldrich and his team held an L-1 review—without involving Abbey or anyone from flight crew operations. The weather team reported that temperatures were likely to drop into the twenties overnight, below the hard minimum for a safe launch—which was thirty-one degrees. But the forecast said temperatures on Tuesday would be thirty-one degrees or better. Aldrich would say later, "There wasn't an immediate feeling that that was going to be unacceptable." And with all other systems ready to go, the L-1 team concluded that a launch would be attempted.

The next morning, Abbey and Scobee's crew went through the breakfast and suiting for the second time. The weather report from astronaut Fred Gregory mentioned record low temperatures—which was obvious to Abbey as the crew did their walk out to the Astrovan.

Abbey exited the Astrovan at the launch control center as Scobee's crew drove on to the pad.

The launch had originally been planned for 9:00 AM, but the management team called a pair of holds, pushing the liftoff to closer to noon, in order to allow the pipes on pad 39A to warm up—and the ice to melt.

Aldrich stepped away to confer with senior management—Bill Lucas, the head of Marshall; Stan Reinartz, the shuttle manager at Marshall; and Jesse Moore, the head of spaceflight at HQ; and twenty others in a conference room one floor above the firing room.

Abbey confronted Aldrich the moment the program director arrived back at his console. "What did you decide?"

"We're still go." He went on to remind Abbey that they needed to get *Challenger* off the ground, then back in time to be ready for the firm May 15 Ulysses-Centaur launch date.

"Did Rockwell say they were go?" Abbey received a positive reply.

Meanwhile, the count had picked up and was proceeding.

Frustrated, Abbey stepped over to the windows and, like citizens and schoolchildren all over the United States, at 11:38 saw *Challenger*'s main engines ignite. The vehicle did its twang, then swung upright as the solids lit.

Up it went. Abbey could hear the callouts between Scobee and capcom Dick Covey, right up to "Go at throttle up."

Scobee replied, "Roger, go for throttle up." Then Abbey saw an orange cloud in the sky as *Challenger* exploded.

Abbey stood there in shock—like everyone else in the launch control center—then he realized that he needed to get to the families, who were watching from the roof of the launch control center.

He called for cars to transport the families back to the crew quarters building. Then, knowing the route they had taken up to the roof, and assuming they would be coming down, he intercepted them in the hallway. All were emotional. Abbey picked up eight-year-old Erin Smith, Mike's daughter, and carried her down to the waiting automobiles.

At the crew quarters, John Young was waiting, as were several other astronauts—Bill Shepherd, Frank Culbertson, Jim Bagian, and Sonny Carter. Abbey gave Erin back to her mother and concentrated on getting an update on the state of events and laying out a plan of action.

Word arrived that Vice President Bush and Senators Garn and Glenn were flying down from Washington to meet with the families.

Abbey and Young met them in front of the building and escorted them up to the fourth floor, where the families were all assembled in the conference room. "The families were all in tears," Abbey remembered. Bush managed to calm them, speaking individually to each family, including the children. "It was a very tough audience that desperately looked for answers that might provide some hope," Abbey said later. "That was not to be." Nevertheless, he was impressed by Bush, and by the way Glenn and Garn assured the families that they would remain in contact.

The families were driven to the Shuttle Landing Facility and put aboard the Shuttle Training Aircraft for their flight back to Houston. Most passengers were by then worn out from the shock and slept. Abbey sat with Mike Smith's sister, Ellen Leonard, and they talked quietly throughout the flight.

Bill Bailey was the constable for the precinct that included the Johnson Space Center. A former popular DJ for KIKK, playing country music, he had connections to the JSC community. "Pete Conrad, Stu Roosa, and Charlie Duke were big country music fans," he said. The astronauts carried recordings of Bailey's selections to the Moon on Apollo 12, 14, and 16.

Learning of the *Challenger* accident on the news, Bailey had sent deputies to the houses of the NASA astronauts, and to the apartments where McAuliffe and Jarvis had lived. He also brought a team to Ellington, where he asked base ops to corral the news media in a location far from the arrival hangar, then turn off the tarmac lights to further degrade any invasive imagery.

After the families had been escorted away, Bailey realized he was alone on the sidewalk next to the hangar. "I felt a presence behind me." He turned and saw Abbey standing alone. "He had lost everyone. He looked as gutshot as any human being I'd ever seen."

All Bailey could think to do was give Abbey a hug and then guide him to his car.

Even before leaving KSC with the families, Abbey had begun deploying astronauts. Many would serve on the inevitable investigating boards, but the first priority was recovery of *Challenger* debris. Mike Coats "flew down to the Cape two hours after the accident and stayed four months" on that awful job.

Overall responsibility was Robert Crippen's. Physician-astronauts Sonny Carter (who had been the last support astronaut to see the crew) and Jim Bagian joined the salvage diving team.

Recovery operations began within minutes of the accident, as the launch recovery director ordered the USS *Liberty* and USS *Freedom*—the two ships that normally recovered the spent solid rocket boosters—into the impact zone.

Search-and-rescue aircraft from Patrick Air Force Base and Canaveral were also called up, though the air force range safety officer kept them out of the zone for an hour due to danger from falling debris. The coast guard began a surface search, reportedly the largest in its history. (The search zone was a rectangle 58 miles wide and 155 miles long.)

Defense contractor Tracor Marine of Fort Lauderdale would soon provide two recovery vessels.

At the end of the day after the tragedy, payload specialist Millie Hughes-Fulford walked over to building 1 and up to the eighth floor. Everyone seemed to be gone except George Abbey.

She barely knew the director of flight crew ops. She had met him during her own astronaut candidate interview in 1977, but had not had dealings with him since her selection as a Spacelab payload specialist two years earlier.

But he knew her. "Millie, come in."

She didn't want to take up much of Abbey's time, so she just expressed her wish, and those of the other PS, to attend the memorials. "Of course you can," he said. The memorial ceremony Friday, January 31, at JSC was solemn. Ten thousand people crowded into the square in front of building 2 as Steve Hawley's father, the Reverend Bernard Hawley, led prayers.

Then President Reagan spoke. Four T-38s whistled overhead, with the number three plane lifting away in the missing man formation.

John Blaha remembered seeing Crippen, Gardner, and other hardened astronauts weeping openly.

As for Abbey, he had already received phone calls and quick visits from Young and others sharing the news that the cause of the shuttle's loss was probably the right solid rocket booster. Knowing that there had been problems with the SRBs, issues quickly dismissed by program managers, especially those at NASA Marshall, Abbey found his mood shifting from numb horror and shock to cold anger. He had to find the strength to cope and somehow move forward.

41 | Recovery

IN THAT FIRST AWFUL week after the *Challenger* disaster, NASA's temporary leadership floundered. Acting administrator Graham made several public appearances, and it was clear that he was out of his depth in answering basic questions about the shuttle, much less the disaster. He also seemed to hope that NASA could investigate itself, a concept that had worked with the Apollo fire in 1967—but would not be tolerated in the post-Watergate era.

The Reagan White House created a Presidential Commission to investigate *Challenger*, and named former secretary of state William Rogers to head it. Neil Armstrong was named vice chair.

Other members—thirteen in all—included Albert "Bud" Wheelon of Hughes Aircraft, Eugene Covert of MIT, air force general Donald Kutyna, physicist Richard Feynman, and former test pilot Chuck Yeager.

Rogers asked for Sally Ride too.

The commission set up shop in DC and began creating a list of interview subjects, from NASA and the shuttle contractors. It had 120 days to deliver its findings on the cause of the accident.

Engineers and other NASA officials already suspected that the solid rocket booster was the cause of the accident. Word of that reached Abbey and Young before the memorial service, but hints and guesses meant nothing. *Challenger* had been lost because of a larger failure, and they needed to hear what the Rogers Commission discovered.

Graham asked Abbey for advice on several occasions. Abbey stated, "I told him to bring Dick Truly back from the navy to run manned flight at HQ.

Dick knew the system, and he had been away from the agency for over two years, so he wasn't going to be part of the investigation."

Graham accepted Abbey's recommendations.

In late January and early February, Abbey was consumed by the investigation. He assigned a dozen astronauts to work with teams that would examine systems and hardware. Over the next several weeks that number would rise to sixty or more.

Always a workaholic, Abbey's post-*Challenger* hours soon alarmed his family. "He never came home," Joyce said.

Trouble was, Abbey was simultaneously trying to be a father to several families, those of the *Challenger* crew as well as his own. Part of that was maintaining constant contact with the recovery effort off the Cape, where Sonny Carter and Jim Bagian were working with the salvage crews.

Press and the public were fascinated by the search. There had been four hundred reporters at KSC for the teacher-in-space launch. According to a report years later, that number quadrupled after the disaster.

There was some hope that components or wreckage of *Challenger* could be found that would reveal the cause of the accident.

USS *Preserver* was a navy auxiliary recovery and salvage vessel. It had been operating off the coast of Georgia at the time of the *Challenger* launch, and was ordered south.

Commanded by John Devlin, it had a crew of eighty-five, many of them qualified divers. (He would rotate to a new command at April's end, succeeded by Robert Honey.)

The work was incredibly difficult. It was winter in the Atlantic, so salvage ships like the *Preserver* battled winds, waves, and cold. The search zone was 480 miles in area; divers and machines had to operate at depths up to 1,200 feet.

Surface operations ended on February 7, but underwater search and salvage work went on using divers and submersibles—remotely operated and manned—methodically examining each target identified by sonar.

An air force vessel, the *Lucy*, with a crew of five, was part of the search team. Two of *Lucy*'s divers made what they thought would be a routine dive on target 67, sixteen miles northeast of KSC, toward the end of that day—and located wreckage containing a shuttle EVA suit at a depth of ninety-five feet.

The news was carried to the nearby *Preserver* in person. The larger vessel motored to the site, and astronaut Jim Bagian performed one dive, which seemed to confirm the news. But it wasn't until the following morning, after another dive, that Bagian found a cockpit hand controller that convinced the team that they had indeed located the cabin and the crew.

Abbey heard the news on March 7. He knew that the orbiter hadn't actually exploded, that instead it had suffered an aerodynamic breakup, but, like most in NASA, he believed that the crew had died instantly.

The next morning, Saturday, before NASA officially released the information, Abbey gathered the families at Scobee's home to tell them their loved ones had been found.

That night, *Preserver* put into Port Canaveral with its running lights doused.

Admiral Richard Truly became associate administrator for spaceflight at HQ on February 20. He found the agency "in utter turmoil. You had the Rogers Commission investigating *Challenger*, and you had all these people—engineers—sitting around not knowing what to do. I was really afraid that NASA was simply going to go away. We needed direction."

Within two weeks, Truly put together his own Tiger Team of Abbey, John Young, Jay Honeycutt, Robert Crippen, and a few others. They developed a plan for returning the shuttle to flight, composing a multipage memo they called the "Manifesto." "It listed a bunch of goals we all agreed on, such as a daytime launch, only experienced crew members, etc."

Truly visited JSC the last week in March and presented the Manifesto to a thousand staffers. "It totally settled down the system." Internally, at least. But Abbey and others still had the Rogers Commission to contend with.

By now they had heard damning testimony from engineers at Thiokol about multiple problems with the solid rocket motors on many shuttle launches before *Challenger*—and about their pleas to postpone the launch on January 28. John Young's internal March 4 memo to Truly, Abbey, and JSC leadership about lapses in safety got leaked to the *Houston Chronicle*, which posted it on

Saturday, March 8—the day Abbey was speaking privately with the *Challenger* families. Truly released it officially two days later.

Another document was released at the same time, a lengthy February 27 report by Steve Bales (onetime hero of the Apollo 11 landing) citing "34 safety items and seven studies to be considered." On the list—fill and drain valves for the orbiter on the pad and on-orbit; SSME turbopumps; the operation of orbital maneuvering jets; the August 1985 launch of the shuttle when there was rain and hail at its emergency landing strip; and so on.

Young attached his own comments, saying, "This proves to me that we have some very lucky people around here"—astronauts who survived flights with compromised systems. "They ought to be fixed so we don't lose any more space shuttles and flight crews."

Abbey was quite familiar with Young's memos. They had been useful to Gilruth and Low in the Apollo era, and had equal value in the world of the space shuttle.

Jesse Moore had been appointed acting director of JSC just five days before the *Challenger* launch, though he did not move into the job until summer (he had been based at HQ as associate administrator for spaceflight) and ultimately wound up holding it for five months.

Moore had told the Presidential Commission that he'd had no knowledge of the continuing problems with the SRB joints or of the waivers that allowed their use on the six flights before *Challenger*. It was largely Moore's testimony that forced the commission to conclude that shuttle program decision-making was flawed. Abbey appreciated Moore's honesty while feeling that he had some responsibility for the accident. Moore left NASA in the summer of 1986.

The Reagan White House, understandably eager to insulate itself from further criticism, wanted an experienced manager to lead NASA through the Return to Flight mission, and had turned to James Fletcher—NASA administrator from 1971 to 1976. (Since that time, the Utah native had been a professor at the University of Pittsburgh as well as a consultant to aerospace companies.) According to news reports, the sixty-six-year-old physicist and manager had been reluctant to accept the appointment but had found it impossible to refuse a presidential request.

Abbey had high regard for Fletcher's technical knowledge, and had considered joining his (and Low's) staff back in 1971.

Meanwhile, the Rogers Commission worked on, and soon it was Abbey's turn to testify.

On the morning of Thursday, April 3, 1986, Abbey, accompanied by John Young, Robert Crippen, Paul Weitz, and Henry Hartsfield, entered the Dean Acheson Auditorium at the Department of State building to face the Rogers Commission.

Abbey led with an overview of the flight crew operations organization, the number of astronauts, and the positions held by Young, Weitz, Crippen, and Hartsfield.

The bulk of the morning testimony came from those four and dealt with their actions, responsibilities, and actions on shuttle missions in general, and on *Challenger* specifically.

It was obvious that Abbey and his team had been largely shut out of the high-level decision-making progress. For example, Weitz had been included on a teleconference on January 26, but not on the fateful January 27.

In the afternoon session, the five continued to explain their positions on training and information flow—or the lack thereof. Young was asked about his memos. Then they were dismissed, replaced as witnesses by Truly, Aldrich, and Charlesworth.

On February 20, 1986, Jesse Moore had officially postponed the two Centaur missions—the shuttle was grounded and the planetary launch windows would not open again for thirteen months.

Gary Johnson's memo on the problems somehow found its way to the press. On May 22, Hauck made a briefing about the program at HQ reiterating his concerns. Three weeks later, on June 18, wisely listening to those who knew the program, Fletcher canceled Shuttle-Centaur. Astronauts and flight controllers celebrated with a "Centaur's Been Canceled" beer bust at the Outpost.

As the hearings wound down in Washington, with Rogers's staff drafting its report, another task remained for Abbey. On Tuesday, April 29, 1986, Abbey and Young flew to KSC, where they joined Truly and others on the runway as seven flag-draped coffins were transferred from hearses to an air force C-141. Two hours later, the process was reversed at Dover Air Force Base in Delaware, as another honor guard removed the silvery flag-draped coffins of the *Challenger* crew: Scobee, Resnik, and McNair. Teacher in space Christa McAuliffe. Hughes engineer Greg Jarvis. And two close friends, Mike Smith and El Onizuka.

The coffins now belonged to the families, who had arranged for disposition. Mike Smith would be buried at Arlington on Saturday, May 3. Scobee's interment would follow on May 19—what would have been his forty-seventh birthday. El Onizuka's final resting place would be in Hawaii, Ron McNair's in Lake City, North Carolina, his hometown. Christa McAuliffe would rest in Concord, New Hampshire. Judy Resnik's remains would be cremated and scattered in the Atlantic off the Cape. Jarvis's remains, also cremated, would be scattered in the Pacific off Hermosa Beach, California.

Abbey attended every memorial, public or private. All he could do was offer prayers and trust that God had embraced the astronauts.

This moment of closure left Abbey weighing his next steps. He never considered quitting, but he wondered about the future of the agency, and of the shuttle program itself. The Return to Flight mission was a long way off, and even then, what type of program would it be? Would NASA still be trying to launch a dozen missions a year? Flying commercial satellites? DOD or NRO payloads?

(Pete Aldridge, as secretary of the air force, would order the mothballing of the Vandenberg shuttle launch site in August.)

If not them, then what would remain? The new Space Station Freedom? Abbey had no faith that Freedom would fly much before 1992—leaving a gap to be filled with Spacelab missions, and little else.

His own future was also uncertain. Investigations inevitably led to firings, and while Abbey felt that he was not complicit in the accident, he knew that blame didn't always fall where it should.

Morale in the astronaut office plummeted after *Challenger*, a natural reaction to the shocking loss of colleagues—and shaken faith in program managers.

Mundane career issues contributed to the dark mood. The military officers selected in the 1978 group were nearing the end of their second details to NASA; the 1980 astronauts were at the end of their first. They had the option of returning to the military, and several did: two-time fliers Bob Stewart (back to the army) and Dale Gardner (to the navy), one-time flier Roy Bridges (back to the air force). Abbey's 1982 astronaut office staffing memo had assumed several losses around this time.

Joe Engle reached the mandatory military retirement age, but secretary of the air force Pete Aldridge, aware of the rush for the exits, asked him to delay his departure, something Engle was happy to do.

Veteran mission specialist and scientist Don Lind triggered his long-planned retirement. Owen Garriott left. So did two-time TFNG astronaut James van Hoften.

Gordon Fullerton retired from the air force and transferred to NASA Dryden as a test pilot. "He could have flown a third mission," Abbey said. "But he didn't want to wait two or three more years. He'd always been interested in aircraft flight tests."

Even those officers who planned to stay with NASA were ordered off for refresher visits to their home services, getting "regreened" or "reblued."

A civilian astronaut asked Maj. Mark Lee of the US Air Force if anything similar would happen to scientists. "Sure," Lee said. "You guys will get re-nerded."

Reinforcements had already arrived. The 1985 ASCANs completed their initial training and were certified in July 1986. They had been working the first technical assignments since just before *Challenger*—Steve Oswald and Rick Hieb had actually been with the astronaut support team at the Cape when *Challenger* was launched. Some had been shifted to the investigation; now they were all back on task, and waiting for the shuttle to resume launching.

Adding to the general gloom was a tragic accident on Saturday, May 24, 1986. Navy test pilot Steve Thorne from the 1985 group was on an off-duty flight with flight controller Jim Simons, a former air force Thunderbirds pilot, in a Pitts 2A acrobatic plane. In the sky near Runge Air Park in northern Galveston County, the Pitts went into an inverted tailspin during acrobatics, plunging to the ground, killing Simons, who was at the controls, and Thorne.

If the astronaut office was a family, it was a family that was still in mourning.

The Rogers Commission officially delivered its report on June 9, attributing the loss of the orbiter to a leak in one of its five-segment solid rocket boosters. Frigid temperatures at the Cape had compromised vital O-rings, sealing one segment of their flexibility, allowing superhot gas to escape and rupture the external tank, causing its destruction.

The report was brutal in its judgment of NASA management, particularly leaders like Marshall's William Lucas and his solid rocket team. The agency had known about the SRB O-rings and their potential failure since 1977, and had failed to correct the problem even though there had been some issue with a dozen different missions back to STS-2 in November 1981.

The *Challenger* disaster was "an accident rooted in history."

Contractors like Morton Thiokol and Rockwell came in for their share of blame too.

Not long after this, Joseph Kerwin delivered his *Challenger* crew pathology report to Abbey and the astronaut office.

Now that the formal hearings were over and the Rogers Commission's report had been delivered, Abbey could begin to look ahead to the Return to Flight mission, which would officially be called STS-26. Truly had discarded the naming system and decided to resume sequential numbering, even if at some point flights launched out of sequence. Since 51-L had been the twenty-fifth shuttle mission, the next would be twenty-six.

Its payload would be a TDRS satellite, next in line after the one lost on 51-L.

Launch was still at least a year and a half away, early 1988 at best. But Abbey had the crew in mind.

Even though Jon McBride had been scheduled to command the next mission after *Challenger*, his flight, 61-E, was a Spacelab mission. Also, McBride had not yet flown as a commander. Abbey felt he needed a veteran commander for this mission and, indeed, a veteran crew. (Truly had asked for this in his March 1986 Manifesto.)

Rick Hauck had been scheduled to command the first of the Centaur missions, 61-F, with Roy Bridges as pilot and Mike Lounge and Dave Hilmers as mission specialists. He had been the first TFNG to fly as pilot, the first as commander, and Abbey considered him one of his stars. He also had great confidence in Lounge and Hilmers, so they would be part of the crew.

Since STS-26 would require a third mission specialist, Abbey told Hauck that he had already contacted Pinky Nelson, veteran of 41-C and 61-B, currently on leave at the University of Washington, and invited him back.

The pilot slot was another matter. Before *Challenger*, Roy Bridges had already let it be known that he would return to the air force after the Centaur mission. With the accident and resulting hiatus, he had brought his departure forward and was already gone.

There were obviously other candidates for pilot, but TFNG Dick Covey had sat in with Hauck's crew on several generic sims after Bridges's departure, so Hauck proposed him. (Covey had been ascent capcom for *Challenger*.)

Abbey had to ponder this. Every other shuttle pilot had flown once as PLT, then moved to command. He would be asking Covey to fly in the right-hand seat a second time.

Covey, however, was happy to put his ego aside, reasoning that while he may have been last in his class to fly, he was getting a fast second assignment, and on a vital mission.

First, though, James Fletcher and Henry Clements at HQ decided to bring Hauck to Washington. There was conflict between Texas and Alabama congressional delegations over the Space Station Freedom program, and these men thought that Hauck would be useful in external relations.

Hauck was reluctant. He told the pair that he had already been informed of his assignment to STS-26, and knowing that Fletcher had the power to remove him, said, "Please don't take this away from me." Fletcher promised he wouldn't, if Hauck would come to HQ for six months.

Sally Ride was another astronaut who had moved from Houston to HQ. At the time of *Challenger* she had been assigned to a mission scheduled for launch in August 1986, with CDR Don Williams and PLT Mike Smith—and the planned journalist in space. Ride believed that this flight, her third, would be a nice way to end her astronaut career before returning to research at some university.

Her assignment to the Rogers Commission had changed all that. When work on the investigation was completed in June, Ride had asked Fletcher for an HQ assignment. (Ride was also in the process of breaking up her marriage to Steve Hawley.)

Fletcher appointed her as his special assistant for strategic planning, and gave her the job of defining a future direction for NASA, especially beyond Space Station Freedom.

Ride had turned to her friend and mentor Carolyn Huntoon, as well as Abbey, for advice. It was obvious that she was never returning to Houston, and Abbey realized that his original group of six women TFNGs was now reduced to three—Shannon Lucid, Rhea Seddon, and Kathy Sullivan. Of course, he had recruited ten more women astronauts. Women were also making progress in mission operations, where a controller named Michele Brekke would soon be selected as a flight director.

As the summer of 1986 turned to autumn, NASA management stabilized. Fletcher was the administrator, Truly head of spaceflight, and Aldrich still in place as director, National Space Transportation System. Former Rockwell Apollo manager Dale Myers became deputy administrator.

What was lacking was a director for JSC. Jesse Moore had been appointed to the job a week before *Challenger*, and after his wrenching testimony to the Rogers Commission had briefly moved to Houston. But he had reconsidered, and offered his resignation to Truly and Fletcher in September.

On October 3, Aaron Cohen was named director of the Johnson Space Center.

42 | Marooned

AFTER THE ROGERS COMMISSION published its recommendations—and even before that—NASA under Fletcher and Truly had begun the process of returning to flight, hoping to launch STS-26 in July 1987.

The primary engineering task was improving the design of the field joints for the solid rocket boosters. Equally important were improved program office structure and communications, with astronauts placed in key positions, and the creation of an office of safety, reliability, and quality assurance at the associate administrator level reporting directly to the administrator.

Other steps included review of the shuttle's ostensible role as national launch system (which led to the removal of commercial communications satellites, among other deployables) as well as all critical safety matters—including brakes, tires, and landing matters as noted by John Young.

Another goal was the creation of a crew escape system. It was impossible to design such a system for the orbiter during the SRB phase—the only useful method would have been a detachable crew cabin like that on a B-1 bomber, a change that would have forced redesign of the entire orbiter.

But NASA engineers did come up with a means of allowing astronauts to escape—in theory—from a crippled orbiter during a controlled glide: a pole that could be extended from the vehicle's side hatch allowing astronauts to hook on and parachute away. (The pole was needed to keep parachuting astronauts from striking the orbiter's huge wing.)

Rockwell was authorized to build a replacement orbiter—to be called *Endeavour*.

At JSC, Aaron Cohen settled into his job. Although he had spent a decade as manager of the shuttle orbiter project office, his last post before his appointment as director had been head engineering.

For support, he reached into Abbey's team and appointed Paul Weitz as his deputy. Abbey replaced Weitz as deputy chief astronaut with shuttle veteran Hank Hartsfield.

Abbey had astronauts working at a variety of levels in the shuttle program, including Hauck at HQ and Bryan O'Connor as chair of an agency-wide Space Flight Safety Panel. Charles Bolden was temporarily assigned as chief of the center's safety division. (Bolden had been pilot on the troubled 61-C mission, the last before *Challenger*.) Pinky Nelson and Steve Nagel were part of the team designing the escape pole.

Shortly after the New Year, on Friday, January 8, 1987, Abbey appeared in the astronaut office to announce the STS-26 crew of Hauck, Covey, Lounge, Fisher, and Nelson. By this time the launch had been bumped to summer 1988.

Abbey then turned to a new issue, the resumption of the suspended 1986 astronaut candidate selection. Some astronauts wondered why, since nobody was flying now and wouldn't be for over a year. But Abbey was motivated by the knowledge that when shuttle launches resumed, missions would be fast and furious as the program worked off a backlog of payloads. He also knew that the office had to replace ten astronauts who had left or retired.

The first group of applicants was scheduled to arrive on February 20 for a week of medical tests and interviews with the selection board. Additional groups of twenty would show up on March 20, March 30, and April 6. A final set of five would come to JSC the week of April 28.

In April of 1987, Cohen told Abbey that he had to replace John Young as chief of the astronaut office. Abbey protested. Young was the most experienced space traveler in the world, a first-rate engineer, looked up to by each class of new astronauts.

Cohen said he was transferring Young to his staff in order not to lose his expertise. After telling Young the news, which was announced swiftly by Cohen on Wednesday, April 15, Abbey appointed Hank Hartsfield as the acting chief.

But Abbey felt he needed an astronaut from the new generation—a TFNG—as the new chief. A pilot, obviously, and one with experience as a commander. That left four candidates: Hauck, Brandenstein, Shaw, Gibson. Knowing they would eventually become office leaders, Abbey had been trying to expose them all to management—Brandenstein had taken Crippen's place as Abbey's deputy in the past few months, once it became clear that Crippen was staying at HQ.

Hauck was going to command the Return to Flight mission; Shaw was still assigned to a vital Department of Defense flight.

That left two potential candidates.

On Sunday night, April 22, Abbey telephoned Dan Brandenstein at home, asking him to meet for coffee. Rushing out for this unusual summons, Brandenstein heard, "I want you to be the new chief astronaut. And I want Steve Hawley as your deputy." Thrilled at the opportunity, and a big fan of Hawley's, Brandenstein eagerly accepted the job and the condition.

The first change the new chief and deputy made was to take charge of the Monday morning pilot meetings. Young had always deferred, leaving the deputy, Al Bean, Paul Weitz, or others, to work through the agenda. Brandenstein and Hawley had never understood that attitude and eagerly took charge.

Abbey flew to Seattle in April 1987 to visit his aging mother, Brenta, in the hospital.

In the years after Sam's death in January 1983, Brenta Abbey had continued to live alone. She started to appear out of sorts, vaguely ill, so Vince took her to see a doctor, who immediately asked how old she was. "I'm not going to tell you," said Brenta, then ninety-three. "You'll just blame everything on that. You should be able to tell me what's wrong without knowing my age." Nothing conclusive was discovered, though Brenta was hospitalized for a short time for tests and observation.

When Abbey visited her in the hospital, she seemed fine. On Wednesday, May 20, Vince telephoned to give his younger brother the news: Brenta had passed away. The woman who had made the journey from a small Welsh town to London to Seattle, who had raised five children, who had been so strong and even fierce . . . now gone.

There was no specific cause of death. Abbey believed it to be a broken heart: "She missed my dad and my sister."

He regretted going back to Houston when he did.

Back in Houston after Brenta's funeral, Abbey had a more pleasant chore—on Thursday, June 4, he telephoned seventeen new astronaut candidates, seven pilots and ten mission specialists. Among the happy recipients were air force test pilot Curt Brown, who would go on to fly six shuttle missions; English-born JSC engineer Mike Foale, who would do tours on Mir and the International Space Station; and NASA physician Mae Jemison, the first African American woman to become an astronaut.

Fifteen were men, two were women.

Others included Bill Readdy, a former naval test pilot more recently with JSC's aircraft ops; Bruce Melnick, the first coast guard officer to be selected as an astronaut; and navy oceanographer Mario Runco—whose wife, Susan, was also a navy oceanographer. Abbey hired her at JSC.

The new group was ordered to report September 17.

That same month, Hoot Gibson and Brewster Shaw decided something needed to be done to raise spirits. What they proposed was a sock hop the following Saturday night. They pitched the idea at the Monday meeting, appointing Sonny Carter as DJ and encouraging others to come up with skits or to be prepared to do karaoke.

The next morning, Shaw had an additional inspiration. He popped into Gibson's office and said, "Hooter, what do you think about putting together a four-man band?"

The two of them, along with Pinky Nelson, had occasionally gotten together at the Shaw or Nelson residence to, in Nelson's words, "drink beer, play guitars, and scream into mikes, annoying our wives."

This trio met up in the astronaut gym that night to rehearse, and resolved the first of many band-related issues. Gibson would recall, "Brewster played rhythm guitar, didn't want to play lead. Pinky could play bass or rhythm, didn't want to play lead. I could play rhythm and I didn't want to play lead, either, . . . but my protest was the weakest, so I became lead guitarist.

"I went out to a local pawn shop and bought a Yamaha amp for $150 and a guitar for $75—you know it was quality stuff."

Then they faced the next major problem—they didn't have a drummer. Shaw and Nelson knew that Jim Wetherbee owned a drum kit, which, it turned out, he had not touched in seventeen years. Nevertheless, their pitch was persuasive. Wetherbee hauled the kit out of his garage and joined Shaw and Nelson—but not Gibson, who was otherwise committed—for a Friday rehearsal.

Somewhere in here Shaw named the group "Max Q," the aerodynamic term for maximum dynamic pressure, or, as Shaw thought of it, "maximum noise."

Max Q went on stage for the first time in the open air of Walter Hill Park in League City, blasting through several Chuck Berry tunes to the general approval of the astronaut audience. "We weren't good," Gibson remembers, "but we weren't bad."

In fact, they were good enough to be invited to appear at the Fajita Festival a month later—which gave them time to rehearse. And to acquire a fifth member, keyboardist Steve Hawley.

Did any of them wonder if they ought to have Abbey's approval? No, Shaw said, "We were doing this on our own time." They need not have worried. The sponsor of the Fajita Festival was George Abbey, for one thing. And the creation of the band showed Abbey that the TFNGs were no longer just astronauts but on their way to being leaders and mentors.

Abbey had immediate need of Hoot Gibson in a more professional capacity. In August he summoned him to the eighth floor to tell him he was going to command STS-27, the second mission after Return to Flight, with a crew largely composed of the original first Vandenberg team. Dale Gardner had left NASA and would be replaced by Bill Shepherd, and Robert Crippen had been promoted to a program management job at KSC. Gibson was essentially replacing Crip. "But it's not my turn," Gibson said, knowing that he had commanded the last shuttle mission before *Challenger*.

"Turns have nothing to do with it," Abbey told him.

This was the kind of decision that baffled and infuriated many astronauts. What about the rotation? Jon McBride had been training to command the mission after *Challenger*. What about Dave Walker's crew? Mike Coats's team? Air force test pilot Dick Covey was enduring a second assignment as pilot, but navy aviator Hoot Gibson—the über-bubba—was getting his second command!

And then there was the matter of Gardner's replacement—Bill Shepherd from the class of 1984. There were still several members of the 1980 group like Jim Bagian and Bob Springer who had yet to fly.

It turned out there was a reason Gibson was assigned "out of turn." The payload for STS-27 was a first-of-its-kind imaging radar system that the National Reconnaissance Office had been developing for years. Special Projects director Maj. Gen. Nate Lindsay had asked Abbey to assign a veteran commander to the crew—someone who had recently served in that role.

Looking back over 1985, the commanders in order were Mattingly (gone), Bobko (two flights that year, now in management), Overmyer (resigned), Brandenstein (now chief astronaut), Fullerton (gone to Edwards), Engle (retired), Hartsfield (management), Shaw, and Gibson.

Hauck was already assigned to STS-26, the Return to Flight mission.

Shaw was already in the training flow for STS-28 (formerly mission 61-N), to carry another vital NRO payload. While it was possible for Abbey to have brought Bobko or Hartsfield out of management, it made more sense—looking to the future and the resumption of a high rate of launches—to assign a qualified TFNG—in this case, Gibson.

By the fall of 1987, as the shuttle program continued its march toward the Return to Flight, management of the Space Station Freedom program was moved to NASA headquarters, which established an office in Reston, Virginia.

Hearing this, Abbey told a meeting of Cohen's staff, "We should get Harvard Business School involved in this right now."

Never one to see a joke coming, Cohen said, "Why would we do that?"

"Because they do studies of failed business plans," Abbey said. "And they'll love this one."

Cohen was not amused. It was time for Abbey to be moved, too.

On Thursday, October 29, he was told he was being transferred to Cohen's staff.

Abbey knew it was pointless to argue. He had no recourse except Cohen's boss in the NASA hierarchy, who happened to be Truly and who clearly supported the change. "I believed that my center directors should have the deputies they wanted," Truly would admit later.

Even Crippen had recently told Abbey directly, "It's time for you to move on."

Hank Hartsfield became the acting director FCOD (flight crew operations directorate), warming the seat for Cohen's choice of Don Puddy, veteran flight director (Skylab, ASTP, and the first two shuttle missions) and friend of Truly and Crippen, who had spiced up his résumé by serving in mission operations management, then as deputy director of NASA Ames. Puddy was completing a yearlong rotation at NASA HQ and headed back to JSC as the new director of flight crew operations.

Abbey disagreed with the choice. Hartsfield, fine; Hartsfield was a pilot, and anyone leading the aircraft ops, not to mention the astronaut office itself, needed to be a pilot.

But there was no way to appeal.

The choice of Puddy wasn't universally welcomed by the astronaut office. Dick Covey recalled a meeting in which Cohen asked his staff if the director of flight crew operations needed to be a pilot, and the universal response was *yes*.

Cohen knew that Abbey had already elevated Bo Bobko to assistant for operations. The veteran shuttle commander and test pilot could advise Puddy on Ellington issues.

When the astronauts held a George Abbey Appreciation Night at Pe-Te's Cajun restaurant on Saturday, December 5, master of ceremonies Mark Lee made the event into something closer to a celebrity roast.

Cohen made Abbey the JSC contact for the National Aerospace Plane, an ambitious joint program between NASA and the Department of Defense whose goal was to create what President Reagan called, in a 1986 speech, "a new Orient Express that could, by the end of the next decade, take off from Dulles Airport, accelerate up to 25 times the speed of sound, attaining low Earth orbit or flying to Tokyo within two hours." There were three prime contractors—and several subs—spending money on the project already, and given Abbey's history with Dyna-Soar and the shuttle, he was far from the worst choice to lead JSC's involvement.

He had real doubts about the viability of the concept. "It was going nowhere." But he accepted the assignment and moved to the ninth floor.

(The NASP would eventually be awarded to Rockwell, becoming the X-30 and taking Vance Brand away from JSC. But it would be canceled in 1992.)

Early in 1988 Dick Truly called Abbey and asked if he would be willing to come to HQ as deputy associate administrator for space transportation systems. The offer left Abbey with mixed emotions. He loathed the NASP job, but he had already turned down two prior chances to go to Washington.

He also knew that he wasn't Truly's first choice for the job—"or even his fourth. A lot of people turned him down."

But at HQ, working for Truly, Abbey felt he could make valuable contributions to returning the shuttle to flight. And he would gain new experience.

He crafted a compromise with Cohen and Truly: he would go to HQ but remain assigned to JSC. His HQ tour, however long it lasted, would be "temporary duty."

So, in late February 1988, twenty-three years after arriving in Houston as a reluctant detailee, George Abbey turned his Houston house over to his daughter Joyce and flew to Washington, DC. He found an apartment in Crystal City, not far from Tom Stafford's office in Old Town Alexandria. It was also close to restaurants—and, with his rental car, no more than a fifteen-minute commute from NASA HQ.

He could stay in touch with his friends in Houston by phone, and during monthly trips home.

He lost his T-38 flying privileges finding himself grounded for the first time in thirty-two years.

He discovered that Truly didn't want him involved in shuttle matters at all. "I was to work on unmanned vehicles and other activities within the Office of Spaceflight."

"At JSC, you dealt with your center and its programs, or other NASA centers. At HQ, you dealt with all these outside elements that affected them, such as Congress and the White House." This was a learning experience for Abbey, and one that ultimately proved worthwhile.

As a student of history, Abbey would visit the nearby Civil War battlefields, sometimes with friends like Ed Pickett (a distant relative of George Pickett, who famously charged the Union position at Gettysburg) and Rick Nygren (another detailee from Houston). Both were Civil War buffs. Nygren said, "It was like having your own personal guide."

Part VI

"A National Quasi-Emergency"

43 | PEPCON

ON THE AFTERNOON OF Wednesday, May 4, 1988, HQ learned of a devastating explosion at the Pacific Engineering and Production Company of Nevada (PEPCON) plant in Henderson, which manufactured ammonium perchlorate (AP), a vital ingredient in the shuttle's solid rocket boosters and, indeed, all of America's solid rocket missiles, including ICBMs and sub-launched Tridents. (There was only one other AP manufacturer, a Kerr-McGee plant located only a mile and a half away from PEPCON, which was fortunately spared.)

Maintenance workers at PEPCON fled when a fire broke out, ultimately igniting a tank and triggering two explosions that were so massive that they sent mushroom clouds into the sky and disrupted the flight of a Boeing 737 at McCarran Airport eleven miles away. Windows rattled in Las Vegas hotels and casinos.

Two PEPCON workers died in the disaster; a hundred others were injured.

The *Challenger* disaster was a contributor to the accident because NASA had continued to order AP for its shuttle fleet—but had not taken possession of it. So storage containers at PEPCON were filled with 4,500 tons of AP that had nowhere to go. Except to burn.

The disaster left NASA and the Department of Defense scrambling. A joint NASA-DOD group was formed to deal with the problem, and Truly assigned Abbey as the space agency rep.

The immediate challenge for Abbey was acquiring enough AP so shuttle launches could resume later in 1988—and to ensure an ongoing supply. The figures made it difficult. Total US military and civilian need for AP was between sixty and seventy million pounds a year. NASA needed close to seven million for 1988–1989, ramping up to over twenty as the space shuttle program resumed its typical schedule. Kerr-McGee was only able to supply half that.

The NASA-DOD team weighed several options, from paying Kerr-McGee to build a new plant, to building up new suppliers, including one in Japan. Ultimately, the panel decided to pay PEPCON to relocate and rebuild, since that was the cheapest and fastest option.

PEPCON never rebuilt its Nevada site, choosing instead to move to Cedar City, Utah, 170 miles away, where a new plant was built in a more isolated area with more stringent standards.

Working with a bank to arrange a loan for PEPCON took a great deal of Abbey's time and effort, too. But it paid off.

He began to make new acquaintances not only at HQ but also in other government agencies, and especially congressional committees. Here he was aided by Tom Stafford and also by Thomas Tate, a lawyer for Rockwell whom Abbey had met in the 1970s. Tate had been on the staff of the House Committee on Science and Technology and was in the process of becoming vice president for legislative affairs for the Aerospace Industries Association. He became an invaluable connection to the world of space-related contractors and their lobbyists.

Abbey also maintained informal conversations with astronauts and other associates in Houston and elsewhere. Steven Hawley, now at NASA Ames, said, "I talked with George more while I was out of Houston than I did when I was in Houston."

STS-26 with a crew commanded by Rick Hauck successfully launched on September 29, 1988, deploying a TDRS satellite and returning the shuttle to flight.

For Abbey, it was a bittersweet moment—the first time since 1975 that astronauts had been escorted to a mission, and greeted upon return, by someone else. (Don Puddy and Dan Brandenstein did the honors.) It was his crew, after all.

But he could only follow the mission on television.

He had similar feelings in December, when Hoot Gibson led his crew on STS-27, a dedicated DOD mission. Although operating in a news blackout like 1985's 51-C, several informed sources claimed that the crew deployed the first ONYX imaging radar satellite for NRO.

And there were problems. The *Discovery* crew had to make a second visit to ONYX after its initial deployment because solar panels failed to deploy. Some heroic work by Mike Mullane with the orbiter's RMS—or possibly a secret EVA by Jerry Ross and Bill Shepherd—saved the vehicle.

Then the crew faced a life-threatening situation. During launch several chunks of foam had fallen off the external tank, gouging the underside of the orbiter. Gibson and team were able to see some of the damage with the RMS camera, but not all of it.

New missions would be launched in March 1989 and May—the first to be flown by crews not selected by Abbey, or not completely. They were the handiwork of Dan Brandenstein and Steve Hawley, who had pulled off an administrative coup. "After George left, and Don Puddy came in," Brandenstein said, "we realized that Steve and I were in a better position to know the strengths and weaknesses of individual astronauts."

Feeling, he said, "like teenagers given keys to the family car for the first time," Brandenstein and Hawley simply "put together a set of crews and presented them to Don for his approval."

One was STS-28, Brewster Shaw's pre-*Challenger* crew—with 1980 astronaut Dick Richards replacing 1984 selectee Mike McCulley. "We thought it was only fair that guys who had been waiting longer got first crack at missions."

George H. W. Bush succeeded Reagan in January 1989, and, as planned, James Fletcher resigned on April 8 to allow the new president to have his own NASA administrator. According to some sources—notably the book *Mars Wars* by Thor Hogan—in spite of Fletcher's advance warning of his departure, the Bush team had no one immediately in mind to replace him.

Dick Truly had submitted a one-page paper to the Bush team urging them to set new goals for the agency, impressing Vice President Dan Quayle and chief of staff John Sununu, who decided that the former astronaut might be a good choice to succeed Fletcher as administrator.

According to Quayle's memoirs, Truly became the default choice because he "didn't have the sort of negatives that might make news and sink the nomination." (The Bush team had just suffered an unprecedented defeat when its nominee for secretary of defense, Sen. John Tower of Texas, was rejected, largely owing to reports of womanizing and drinking.)

When Truly was going to be sworn in, he asked his old friend Bill Lenoir, who was now working for Booz Allen on a variety of projects, including Space Station Freedom, to come along. Lenoir held back the information that he and Sununu used to play lacrosse at MIT, so Truly was surprised to see the White House chief of staff walk over to Lenoir and shake hands, like an old pal.

In the car after the ceremony, Truly revealed his own surprise: he wanted Lenoir to come to NASA HQ as associate administrator for spaceflight. Lenoir was reluctant, telling him, "I'm just getting to the point now where I'm starting to make some money, and save."

So the answer was no. But Truly kept calling, playing the "national service" card. Lenoir eventually relented, becoming the associate administrator just as Space Station Freedom was facing a critical vote in the House of Representatives. (Rep. Tim Roemer, D-IN, had offered an amendment to the NASA authorization that would terminate funding for Freedom. The amendment ultimately failed but would not be the last attack on space station funding.)

The Bush administration also revived the National Space Council, announcing it in a ceremony at the Old Executive Office Building on April 20. Vice President Quayle would be the chair; there would be members from the Departments of State, Defense, Commerce, the Treasury, and Transportation, as well as the ever-present Office of Management and Budget, the director of the CIA, the White House assistants for the National Security Agency and for Science, Space and Technology, Sununu himself, and the incoming administrator.

The executive secretary would be a young staffer named Mark Albrecht, thirty-eight years old with graduate degrees from UCLA and from the Rand Institute.

44 | The Space Exploration Initiative

ON SATURDAY, JUNE 17, Abbey learned that an astronaut had died. TFNG Dave Griggs, a former naval aviator and test pilot, was killed in the crash of a North American T-6 in Earle, Arkansas. He had been rehearsing acrobatics for an upcoming air show in the venerable trainer—the same type Abbey had flown in 1955.

Griggs was a good friend; Abbey had recruited him for JSC's aircraft ops in the summer of 1970, had worked with him closely on the Shuttle Training Aircraft, and had selected him as an astronaut in 1978. Griggs had flown as a mission specialist with Bo Bobko and Jake Garn in 1985. More to the point, at the time of his death he was assigned as pilot for STS-33, scheduled for launch in November 1989.

As head of flight ops and flight crew ops, Abbey had instituted strict rules for astronauts on flight crews—no skiing, for example. They couldn't even play *softball.* Why was Dave Griggs off doing acrobatics on a weekend? Yes, astronaut candidate Steve Thorne had been killed in 1985 in an off-duty accident, but Thorne was not in training for a mission.

He learned later that Griggs had not been given permission for his off-hours flying, but even that reflected badly on Puddy and Brandenstein. Didn't they know? Or suspect?

Abbey had been criticized for his restrictions on the astronauts, but here was justification. Some of them were just not willing to follow the rules.

July 20, 1989, was the twentieth anniversary of the Apollo 11 landing. With astronauts Armstrong, Collins, and Aldrin in attendance on the steps of the National Air and Space Museum, President George H. W. Bush proposed a new Space Exploration Initiative, a long-term effort to build Space Station Freedom, return American astronauts to the Moon, then press on to Mars.

The effort would be led by Vice President Dan Quayle and the National Space Council.

The proposal was the result of weeks of work by Truly and the HQ staff proposing new steps for NASA in the next decades. The rough outline of a return to the Moon, then to Mars carried a projected cost of $400 billion.

One reason for Bush's speech was to give his new administration a signature program in space, à la John F. Kennedy's Apollo.

The other was more pragmatic. According to Albrecht, "The Soviet Union was falling apart. Defense budgets were falling by 30 percent. Don Rice and Dick Cheney, everyone was looking at NASA's programs, too, thinking of carving up the agency. The study was a way for NASA to justify itself." And save itself.

Tom Stafford, who was present, approved of the message, but wondered, "Where is the money going to come from?"

More the point, what would be the steps? Weeks before Bush's announcement, Truly and Albrecht had asked Aaron Cohen to form a working group at JSC to provide more technical details. Albrecht had told Cohen, "Give us a Cadillac option, then give us the el cheapo alternative."

As many as 450 people worked on the 150-page 90-Day Study, which arrived at HQ on November 20—and landed with a splat. The big takeaway was a reported price tag of $541 billion over twenty or thirty years, far more than the Bush administration or the Space Council wanted. Historian Dwayne Day would later describe it as "like selecting five different ways of paying for a Rolls Royce, rather than looking at cheaper cars."

Albrecht was blunter. "NASA blew it. They came back with one option—to Mars via the Moon and still using Space Station Freedom as the centerpiece, the facility that would be all things to all people."

As always happened with big new NASA initiatives, the National Research Council commenced a thorough review led by Guy Stever and including Tom Stafford.

The report was also turned over to a council-chosen review panel that included Buzz Aldrin, novelist Tom Clancy, Carl Sagan, space policy academic John Logsdon, and physicist Lowell Wood. The panel found fault with the 90-Day Study in a variety of areas, both technical and managerial.

Stever's NRC response arrived in late February 1990. It found NASA's proposal to be flawed—and also criticized the White House for failing to provide proper guidance and support.

In January 1990 NASA announced the selection of a new group of astronaut candidates, the first since 1978 that Abbey had not supervised.

There were twenty-three, sixteen mission specialists and seven pilots. Among the pilots was air force major Eileen Collins, the first woman selected for that job. Collins had been a T-38 instructor, then a C-141 transport pilot with combat experience in Grenada. She was also a graduate of the Air Force Test Pilot School—the second woman to attend, in 1989.

Air force captain Susan Helms was a new mission specialist ASCAN; she had been in the first Air Force Academy class (1980) to admit women.

There were two other women in the new group: army helicopter test pilot Nancy Sherlock (later Currie) and physicist Ellen Ochoa from NASA Ames, the first Hispanic woman to join the astronaut team.

Among the male mission specialists were Bernard Harris, an African American NASA doctor, and Leroy Chiao, a Chinese American chemical engineer from Lawrence Livermore Labs.

Abbey was pleased to see that his original outreach was paying dividends. Women and minorities were now routinely included in ASCAN selections.

It was a small consolation.

William Lenoir was an MIT-trained electrical engineer with a sly sense of humor—he was the astronaut who carried the NASA OFFICIAL GEORGE ABBEY sign. As he had with all the holdover scientist-astronauts from the 1960s, Abbey had felt obliged to give Lenoir a chance to fly in space, and he had, on STS-5.

Now he worked for the former astronaut, who had become the associate administrator in charge of both shuttle and space station Freedom.

From the summer of 1989 and into the winter and spring of 1990, Abbey continued to follow the construction of the new PEPCON facility in Utah. He also found himself evaluating the sources of vital materials like rayon, used in the nozzles of solid rockets and increasingly unavailable due to reduced American manufacture.

The shuttle continued to fly. Brewster Shaw led a crew of five on a DOD mission, STS-28, in August. The deployment of an NRO payload (reported to be a relay satellite, code-named QUASAR, for NRO's imaging reconnaissance constellation) went well.

In October, STS-34 carried the long-delayed Galileo probe into orbit, where it was sent on its mission to Venus and then Jupiter by a crew of five under Don Williams.

On Thanksgiving 1989, STS-33 lifted off with a crew of five commanded by Fred Gregory, becoming the first African American astronaut to lead a mission. The crew—with pilot John Blaha replacing the late Dave Griggs—deployed a payload for the National Reconnaissance Office.

Dan Brandenstein, Young's replacement as chief astronaut, launched on STS-32 in January with a crew that included James Wetherbee, Marsha Ivins, Bonnie Dunbar, and George Low's son Dave on his first mission. After deploying the Syncom IV-5 communications satellite, Brandenstein's crew successfully retrieved the Long Duration Exposure Facility pallet, originally orbited in April 1984 and intended for return to earth in March 1985.

Another NRO mission followed in March, after several delays owing to the health of commander John Creighton—who came down with a serious cold in the last week before launch. For the first and only time in shuttle program history, STS-36 used a sixty-two-degree inclination (before this the maximum allowable had been fifty-seven degrees, used by STS-27) in order to deploy an NRO payload originally intended for launch into polar orbit from Vandenberg.

STS-31 lifted off on April 24, 1990, with a crew of five commanded by Loren Shriver, with Charlie Bolden as pilot, and mission specialists Bruce McCandless, Kathy Sullivan, and Steve Hawley. Their task was to deploy the precious Hubble Space Telescope.

Originally developed as the Large Space Telescope beginning in the mid-1970s (and named for pioneering astronomer Edwin Hubble in 1983), the Hubble's primary instrument was a 2.4-meter diameter reflecting mirror that delivered light to five different detectors: the Wide Field and Planetary Camera, the Goddard High Resolution Spectrograph, the High Speed Photometer, the Faint Object Camera, and the Faint Object Spectrograph.

The mirror, instruments, guidance, propulsion, and other systems were housed in a giant cylindrical bus that bore a strong resemblance to the NRO KH-11 KENNEN satellite, which also carried a Pekin-Elmer imaging system.

The construction was managed by NASA Marshall, with the scientific payload under NASA Goddard.

At one time the launch was planned for April 1985, but Pekin-Elmer continued to suffer delays, and the launch slipped to the right, ultimately landing in August 1986.

The *Challenger* disaster forced another, longer delay, to April 1990, by which time the price for the spacecraft—originally budgeted at $400 million—had reached $2.5 billion.

Soon after Hubble was deployed at an altitude of 380 statute miles, however, operators at NASA Goddard discovered that the telescope's optics were flawed. Yes, the big telescope was near-sighted, or so a host of late-night television comics claimed.

Then, in late June 1990, a hydrogen leak was detected in an external tank umbilical disconnect with the orbiter *Columbia* as it was stacked for STS-35, a long-delayed Spacelab mission.

The shuttle team at KSC looked at *Atlantis*, already on pad 39A and scheduled for launch as STS-38, and found the same problem. *Atlantis* had to be rolled back to the Vehicle Assembly Building for repairs—and while it was outside the VAB, allowing *Columbia* to move to the Pad, the Cape suffered a hailstorm . . . and *Atlantis* was damaged.

All of this forced the program to suffer another few months of delays at the same time the flawed Hubble was becoming a national joke.

NASA established a board to determine what had gone wrong—it was headed by Lew Allen, director of the Jet Propulsion Laboratory, and luckily enough, a former head of the National Reconnaissance Office's Air Force Element in Los Angeles. Allen was quite familiar with the "black" side of Pekin-Elmer's operations.

What the board discovered is that Pekin-Elmer had incorrectly assembled a device called the main null corrector, which was designed to judge the precision of the grinding and polishing of the main mirror. The company also skipped some final tests, and ignored data that showed a spherical aberration.

While most of the blame landed on Pekin-Elmer, NASA came in for its share of criticism for failing to exercise proper supervision of the contractor. Fortunately, Hubble had been designed to accommodate servicing missions. It operated in an orbit that could be reached by the shuttle, and had external grappling fixtures.

45 | Synthesis

IN 1990 TOM STAFFORD was a sixty-year-old retired air force lieutenant general as well as a veteran astronaut. He had been a Beltway bandit for the last eleven years as well as a board member for such aerospace giants as Boeing and AlliedSignal and his personal favorite, the watch company Omega.

Like Abbey, he was brilliant, with near-total recall memory (not, he would always insist, as good as Abbey's). He had demonstrated a useful ability to get along with powerful bosses like Deke Slayton and Al Shepard, and then to be an equally strong leader himself—commanding the Air Force Flight Test Center at Edwards.

Unlike Abbey, Stafford's maneuvers were out front, blunt, frequently take it or leave it. He had strong opinions about military technology as well as NASA and the human spaceflight program, and wasn't shy about expressing them.

Abbey and Stafford had known each other since the late 1960s, crossing paths on Apollo, but had only become close friends during the Apollo-Soyuz days in the mid-1970s. Since that time they had remained in touch, as Abbey largely stayed in Houston while Stafford moved to California, then to the DC area.

In late 1989 Stafford had been asked by Truly to review Cohen's 90-Day Study, and the former astronaut's board had found the study flawed and too expensive. That verdict made Quayle's office and the Space Council unhappy. Albrecht wanted a more comprehensive roadmap going forward, and suggested as much to Truly.

Truly invited Stafford to HQ on May 30. There he and deputy administrator J. R. Thompson asked Stafford to "volunteer" to head this comprehensive study, which would be known as the Synthesis Group. Truly wanted it based in Los Alamos or some equally remote tech center, but Stafford knew better: this would be a Beltway team, and he wanted to be located within the Beltway. If nothing else, it would be easier and cheaper for team members and subjects to travel to and from DC. (Stafford also wanted to be near his consulting firm, Stafford, Burke & Hecker, which had its office in Alexandria, Virginia.)

Truly relented, and NASA rented the entire floor of Crystal City Gateway 2, an office complex near the Pentagon and Reagan National Airport.

Knowing that Abbey was not being fully utilized at HQ, Stafford asked Truly to assign him as his deputy for operations. Truly agreed, since the move would create a vacancy for Lenoir to fill with his own choice as deputy. Abbey was pleased to work with Stafford. He and the general spent the Fourth of July weekend planning their activities and schedule.

Then it was time to staff the group, and Abbey reached into his Rolodex and memory. He had Stafford hire Spence Armstrong, recently retired from the air force. Like Stafford and Abbey, Armstrong was a graduate of the Naval Academy who went on to become a pilot in the air force. He attended the Aerospace Research Pilot School; following a combat tour in Vietnam, he would become commandant of the school. He had also worked under Stafford in the Pentagon.

Armstrong had been looking forward to retirement and was literally painting his home in South Carolina when Stafford telephoned. Not only would he serve on the Synthesis Group for a year, he would then join NASA—where he remained until 2002.

Another key hire was Col. Michael "Mini" Mott, a Marine Corps aviator and test pilot then stationed at Andrews Air Force Base, where he commanded a marine unit. Mott had applied for the 1984 astronaut group and been one of the finalists. But a minor yet disqualifying physical issue cropped up and he was not selected.

Abbey had been impressed with him. And as Abbey had done in the past with promising talents like Jim Adamson, he had stayed in touch.

Mott found Abbey's offer of an assignment to the Synthesis Group so attractive that when the Marine Corps declined to give him a detail, he submitted his resignation papers, impressing Abbey and Stafford with his commitment.

Another marine was Lt. Col. David Lee, son of NASA's Chet Lee, who had been an Apollo mission director. The younger Lee had just completed a tour in the navy's space program and was about to retire, but the Marine Corps wasn't ready to let him go.

Stafford took to the phone, making a personal plea to Jack Dailey, the assistant commandant of the Marine Corps—which was successful.

Abbey recruited staffers from Lockheed, Rockwell, and other contractors; from NASA JPL, from NRO, from various national labs like Sandia. He created a separate team of tech advisers that included astronauts Jim Adamson, Sonny Carter, Steve Hawley, and Bill Shepherd, along with individuals from the Space Telescope Institute at NASA Goddard, JSC, and Brookhaven National Labs, as well as several private consultants.

Douglas Beason, a young air force major with an advanced degree in plasma physics, was called in to see his boss at the Air Force Weapons Laboratory (AFWL) at Kirtland Air Force Base, New Mexico, Col. John Ott, who told him, "You're being assigned TDY [temporary duty] to this Moon-Mars group in DC."

"But I don't know anything about human spaceflight," Beason said. (He was a huge science fiction fan, though, and later published SF and high-tech thrillers.)

No matter, Ott said. "We want to push you, get you outside the lab, broadening your background." Beason would say later that his tour on the Synthesis Group "changed my life for the better. I showed up in Crystal City with Colonel Rich Davis, also from the AFWL. There were about ten active duty military—four air force, a couple of marines, a couple of army and navy."

"Pete Worden was on the Space Council with Jim Beale, and they represented different sides of the space world, human vs. unmanned intel. The belief was that there was a finite amount of space money, and that an increase in dollars spent on humans meant less for NRO.

"We were supposed to live in a hotel for nine to ten months, but Davis said to hell with that, I'm renting an apartment. 'Can you do that?' 'I'm a colonel, I can do anything.' So there we were, signing these lease agreements and driving the accountants back in Kirtland crazy—this turned out to be the Synthesis Way . . . better to ask forgiveness than permission."

To Beason, everyone he met seemed to have some connection. He knew Don Pettit from Los Alamos. Worden and Pettit had both attended the

University of Arizona. Several team members had come to Abbey's notice because they'd applied to be NASA astronauts, others because they had worked on programs Abbey knew.

Even before Stafford and Abbey's team got started, NASA asked the RAND Corporation to conduct a public survey called Project Outreach, to, in Vice President Quayle's words, "cast the net widely." All federal research would be reviewed, the American Institute of Aeronautics and Astronautics would be consulted, but what was really sought were ideas from space contractors to universities, from government agencies to individuals.

RAND had collected 1,697 concepts for space technology and operations. The Association of Aeronautics and Astronautics submitted its own list of 542. Stafford and Abbey's team reduced the list to 215, adding a few of their own. And then they got to work.

In the midst of a divorce, and knowing that Jake Garn was not going to run for reelection, Jeff Bingham, Garn's space staffer, had decided to search for something else to do. He had asked Truly for ideas, and the administrator forwarded his name to Stafford for the Synthesis Group.

A meeting with Abbey resulted in an offer to join up. Bingham jumped right into what he described as a "non-stop succession of hour-long sessions with experts in every aspect" of future space missions. The days ran from 7:00 AM to 7:00 PM.

"People from Bechtel came in and showed us how they would move regolith around on the surface of the Moon with tractors and specially designed vehicles, and build habitats and protect them from radiation," Bingham would recall. "We had every rocket designer, every propulsion system; we had people talk about the psychology of long-duration spaceflight. It was just nonstop and endless, and we divided up in teams and came up with four different architectures."

At one point, feeling overwhelmed and technically out of his league, Bingham went to Abbey and said, "What am I supposed to do here?"

"You're supposed to ask the dumb questions."

Doug Beason had no idea who Abbey was when he arrived at Crystal City. "But you quickly realized who was running the show. Not General Stafford, but George Abbey."

He was also amused by Mini Mott and David Lee, who "liked to present themselves as dumb marines" but quickly established themselves as leaders and doers.

Beason says that "the real Synthesis Group—we forty were the staff—probably only met three or four times, and everything they saw came from us, and George."

In addition to the formal briefings, Abbey continued his unofficial information gathering and spreading. Beason says, "You also saw that one of the keys to operating in the group was getting to know George socially, which meant being part of beer call or the George Watch . . . which meant going out with him in a group." The size of this group varied but usually numbered ten.

"It wasn't automatic. George had to get to know you a bit. Pete Worden would spec you out. There was a sort of inner circle, and I wasn't part of it, though Worden was. (Mott wasn't yet, either.) And there were some who just didn't make the cut."

And beyond these informal, unscheduled social gatherings, Abbey soon established a regular Friday evening event he called "Vespers." Like something out of Victorian tales of gentlemen's clubs, that day's guest briefer would essentially sing for his or her supper—would make conversation with staffers that would, in classic Abbey fashion, illuminate or give a different perspective on the official subjects.

It wasn't healthy or expensive. Beason says, "Stafford paid for the beer and wine and food. We were just junior officers, most of us, with not much money, and he would give out cash or reimburse us. We would run out, buy the booze and the not-healthy chips and dip."

But it was useful.

Vespers became so popular that Stafford once skipped a flight to California and a paid board meeting so he could linger. (Hearing that this gesture cost the general something like $40,000, Mott offered to take Stafford's place at any future board meetings.)

Visiting cosmonaut Aleksey Leonov asked to attend.

Vespers opened another door for Abbey, too. Mark Albrecht said, "Sometime during the Synthesis Group's work, I heard from Pete Worden, who was the Space Council representative to Synthesis, that there was this frequent Friday evening event called 'Vespers.' I met George at one of these and it was immediately apparent that he knew everything about NASA." Until that time, Albrecht only knew Abbey as "the man from NASA running Tom Stafford's committee.

"Talking to him was like having a football coach explain the Zone Defense to you for the first time."

Albrecht couldn't attend Vespers often, nor did he take part in the other casual Abbey dinners, but he began to treat Abbey as a source of informed information about NASA and human spaceflight. "As I got to know him, there was one mystery that always puzzled me: how does a man who knows so much, who does so much, who feels so strongly about the importance of human spaceflight. . . . How does he have no ego?"

The group met through the winter. By spring 1991 it was time for Stafford, Abbey, Mott, and the staff to evaluate the proposals and begin drafting the actual report.

On the morning of Monday, April 5, 1991, NASA launched the orbiter *Atlantis* on mission STS-37, aimed at deploying the Gamma Ray Observatory. The crew of five was commanded by TFNG Steve Nagel. In Washington, Abbey could only watch on television as the crew reached orbit safely.

Meanwhile in Houston the weather was terrible, a spring storm so strong that astronaut T-38 flights were canceled. Astronaut Sonny Carter, committed to speak on Brunswick Island, Georgia, boarded a commercial flight from Houston Hobby to Atlanta Hartsfield, where he connected to Atlantic Southeast Flight 2311, which took off at 2:23 PM local time. Also aboard this flight were former senator John Tower of Texas and Tower's daughter, Marian, and eighteen other passengers in addition to a crew of two.

Shortly before 3:00 PM, the flight (which used an Embraer 120 aircraft) was on approach to Glynco Brunswick Airport when it suddenly rolled left and crashed, killing everyone on board.

Abbey was shocked by the accident. A navy flight surgeon and test pilot—and a former professional soccer player—Carter had been not only one of his personal favorites but also one of the most popular and capable astronauts in the office. He attended the memorial for Carter in Warner Robins, Georgia.

By now it was time to make conclusions and to write the Synthesis Group report.

Beason says, "I believe that George Abbey knew what the report was to say from the beginning. As far as the next steps in space . . . He was a big believer in, you don't go SCUBA diving in the ocean until you've tried it in a pool. In other words, don't try a Mars mission unless you've tested all the technologies within three days of home. He was interested in what the survey groups had to say."

It's not as though the conclusions were preordained, though they were shaped by what Abbey had learned over several months of reviews.

The report called for several architectures, all of them in one way or another treating a return to the Moon as a stepping stone to Mars. Published as *America at the Threshold*, the document was delivered to the Space Council and Vice President Quayle on May 3, 1991. Mott wrote the final text, under Abbey's direct supervision.

Stafford led a press conference on June 10 to present the report to the public.

46 | Space Council

MARK ALBRECHT HAD DECIDED that he needed George Abbey on the Space Council, and the completion of the Synthesis Group report freed Abbey for a new assignment.

As for Abbey, he was eager to go with Albrecht. He had no desire to go back to NASA HQ.

Richard Truly objected to Abbey's move to the Space Council. Albrecht said, "While he had his problems with George, he knew how valuable he was. He kept suggesting X or Y alternatives, but I kept pushing and finally said, 'Should I have the vice president call you to make the request'"?

"We knew—and I think Truly knew—that with George at the Space Council, the White House couldn't be flimflammed by NASA.

"Talking to him was like having the decoder ring for NASA. He was the Indian guide. I was fairly animated, and I would be waving my arms and saying, 'What the fuck are they doing?' And he would shuffle and shift and mumble, 'This is what's going on.' And he was always right.

He was loyal. He was smart."

Abbey felt that he was rehabilitating himself, finding new ways to contribute. NASA's Frank Martin, originally a bit of an Abbey skeptic, had concluded by this time that, "here was a guy who couldn't help being effective. They kept shunting him off to dead-end jobs to hide him or get him out of the way, and still he managed to do productive work.

"He just had a passion for NASA and its mission."

By this time, the fall of 1991, Albrecht had grown frustrated with the NASA administrator. "I would be ranting about Truly, how I was just going to replace him."

In order to get rid of Truly, Albrecht needed several planets to align. The first was the November 1991 resignation of J. R. Thompson as deputy administrator. Albrecht had the job of finding a replacement—and got nothing but turndowns, primarily because none of the potential appointees wanted to work for Truly. That led to a conversation with Quayle, who urged Albrecht to find out if Truly was really that unpopular.

This set of queries led to a meeting with Albrecht and three former NASA administrators—Paine, Fletcher, and Beggs—in the office of William Kristol, Quayle's chief of staff. The verdict was unanimous that Truly ought to be replaced.

Truly still had a powerful supporter inside the Bush White House—Chief of Staff John Sununu. But Sununu abruptly resigned on December 4, leaving Truly unprotected.

It was left to Dan Quayle to deliver the message to Truly that it was time to step down, offering him a graceful and face-saving transition to an ambassadorship. Truly promised to consider it; over the weekend, he did. And told the vice president that he wouldn't go.

Two weeks of paralysis followed.

Then Truly arranged a meeting with Samuel Skinner, Sununu's replacement as White House chief of staff. Truly told Skinner he wanted to meet with President Bush himself, and on Wednesday, February 12, he did.

Nothing changed. Truly was out, effective April 1.

According to the account in Bryan Burrough's book *Dragonfly*, shortly after the meeting Albrecht was unlucky enough to encounter President Bush himself in the halls of the West Wing. Stung by the unnecessary confrontation with Truly, the president informed Albrecht that his primary job was to find a new, qualified, fantastic NASA Administrator who could be confirmed by the time Truly formally departed on April 1.

That was forty-five days away.

Albrecht asked another White House staffer: What's the fastest confirmation this administration has had? James Baker as secretary of state, he was told, in forty-three days.

Hoping to give the agency some continuity, Truly appointed JSC Director Aaron Cohen as acting deputy administrator—that job had been vacant since J. R. Thompson's departure the previous November. The move left Paul Weitz as acting head of JSC.

Meanwhile, Albrecht searched for a new administrator, someone who not only had the requisite political skills and technical knowledge, but who was confirmable on short notice.

There were the usual suspects, aerospace professionals whose names always emerged as candidates to lead NASA. Former astronaut Tom Stafford. Lockheed president Norman Augustine. Planetary scientist Carl Sagan.

Fortunately, Albrecht had a candidate in mind, a former NASA engineer who had gone into the contractor side of space and had worked on both NRO and NASA programs. This candidate shared Albrecht's frustration with NASA, having complained about the agency's management of a Mission to Planet Earth program.

On Monday, March 23, 1992, Abbey arrived at the Executive Office Building to find Albrecht with someone new. "Meet Dan Goldin," Albrecht said. "He's going to be the next NASA administrator."

47 | Dan Goldin

YEARS LATER, MARK ALBRECHT would say, "I know there's a mythology that George Abbey somehow arranged for Truly's downfall and handpicked Dan Goldin as his successor because he could control Dan. Not true. The entire Bush Administration was fed up with Truly. And I was the one who found Dan Goldin—he was my number one pick to head NASA in the spring of 1992. I was the one who introduced George to Dan about a week before Dan was nominated."

Who was this new NASA savior?

Born in the Bronx to an Orthodox Jewish family, and educated at City College, where he received a BS in mechanical engineering, the fifty-one-year-old Daniel Saul Goldin had a prior tour with NASA, having joined the agency's Lewis Research Center in Cleveland in 1962, working on electric propulsion systems.

Five years later, seeing that NASA had few realistic plans for missions beyond the Apollo lunar landing, he joined TRW in Redondo Beach, California, where he spent the next twenty-five years, rising to vice president and general manager of its Space and Technology Group.

Although his primary work was in signals intelligence satellites for the National Reconnaissance Office, Goldin also managed civilian programs such as the Compton and Chandra observatories as well as advanced communications satellites.

Dealing with NASA as a contractor, he found the agency difficult to work for, feeling it showed too great a preference for very large, expensive, and one-vehicle-does-everything projects. He also felt that NASA tended to threaten to eliminate competing programs. (It was a Goldin complaint about NASA to Albrecht of the National Space Council that earned him a place on the shortlist of potential agency administrators.)

He would be taking over an agency that burned through $17 billion of federal funds each year, that had eleven separate (and sometimes competing) centers, including the politically powerful Johnson, Marshall, and Kennedy operations.

It was an agency with a history of genuine accomplishment marred by recent failures and now judged to be adrift.

Goldin would proclaim his support of and interest in monitoring Planet Earth and cutting-edge aeronautics, but he was well aware that NASA's core mission was the exploration of space. Shuttle, space station, those possible returns to the Moon, voyages to Mars, unmanned planetary probes—how to do them all with less money?

His solution, using a catchphrase from the world of military tech, was to make NASA programs "faster, better, cheaper." One step was to downsize and reorganize the agency to meet upcoming budget cuts.

For a new leader with no allies in the organization, and no real insights into the way it worked, implementing these changes would be a huge challenge—the very issue that scared off other candidates for the administrator job. Goldin discussed this with Stafford and Albrecht, who both told Goldin that he needed George Abbey at NASA Headquarters.

Goldin asked Abbey to be his special assistant, specializing in day-to-day agency matters while Stafford would advise on international and policy questions. Abbey agreed, though he would remain on the staff of the Space Council. Over the rest of 1992, he would spend half his time at NASA.

By the time Goldin was officially confirmed on March 31, 1992, Albrecht and Abbey were already meeting every evening in the Space Council offices, plotting changes. Stafford would admit years later that he "had a little list" of NASA leaders who needed to go. Abbey had his own.

One of the first out the door was Lenoir, who announced his resignation on March 30, the day before Goldin's confirmation. It was effective May 4.

What kind of new leaders? Abbey had learned things over the years, traits of leadership that he was able to codify. He actually wrote them down:

> Must have experience managing or leading major projects or organizations
> Must have demonstrated ability to work with people
> Must have demonstrated ability to work successfully as a team member
> Must have technical background
> Must have experience successfully managing projects or activities with significant budgets to tight schedules
> Must have demonstrated ability to perform under stress
> Must have demonstrated ability to present or convey ideas and technical and/or cost analysis
> Must have ability to recognize and select people for their technical and managerial skills
> Must have demonstrated creativity
> [And especially—]
> Must have demonstrated ability to make tough decisions when required and treat people with dignity and respect

Who had the requisite experience? Who was a proven team member who became a leader? Who was creative and yet, if necessary, could make the needed tough decisions? Abbey had identified a few within the agency, especially veteran astronauts he had selected and mentored. But he was ready to look outside.

The US military provided not only a ready source of educated, experienced leaders but also new ideas and ways of doing business. Abbey's first source for recommendations was former marine Mike Mott, who after the demise of the Synthesis Group had been hired by General Research Corporation, a Dayton, Ohio-based firm with offices in DC.

During his marine aviation career, Mott had flown with Jed Pearson, a Gulf War veteran and test pilot who oversaw the introduction of the F/A-18 to the Marine Corps.

Pitched about a possible NASA job by Goldin and Abbey, Pearson—an unsuccessful applicant for the 1978 astronaut group—signed on, soon to replace Lenoir as associate administrator for spaceflight. Since Pearson had zero experience at NASA, Abbey sought to support him by luring former astronaut Bryan O'Connor back to the space agency.

Following shuttle flights in 1985 and 1990—with intervening work on the *Challenger* recovery and in shuttle safety jobs—O'Connor had returned to the Marine Corps and was in the middle of a tour at Patuxent River as commander of the Marine Air Detachment there. He had just realized that he had no vital role to play in flight test programs—but he might be able to help NASA, and accepted a post at headquarters.

Yet another Mott connection was Randy Brinkley, then working for McDonnell Douglas on the F-18 fighter program in St. Louis after twenty-five years of active service.

The biggest hire was yet another Mott associate from his time in weapons development, when his boss was the future four-star assistant commandant of the Marine Corps, Jack Dailey. Dailey had been a third-string All-American football player at UCLA under coach Red Sanders. He had joined the Marine Corps after graduation, became an aviator, then rose to the second-highest ranking position in the service. In July Abbey went after Dailey as a possible deputy administrator. (As Truly was departing, he had asked Aaron Cohen to come to HQ to serve in that role, but everyone knew that was short term.)

Hence Dailey, who remembered, "I think they wanted somebody who was not a former NASA person," but yet had senior management experience.

Dailey still had two years remaining in his post with the Marine Corps. He assumed that NASA could accelerate his separation from the service, a process that typically required six months. (Dailey hoped to become commandant of the corps, but his background as an aviator made that impossible.)

The first step went quickly. Dailey retired, angering the marines because there was no immediate plan for the transition. But then the process stalled. Civil service regulations forbade the direct appointment of someone at Dailey's level into the Senior Executive Service. Dailey moved into an office in Crystal City, Virginia, and waited.

Meanwhile, Abbey's recruitment continued with other targets: On April 2, 1992, the orbiter *Atlantis* landed at the Kennedy Space Center, completing STS-45—the Spacelab ATLAS mission.

Col. Charles Bolden, the mission commander, had made "an unpopular decision" for his crew. Previous shuttle astronauts—some of them in terrible physical condition—had exited their orbiters on foot, the pilots performing the ritual "walk-around" of the vehicle, before being turned over to the flight surgeons. The medical community wanted astronauts to be carried off the orbiter, so that the first stages of readaptation to gravity could be studied. Bolden insisted that his crew would be "carried off on gurneys," and they were.

"I was laying there with my eyes closed, relaxing, and all of a sudden I hear this voice that I didn't recognize, and I look up and it's this guy I'd never seen before. And it was a gentleman by the name of Dan Goldin, who on that day had become the new NASA administrator. So this was his first official act was to come to the Cape and greet the crew of STS-45 as they came back. And so I'm sitting there, and he says, 'I want you to come to Washington and work, and I'd like to talk to you about it.'"

Abbey was present and wouldn't remember the scene quite the same way, but he was behind the Goldin offer to Bolden—Abbey had supported the marine test pilot's assignment as head of the JSC Safety Division after *Challenger*.

Bolden would come to headquarters as deputy to Aaron Cohen.

Why was Abbey "raiding the Marine Corps" for new personnel? Three basic reasons: they were aviators, they had useful experience as managers of high-tech programs, and they had topped-out, career-wise: the Marine Corps had a small number of high-level billets, and its highest post, commandant, always went to an officer from the traditional ground community, leaving aviators frustrated.

It was also relatively easy to convince the Marine Corps to allow officers to take early retirements. (Stafford had demonstrated this in getting David Lee and Mini Mott onto the Synthesis Group.) The US Navy was almost as open, while the US Air Force—which had a great number of senior space-related billets—was not as inclined.

And so the wheel turned, back to the early 1970s, when Abbey worked for Gilruth and Kraft. He had no staff, no direct reports, no team.

He just possessed . . . knowledge. And now, knowledge with twenty years' additional experience.

Those at JSC who had disliked Abbey or chafed under his leadership were alarmed: he was back from the dead, and he was helping Goldin to transform the agency. It wasn't easy. Abbey arranged for Goldin to meet with Chris Kraft at Goldin's apartment in the east tower of the infamous Watergate complex. Goldin demonstrated his lack of practical skills when Kraft showed him how to operate a washing machine in the complex's laundry. Goldin had run out of clean clothing and faced a rapidly growing pile of dirty items. The experience prompted Kraft to take Abbey aside. "Is he for real?"

Abbey shrugged. In the few weeks he had known Goldin, he had concluded that the new NASA administrator was a brilliant and visionary engineer, but like many brilliant and visionary engineers, sometimes lacking in common, everyday skills. When it came to simple survival on his own, Goldin was simply unfamiliar with such basics as shopping in supermarkets and using appliances like microwaves and, yes, washing machines. (Goldin had gone from the care of his mother directly into his wife's household and had obviously spent no time on domestic chores.)

Yet Abbey was happy working with Goldin—and was so tight with him that NASA insiders were calling them "Chief Dan George," a name associated with an actor from the Dustin Hoffman movie *Little Big Man.*

And some of those insiders had knives out. There was increasing gossip—some justified—about Goldin's manner and behavior.

Len Fisk, head of space science at HQ since 1988, worked with Goldin when the administrator was in charge of the Chandra orbital observatory at TRW, and "thought he was wonderful . . . the most responsive contractor that I had."

(Abbey had first seen Goldin, though hadn't said hello, when the TRW manager visited DC to do a presentation on mirror technology.)

But after Goldin's arrival in Washington, Fisk decided that the new administrator had two sides, a deferential one he showed to people he worked for, and an unreasonable, demanding one for those under him.

After several weeks, Abbey concluded that while Goldin had his idiosyncrasies, he was sharp and intelligent, and . . . he was the administrator. And the administrator began to move NASA physically as well as culturally.

In 1988, when Abbey first arrived in Washington, the HQ staff numbered 1,500 spread over three main buildings, with outlying offices in L'Enfant Plaza N and E, and the Freedom program office in Reston, VA.

NASA and the Government Services Administration signed an agreement for a new central office in 1987 at Two Independence Square, E Street SW.

Construction began in 1990, and first staffers were moving in by 1992, though total move-in wouldn't be complete until 1995.

During the transition, 170 staffers were housed at 250 E Street SW, a building shared with the comptroller of the currency.

As for cultural changes, Abbey had always hated the new agency logo—better known as the worm—adopted in 1975 under James Fletcher in advance of the bicentennial. The worm was a stylized four letters—N A S A—meant to replace the classic "meatball," what the designers from Danne & Blackburn called "contrived by test pilots" and about as sophisticated as a Buck Rogers comic.

Abbey loved the meatball, however. To him it symbolized Apollo, Moon, success.

When he and Albrecht flew down to Langley with Goldin for their first visit, and the chat happened to touch on ways to inspire the NASA team, one of them said, get rid of the worm and bring back the meatball.

Goldin was intrigued enough to raise the subject with Langley director Paul Holloway, pointing at a huge worm logo on a hangar. Holloway was excited about the idea of bringing back the meatball, too.

In his speech to the Langley team, Goldin announced that he would, to wild applause. Goldin allowed for the gradual depletion of stationery and other materials that displayed the worm but wanted it removed from buildings and other locations where it was prominently displayed. When he found it still visible, he showed his displeasure.

The worm would never vanish entirely. It was painted on the side of the orbiting Hubble Space Telescope, and on the shuttle orbiters, at least until they underwent refurbishment in 1998. *Enterprise* would always keep the worm.

And it was carved in stone outside the new NASA HQ.

Meanwhile, Jack Dailey was waiting in the former Synthesis Group offices in Crystal City, making use of his time by becoming familiar with the NASA organization and its ongoing programs.

The process of appointing a NASA deputy administrator took time because the post required confirmation by the US Senate.

Then another complication in making Dailey the deputy administrator: His value in the business world had to be demonstrated in order for him to be entitled to the pay and benefits of the Senior Executive Service that went with the position. Enter Tom Stafford. "I got a call from George at 10:30 one night telling me he needed three substantial job offers, in the $250,000 to $300,000 range—for Jack Dailey from the aerospace world by noon the next day.

"I knew Dailey and knew that companies would want him, but it was 10:30 at night!"

Fortunately, in Pacific time it was just 7:30 PM, so Stafford got on the phone to the presidents of two West Coast companies whose boards he sat on, collected two expressions of interest in Dailey, and then a third. "I was able to give George three faxes in time."

Stafford judged this to be the first of several "Hail, Mary" plays he would execute on Abbey's behalf in the next few years. The effort remained strong in memory, too, since, according to Abbey, "Tom mentioned it to me almost every time we met for the next twenty-odd years."

Under Truly, NASA had planned to move shuttle program management to the Cape. Leonard Nicholson, a long-time shuttle manager who was running the program at JSC, had already picked out a home in Florida when Goldin, urged by Abbey and Stafford, halted the process and left the shuttle leadership in Houston.

When Goldin became administrator, he felt that NASA headquarters was too involved in the day-to-day management of shuttle operations. He also believed there was no need for meetings with flight crews. Or so he believed in 1992. "Within a few years," Albrecht said, "Goldin was personally contacting each crew prior to launch." Goldin's lack of operational experience in the world of human spaceflight, and willingness to micromanage, was obvious from the first shuttle mission launched after he became administrator.

On March 14, 1990, an Intelsat communications satellite was launched from the Cape by a commercial Titan III rocket and aimed at GEO. The Titan's second stage failed, however, and the Intelsat had to jettison the Orbis kick stage intended to lift it to GEO. A $157 million satellite still capable of performing its mission was left stranded in low Earth orbit.

Knowing that NASA and the shuttle program had already retrieved a pair of errant communications satellites in 1984 and 1985, Intelsat and Hughes approached the agency with an offer: capture the satellite and attach a new Orbis booster, then rerelease it.

In early May 1990, NASA offered to retrieve and relaunch the satellite if Intelsat would cover the costs, projected to be $130 million. (That figure was in 1998 dollars, the last time the Office of Management and Budget fixed the costs of a "nongovernmental" shuttle mission—the real price in 1992 was over $148 million.)

Following Office of Management and Budget approval of the offer, Intelsat's board of governors accepted it on June 13, agreeing to pay most of the costs. (The retrieval mission would carry other paying payloads.) NASA scheduled the mission for spring 1992 aboard STS-49, the first flight of the new orbiter *Endeavour*.

The retrieval mission—which would also see the maiden flight of OV-105, *Endeavour*—was to be commanded by chief astronaut Dan Brandenstein, who had assigned himself to this sought-after flight, and handpicked a crew of pilot Kevin Chilton, RMS operator Bruce Melnick, and EVA mission specialists Kathy Thornton, Rick Hieb, Pierre Thuot, and Tom Akers.

Launched on the evening of May 7, 1992, *Endeavour* and its crew settled into a two-day chase, reaching Intelsat on May 10. Thuot was to be carried from *Endeavour*'s payload bay to the slowly rotating satellite by the orbiter's robot arm, operated by Melnick. Then he would attach a specially designed capture bar that would allow Melnick to grab the satellite itself.

The latches on the bar failed to catch, then failed again. After the third try, Brandenstein noted that the huge satellite was beginning to oscillate. He fired *Endeavour*'s attitude rockets to move the orbiter away, planning to shut down this attempt and try another day. (Intelsat controllers were able to stabilize the satellite.)

After the failure of the first attempts at capture, Goldin told Pearson and Abbey to end the mission. "Bring them back."

Recognizing the need for the mission to continue, and to give operators a chance to do their work, Abbey convinced Goldin to establish a "red team" that included Kraft and Faget. Without taking more than a day, and requiring no guidance from Abbey, these two old hands were able to reassure Goldin and the other managers that there was still a chance the retrieval would succeed.

On May 11, Hieb and Thuot tried again. Hieb appeared to have the capture bar latched, but no—it came off and Intelsat started to wobble again.

Knowing that a third re-rendezvous and try at capture was all he was going to get, Brandenstein called for a day off while flight controllers and the crew tried to solve the problem.

It was Melnick who came up with this idea: put three astronauts into the payload bay at the same time, and have them grab the errant satellite. There were challenges to this novel idea, including the sheer difficulty of fitting three astronauts in bulky EVA suits into the airlock. (Astronauts Rich Clifford, Story Musgrave, and Jim Voss were dispatched to the WET-F to test the concept.)

On May 13, Thuot, Hieb, and Akers ventured out again, first constructing a makeshift cradle for the satellite out of a Space Station Freedom truss that was to be tested later in the mission. Positioned at equidistant points around the base of the slowly spinning satellite, they managed to grab it and stop it. Melnick was able to grapple it with the arm. The trio then attached the Orbis stage, and after a few misadventures there (a switch setting had been changed by Hughes but had not made it to the crew checklist), Thornton triggered Intelsat's release.

It floated out of the payload bay to a safe distance. On May 21, its new kick motor fired, and Intelsat rose to its geosynchronous orbit, where it operated for eleven years of its planned thirteen-year lifetime.

After three years on the job, Mark Albrecht left the Space Council at the end of May 1992 and was succeeded by Brian Dailey, from the Senate Armed Services Committee, just ahead of a long-planned US-Russia Summit on June 16–17 that would include discussion of future joint technical programs.

Brian Dailey recalled that when he arrived, Abbey and Liz Prestridge, another council staffer, already had several ideas they were trying to place on the summit agenda, notably the flight of a Russian cosmonaut on the shuttle and an astronaut on the Mir space station.

But an actual joint flight, a sequel to Apollo-Soyuz, represented a far greater challenge.

"George and I were downstairs about to leave the old Executive Office Building when he came up with the recommendation." Dailey liked the idea, though he was concerned about what the astronaut office would think, especially about safety and Mir.

Goldin was enthusiastic. But there was considerable pushback from other places—Dan Quayle's staff was cool to the idea, though Dailey said, "it wasn't so much the idea as the lateness." (Quayle himself was apparently quite enthusiastic.)

The State Department also balked.

Dailey happened to attend an embassy reception around this time, and ran into Kathy Sawyer, a reporter for the *Washington Post*, and he spoke to her on background about the possible US-Russian efforts.

Her article, citing the real possibility of a second joint manned mission, was published two days later, on June 14, and triggered the creation of a space-only breakfast at Blair House on the morning of June 16, with Quayle and Yeltsin.

Prior to the collapse of the Soviet Union and its evolution into the Russian Federation, Soviet space programs were managed by several ministries, notably the Ministry of General Machine-Building. When those dissolved, former Soviet contractors, organizations named Energiya (builder of Mir and Soyuz) or Krunichev (builder of the mighty Proton rocket), had stepped into the vacuum, scooping up or claiming assets.

NASA, through Goldin, had insisted on dealing only with another governmental organization. So the Russian government created the Russian Space Agency under former Energiya official Yuri Koptev, with a staff of perhaps ten people, a limited budget, and even more limited authority—though it was based in the palatial offices of the former Ministry of General Machine-Building.

Koptev was included in the Yeltsin party. He had wanted to meet with Goldin, but the NASA administrator was unavailable, so Goldin and Dailey asked Abbey to meet with Koptev at Goldin's Watergate apartment. There he pressed Koptev about Russian involvement in the NASA space station. Koptev said, "Why would we do that? We already have Mir!"

"What would it take?"

Koptev thought for a moment. "If America would promise to involve us in a return to the Moon, or a flight to Mars, then we would join your space station."

Abbey reported the discussion to Goldin, and the next morning Koptev was present for the Blair House breakfast. There it was agreed that Russia and the United States would start negotiations on the joint flights, and the possible use of Soyuz as rescue vehicle for the space station.

In less than a month, the first week in July, Brian Dailey, Goldin, Abbey, and Stafford left for Moscow. It was Abbey's first visit to Russia; having Stafford as guide made it easier, since the former astronaut was a genuine celebrity there. Abbey was also able to reconnect with friends he had made during Apollo-Soyuz, and to make new acquaintances like Maj. Gen. Yuri Glazkov, deputy director of the Gagarin Cosmonaut Training Center (GCTC).

At the age of forty-four, Glazkov had made one spaceflight, an eighteen-day mission to the Almaz military space station. Before becoming a cosmonaut in 1965, he had worked in missile and rocket development for the Soviet Rocket Forces at the Baikonur launch center in Kazakhstan. Glazkov also worked with GCTC, which was part of the Soviet air force, and the civilians at Energiya. There he had met Sergei Korolev, the Soviet Union's leading rocket engineer and spacecraft designer. It was Korolev who personally encouraged Glazkov to apply for selection as a cosmonaut.

Abbey would judge Glazkov to be very knowledgeable, competent, and honest; within two years they were close friends and looked upon each other as brothers.

It became clear to Abbey during his visit that Tom Stafford was fondly remembered for his role in Apollo-Soyuz. He was recognized by passers-by on the streets of Moscow. During discussions with the Russians Stafford learned that Yuri Semenov, Energiya's hardcase general director, wanted NASA to deal directly with Energiya rather than through Koptev.

Eventually, NASA and Energiya/RKA agreed to study the Soyuz as the ACRV and, more important, to fly a cosmonaut on the shuttle, and a NASA astronaut on Mir. Within a few weeks, veteran cosmonauts Sergei Krikalev and Vladimir Titov had been assigned to JSC.

There Cohen and Weitz established a US-Russia program office under Don Puddy.

Chief astronaut Dan Brandenstein had briefed the corps about a possible flight aboard Mir as early as November 1991, following the initial announcement from presidents Bush and Gorbachev. Several astronauts had volunteered, among them Shannon Lucid and the always eager Story Musgrave.

But no assignment had been made. With the new agreement in place, Brandenstein heard via the grapevine that four-time shuttle veteran Norm Thagard, recently returned from a Spacelab mission in January, was interested in the Mir flight. Brandenstein told Thagard he could have the assignment.

In Houston, another new astronaut candidate group was selected, twenty-three in all—nineteen mission specialists and only four pilots. The mix reflected JSC's belief, or hope, that these astronauts would be staffing Space Station Freedom. Among the twenty-three were three women, an African American navy pilot named Winston Scott, and another navy pilot, Michael López-Alegría, born in Spain.

In Houston, Don Puddy had taken a three-month leave in early 1992 to attend graduate school, leaving his FCOD job. Aaron Cohen and Paul Weitz decided to compete for the position, and astronaut David Leestma was selected as the new director.

With the urgent need to repair the deployed but nonfunctioning Hubble Space Telescope, Goldin named Randy Brinkley to serve as mission director for STS-61, providing oversight of the upcoming repair mission. Brinkley would be assigned to the NASA Goddard Spaceflight Center since it was responsible for Hubble, but would be stationed in Houston at JSC. Goldin also asked Tom Stafford to review the effort.

Other decisions on the Hubble repair were taking longer. Before and after their Russia trip in July, Abbey, Goldin, and Stafford were pressuring JSC to assign EVA astronauts to the mission in August 1992, almost eighteen months ahead of the planned December 1993 launch. Abbey and Stafford felt that the astronauts would need the extra time to work with crew equipment teams and Goddard engineers on EVA procedures. The timeline for the planned EVAs on orbit needed to be coordinated with Goddard in order to arrive at the number of spacewalks required.

Astronaut Story Musgrave was assigned as payload commander for the mission. After STS-6 in 1983 Musgrave had gone on to make three more successful flights. The detail-oriented work required on the Hubble repair matched up well with Musgrave's temperament.

Soon thereafter, four more mission specialists were assigned, too—Kathy Thornton and Tom Akers, fresh from the Intelsat rescue, and veteran space walker Jeff Hoffman. ESA astronaut Claude Nicollier was tasked with the challenging job of operating the remote manipulator arm.

An experienced commander and pilot were still needed to complete the team. Based on availability and reputation, there were two leading candidates for commander: Hoot Gibson, veteran of four missions including the recent STS-47, and Dick Covey, veteran of three and, most recently, acting chief of the astronaut office.

Indeed, with Leestma's appointment as director FCOD, Brandenstein was retiring, leaving an opening for chief of the office. Covey met with Cohen, Weitz, and Leestma and was presented with two options: succeed Brandenstein, or fly the Hubble repair mission. He was told that Gibson was the other candidate for both.

"Let Hoot be chief," Covey said. "Let me go fly."

All the new assignments were announced on December 1: Leestma as director of flight crew operations; Gibson as chief of the astronaut office; and Loren Shriver as Gibson's deputy.

On Tuesday, November 3, 1992, Democratic governor Bill Clinton of Arkansas defeated incumbent Republican president George H. W. Bush, becoming president elect.

With the change of administration, the Space Council would not be continued. Abbey knew that he would once again return to full-time at NASA working for Dan Goldin.

48 | Space Cowboys

AROUND 10:00 PM ON the evening of Friday, February 5, 1993, Tracey Pate, a thirty-two-year-old secretary, was awakened from sleep by a telephone ringing in her apartment in northern Virginia. Pate's husband, Jay, was an Alexandria police officer on duty—a late-night call was always troubling.

But the caller was Pate's boss, Tom Stafford, from his home in the Florida Keys, with an unusual request. "We're facing a national quasi-emergency," he said. He wanted Tracey to open up Stafford, Burke & Hecker's townhouse office the first thing tomorrow morning. The keys were to be delivered to a Mr. George Abbey of NASA.

The next morning, Tracey's husband took the key to 1006 Cameron Street, where he met Stafford's chief of staff, Sid Rodgers, waiting in the cold and wind with Goldin, Mini Mott, and Abbey.

By the time Tracey arrived an hour later, with food, she learned from Rodgers that several other men were expected, too.

"Who?" she asked.

"A couple of astronauts, some NASA guys, friends of General Tom's."

And what are they doing in Old Town Alexandria on a cold Saturday morning in February?

"We're going to have a meeting," Mott told Tracey. "We're going to save the space station."

On this Saturday in February, there was a space station in orbit—Russia's Mir—orbiting at an altitude of 170 miles, its six modules occupied by cosmonauts Gennady Manakov and Aleksandr Poleshchuk, who were launched aboard Soyuz TM-16 on January 24 and who were to live and work aboard the station until June.

On the American side of human space exploration, the shuttle *Endeavour* had just returned from the six-day STS-54 mission, deploying the sixth Tracking and Data Relay satellite. Five more shuttle missions were scheduled in the coming year, including two Spacelab missions as well as another carrying the first Russian cosmonaut to fly aboard the American vehicle.

And, while no one in the West knew at the time, China had just approved program 1092, a human spaceflight program using a craft based on the Russian Soyuz.

The subject of this secret, off-site, off-the-books meeting was NASA's future, and that of Space Station Freedom, the estimated price of which had just reached $30 billion for initial assembly, with the first launch still four years in the future.

That Friday, February 5, 1993, Goldin had been summoned to the White House to meet with science adviser John Gibbons.

And there he heard bad news. "We're going to cancel Freedom," Gibbons said. The new Clinton administration had done a quick review of NASA programs and concluded that the space station was going nowhere. (Gibbons's team was also ready to cancel the Super Collider in Texas, and the next day's *Washington Post* would report that both programs were headed for the axe.)

Goldin was not a fan of Freedom. Within two months of his appointment as administrator, hoping to redirect the program, he had convened a meeting of his senior staff—Abbey plus Aaron Cohen, associate administrator for legislative affairs Marty Kress, and several others.

Kress, a former house and senate staffer who was NASA's interface with Congress, had been adamantly opposed to *any* changes in Freedom. Cohen was also opposed and responded with great emotion to such plans, too.

Kress and Cohen insisted that if NASA put Freedom through yet another redesign, Congress would simply cancel the whole thing. Seeing Cohen's emotional reaction, and hearing Kress's dire warnings of Congressional doom, Goldin had backed off the Freedom issue, kicking the can down the road . . . to this Friday night in February.

"Give me some time," Goldin pleaded.

The administration needed to submit a budget to Congress within a week, Gibbons told him. Goldin was sure he could come up with something.

"You have the weekend," Gibbons said.

A weekend! Without Freedom there would be a very limited manifest of missions that would justify the continuation of the space shuttle program. No shuttle meant that the United States would be *out* of the human space business. In practical matters, that meant the loss of thousands of jobs. In larger terms, there would be fewer technological innovations and spin-offs, achievements that space enthusiasts—and Dan Goldin—believed to be major contributors to America's global technical dominance.

There would be no astronauts, no inspiration for future engineers and space scientists.

Freedom was not only intended to be a destination for the space shuttle, but to serve as a laboratory for life science research that would enable missions beyond Earth's orbit.

No, Freedom had to be saved.

Goldin rushed back to the new NASA Headquarters Building. As soon as he got to his office he called in George Abbey.

Abbey's skills would be tested this weekend. He had to not only pull together the right group to sell Dan Goldin on a practical new version of the space station but also prepare Goldin to make a winning pitch to a skeptical Clinton White House.

Abbey needed a "tiger team" of experts, people who knew the engineering, the systems, the operations, the politics . . . people who had *done* this. And he needed them immediately, as in tomorrow, Saturday, February 6.

But who? What skills?

A team of technical experts was the answer, individuals who had been through the trials of successfully designing, developing, and operating human spacecraft.

There was also the question of where they might meet. Any get-together at NASA headquarters, even on a weekend, would leak immediately, triggering inquiries and action from other NASA staffers as well as contractors and political people. What was needed was some location close to Washington where this "tiger team" could gather in private, and come up with a solution.

As the February skies darkened and thousands of bureaucrats and secretaries headed for their homes, Abbey reached for his phone and called Tom Stafford, currently at one of his three homes, this one at Islamorada in the Florida Keys. There was no way Stafford could get back to Virginia in time for a meeting early Saturday morning, but he was happy to offer Abbey the use of his offices in Alexandria and his staff, beginning with the alert to his secretary, Tracey Pate.

Abbey, Mott, Goldin, and Rodgers opened up the townhouse. On the ground floor were offices for Stafford and his partners. On the third floor was a large conference room, and that's where the team would gather.

They would have no high-tech tools. For one thing, there was no need for serious calculations and there was no time to prepare PowerPoint presentations. This discussion would be as much philosophical or political as technical, and its tools would be legal pads and pens, not computers.

Arriving first, by commercial aircraft from Boston, was Joseph Shea, who had served as Apollo program manager in the 1960s.

Shea had left NASA in 1967 after the Apollo fire to become vice president at the Polaroid Corporation. In 1968 he took a position at Raytheon, serving as senior vice president for engineering from 1981 through 1990. After leaving Raytheon, Shea became a professor of aeronautics and astronautics at MIT. He also became president of the American Institute of Aeronautics and Astronautics, was a member of the National Research Council, and also served on a number of technical advisory panels.

In spite of what happened with Apollo, Abbey still thought of Shea as one of the most brilliant systems engineers he had ever known.

Arriving soon after Shea, still wearing his blue flight suit, was astronaut John Young, fresh from piloting a NASA T-38 jet. At sixty-two, Young was serving as technical assistant to Aaron Cohen, the director of the Johnson Space

Center. He continued to follow the space shuttle and space station programs and was very much aware of the issues with both programs.

The third member of the team—spacecraft designer Max Faget—arrived last, flying in commercial from Houston.

Rounding out Abbey's group was Mike Mott.

Getting to business took several minutes. There were the usual hellos from Young and Faget to Shea, whom they had not seen in years. There were introductions, since Faget and Shea didn't know Goldin at all. Faget *had* met Mott, during the time the marine worked with Abbey on the Synthesis Group, but apparently didn't remember—except for taking pleasure in the obvious fact that "he's shorter than I am!"

Soon they were all seated around a conference table, just as Abbey had seen in so many Apollo-era meetings. Mott kept handwritten notes.

The challenge facing them was complicated. Space Station Freedom was originally envisioned as a "dual keel" rectangular truss one football field long by half a football field wide. Solar arrays, life-support gear, and cylindrical modules would be attached to this truss. Some would be built by NASA. Canada would supply the station's robotic arm.

This design emerged at the time of the *Challenger* disaster in 1986. Within a year, as NASA adjusted future shuttle schedules to a more realistic launch rate, Freedom shrank. The dual-keel became a single one.

In 1988, contracts were signed and the real work began with so-called Work Packages—specialized collections of components and hardware—doled out to four major NASA centers, from the Marshall Space Flight Center to Goddard in Maryland, to Lewis in Cleveland, and, of course, the Johnson Space Center in Houston.

This division of responsibilities proved to be a major problem for the program, with different centers—none of them with much experience in managing such a complex development—fighting over money and authority, and then *four different* commercial companies that had no more experience with space stations than NASA did . . . and that were used to operating as competitors.

That wasn't the end of Freedom's problems.

After the *Challenger* disaster, the Rogers Commission recommended that NASA take primary responsibility away from centers and move it to NASA headquarters, which in Freedom's case required the establishment of an office in Reston, Virginia. Reston had since compounded the program's problems by selecting a *fifth* contractor, Grumman, to "integrate" the program—putting one commercial firm in a position of power over its competitors.

Freedom's program director was Richard Kohrs, a former shuttle engineer. Abbey had known Kohrs since arriving at Houston in 1964, had even played basketball with him. He thought of Kohrs as a decent engineer in an impossible situation.

Abbey knew that what made Mercury, Gemini, Apollo, and the space shuttle orbiter such a success was co-locating the design engineers and operators at one NASA center. The Johnson Space Center was unique in this regard. Spacecraft designers, flight controllers, astronauts, and their trainers were all in the same organization. Under the Freedom management structure, JSC engineers, flight directors, and astronauts had fewer ways to get their wishes and needs heard or acted upon.

Since Young, Shea, and Faget hadn't heard more than a general outline of the challenge facing them, Goldin gave the team the specifics of the bad news—$2 billion a year, or no space station.

Abbey had a clear vision of what might replace Freedom, but he did not want to lead the discussion with Goldin. And with Faget, he didn't need to. The veteran engineer was never shy.

For a decade after his retirement from NASA, Max Faget worked on a project called the Industrial Space Facility (ISF)—a module that could be carried into space by the shuttle, left there with scientific or industrial equipment operating autonomously, then visited periodically by later missions. The ISF had a small pressurized module that could accommodate several astronauts for days at a time. Additional modules could be linked together, too. Abbey always liked the concept. And George Low was also a supporter of Faget's approach and had tried to help Faget find private investors for the ISF. Faget had also approached NASA—without success.

The Russian space station Mir was conceptually very similar to the one proposed by Faget. Launched from the former Soviet Union almost seven years earlier, on February 20, 1986, it was originally a single module known as the "base block," 13.3 meters long, 4.5 in diameter, powered by solar arrays and capable of independent propulsion. It had ninety cubic meters of habitable volume.

Over the next six years, three different modules were attached to Mir's radial nose docking port, with two others yet to come. One of the new modules carried experiments in astrophysics and materials science. The second carried new science equipment, as well as an airlock and improved life-support systems. The third was an orbital space manufacturing module.

Mir, like its predecessors Salyut 6 and Salyut 7, could be resupplied by unmanned progress vehicles.

This was the type of station Abbey, Faget, and Shea believed that NASA should have—a vehicle that allowed for a man-tended phase quickly, with full-time occupation to follow, using the shuttle for delivery—and more important, costing under $2 billion a year.

John Young appreciated the simplicity of the Russian approach and was in agreement with the concept being proposed by Faget, Shea, and Abbey.

In order to illustrate his concept to Goldin, Faget, who was still spry at the age of seventy-one, jumped up and started rummaging through a desktop. Eventually, with Tracey Pate's help, he found a box of old-fashioned wooden matches. The brain behind Mercury, Apollo, and the space shuttle—who also built balsa-wood models as a hobby—smiled as he broke the matches into smaller pieces and arranged them on the tabletop. "Here's your hub," he said. "Here's your first module." He proposed a hub-base block (like Mir) with four additional modules attached at ninety-degree points around it. "Shuttle docks to the X-axis," he said. The nose of the hub.

Shea, largely silent until this moment, spoke then. "You need to take a hard look at simplifying the subsystems and streamlining the management of the program. They're both contributing to increased cost, and with a two billion dollar a year limit, both steps are essential." Faget added, "You also need to go back to the approach that was key to our past success and manage the program from Houston."

The redesign of the station would affect the Freedom contractors. The group knew, however, that the alternative—no station at all—wouldn't be welcome, either.

By 4:00 PM Goldin and the team had agreed on the new design and the selling points. Mott had handwritten a set of bullet points for Goldin to carry—the pitch. Goldin needed to get to Gibbons's office but had no car. (He left his in California and took cabs around DC.) He asked to use Abbey's Honda Prelude. Abbey handed over the keys.

Goldin jumped in and drove off to the White House. He met with Gibbons for an hour, laying out the new design just as Faget had, with matchsticks.

Gibbons told him, "If you can do it on $2 billion a year, it will work."

Goldin returned to Cameron Street in triumph—and pique.

The power steering in Abbey's Honda had failed some time earlier. This hadn't been a problem for Abbey, but he had forgotten to tell Goldin, who had struggled to guide the car throughout his round-trip journey to the White House. He snapped, "You almost got me killed." Goldin never again expressed a desire to borrow Abbey's car.

Abbey's team went out for a celebratory dinner that night at a nearby restaurant called Le Gaulois Cafe on King Street in Alexandria. They knew that proposing an alternative station to the White House was one thing . . . lining up the political support quite another.

Nevertheless, Abbey's Space Cowboys suspected then, and Tom Stafford would proclaim years later, "That weekend saved human spaceflight for the twenty-first century."

49 | Redesign

THE BEST WAY FOR Abbey and Goldin to prove their point to NASA and the administration was, in the classic agency manner, to create teams that would wirebrush several options and settle on one, ideally the one they preferred.

They developed three primary pathways: a revision of the Freedom design (Option A); their new modular design (Option B); a more radical design in which an even more scaled-down station, à la Skylab, would be launched as a single vehicle (Option C).

The option teams would present their designs to an outside panel headed by Charles Vest of MIT.

Vest's committee would have three dozen members, including representatives from each NASA center and of international space agencies. There would be a ninety-day deadline for making the decision.

Their first choice to head it was Joe Shea, and Shea was willing—but then . . . along came the Russians.

Rose Gottemoeller of Clinton's National Security Council had been busy searching for a way to keep Russian entities from selling tech knowledge and hardware to enemies of the United States.

The preferred method of preventing that was to enlist the vast space enterprise of the former Soviet Union in a joint project. Cosmonaut and astronaut exchanges weren't enough, nor was the possible use of Soyuz as a space station rescue vehicle.

Gottemoeller had been in contact with Abbey about the matter, and Abbey shared information from his February 1992 meetings with Koptev. He remained intrigued by Russian involvement in the various options, but wasn't sure, in February 1993, that it was time to pursue it.

Nevertheless, prompted by the announcement of a new direction for America's space station, and a redesign, Koptev traveled to DC in March to make a new pitch: to make Mir 2 a key element in an international space station.

Abbey and Goldin welcomed the idea—here in one package was that new space station hub they sought—but they kept their discussions private. For one thing, they weren't entirely sure how much authority Koptev had—was Semenov of Energiya onboard with this? (Another summit, between President Clinton and President Yeltsin in April, would clarify matters.)

Negotiating a partnership with the Russians would also take months, with no guarantee of success. Congress was restless; under control of the Democrats, there was a greater chance that the next budget might still eliminate the station.

Finally, they didn't want to bias the redesign teams.

Assuming they even got started.

Bryan O'Connor was still in his headquarters office at 6:00 PM on March 7 when Abbey phoned from the ninth floor, telling him that Joe Shea was in the hospital. The public announcement of the redesign—which Shea was to lead—was scheduled for tomorrow and would be attended by the press, international partners, and contractors.

Abbey wanted O'Connor to "stand by"; he might be needed tomorrow. He said that the agenda was all laid out, that at the worst O'Connor's job would be to merely introduce the other speakers, like Max Faget, who would then carry the event.

He promised to call back soon with definitive word, but didn't want O'Connor to leave. So, the former marine and astronaut sat there "wondering what this all meant." He had been working on the shuttle, not the station. What if someone asked him about the specifics of the redesign?

Abbey called back. Yes, things with Shea were serious enough to prevent him from leading the redesign. O'Connor would be needed.

Then Faget arrived. O'Connor barely knew the legendary NASA engineer, so he was happy to linger, especially when Faget replicated his original sales pitch to Dan Goldin by sketching the new concept on a napkin. And then he was glad to have Faget at his side as he faced the press, and wound up the de facto head of the redesign effort.

On March 16, 1993, Goldin was told by the White House that he would be retained as NASA administrator. (He had recently given a speech in which he declared that he was "tired of Apollo stories," that he wanted NASA to "tell new stories.")

The redesign teams, with a staff of forty people, set to work in Crystal City.

Jack Dailey finally took up the deputy administrator job.

And the shuttle continued to fly.

The first half of 1993 was a fruitful time for Abbey. Now working full-time at the agency, he had Goldin's total confidence and support, and was able to use all his knowledge of NASA assets and technology, people and politics to shape the agency's future.

Many if not most of those tapped for the station redesign were his choices. Some, like Mike Griffin, came from the Synthesis Group, others from JSC. So Abbey not only affected the information that went into the reviews, he heard reports on the internal debates.

He wasn't putting a finger on the scales—he fervently believed that any experienced space professional, looking at the same budget, facilities, technology, would reach the same conclusion . . . a modular station that could be built up over time was the best option.

The Vest Committee made its formal recommendations to the White House on Monday, June 7, identifying Option A, a greatly modified Freedom, and Option C, the single-launch concept, as the best. Because of its modular structure, Option A had the potential for growth—but Option C was likely to be simpler and cheaper.

Nevertheless, on June 17, the Clinton administration announced its support for Option A, a modular station that would almost certainly include Russian hardware in addition to elements from Europe, Japan, and Canada.

An executive decision was one thing—Congressional approval was another. Freedom had avoided one attempt at cancellation in 1992.

The first vote on the new station came a week after Clinton announced his decision, and thanks to personal lobbying by Vice President Al Gore and the forty-two-year veteran of the House, Congressman Jack Brooks of Texas, was a real nail-biter: 216 for, 215 against.

The space station lived on. Barely.

As more senior diplomatic teams met—those involving Gore and Russian premier Viktor Chernomyrdin—Abbey, with Goldin, O'Connor, Lynn Cline, and William Schumacher, headed to Moscow on June 30. Out of this visit came a new program, Shuttle-Mir, in which ten shuttle missions would launch to the Russian station, one to rendezvous, the other nine to dock.

O'Connor would recall that the original number was four, and that Goldin and Abbey arbitrarily raised it to ten during the flight to Moscow. Abbey disagreed: "I always wanted ten, with eight or nine to deliver NASA astronauts for long-duration missions."

Why that number? The first Shuttle-Mir docking could not take place before the summer of 1995, and the first space station element was not likely to be launched until summer 1998. Three years—thirty-six months—at four months for each visit equaled nine long-duration NASA visits.

The number wasn't just to keep NASA astronauts occupied, it was to force—there was no other word—mission controllers and astronaut trainers and crew equipment teams to deal with the challenges of long-duration flight such as logistics and resupply, and basic operations. Nine missions would also allow NASA to accomplish essential life science research in order to better understand the effects of long-duration flight on the human body well in advance of International Space Station missions.

The Russians would receive $400 million from the United States.

It was understood now that Shuttle-Mir might serve as phase I of an ongoing US-Russian space effort, with phase II involving the use of Russian hardware in the new NASA facility.

Abbey and Goldin returned to the United States for different negotiations.

Aside from the engineering and political challenges of transforming Freedom into a new space station, there were also business matters: Freedom had been approved by Congress in 1984 and contracts signed between NASA and four major contractors and dozens of smaller ones. Nine billion dollars had been spent; billions more were owed for work in progress. Each contract dealt with different hardware, schedules, payments—and completion. (Terms and mechanics of cancellation were also in the contracts.)

How would these be managed? Was Freedom being canceled? Or was it being modified?

On July 22, Goldin met with the CEOs of Boeing, McDonnell Douglas, Rockwell, and Grumman at headquarters and told them that the space station had been redesigned and would be moving forward with a firm financial cap.

Changes had to be made; Goldin intended to select a new "prime contractor" for the new program, with other contracts being "realigned." He asked the execs for their support.

Recognizing the need to keep the station moving forward, they agreed.

On August 17, NASA announced that Boeing would be the prime for the new station. Its subsequent acquisition of both Rockwell and McDonnell-Douglas enhanced Boeing's ability to function successfully in that role. Grumman was dissatisfied with the decision and the subsequent contract realignments, and filed a lawsuit that would go on for several years—until, in fact, the first elements of the new International Space Station reached orbit in 1998.

As NASA was wrestling with the prime contractor question, a Russian team traveled to the United States in late July to begin the tedious and delicate business of adding Russian modules to the new station—and deciding how international modules from Japan and Europe would be affected.

As agreements were being made with the Russians, there were external complications. Goldin, Abbey, Young, and a team of engineers had just arrived in Moscow when a coup was staged by Russian hardliners against President Yeltsin on October 3. The NASA delegation hunkered in their hotel rooms with gunfire echoing in the streets around them. The gunfire continued until early the next morning, when Russian troops and tanks entered Moscow and put down the attempted coup.

Goldin, Abbey, and the others were able to follow the events on CNN even as they saw smoke from fires and observed exploding shells from the higher floors of their hotel.

The next morning, October 4, their meetings with their Russian counterparts commenced as originally scheduled.

On December 9, 1993, the Clinton administration formally invited Russia to take part in what was now going to be the International Space Station.

But the future of NASA was riding on the next mission—already in progress.

50 | Hubble Rescue

ABBEY FLEW DOWN TO the Cape to be on hand as, on December 2, 1993, *Endeavour* was launched with its crew of seven, commanded by Dick Covey. The early hours of the mission went flawlessly as the orbiter maneuvered to Hubble's altitude, then closed in.

Abbey returned to NASA headquarters to be with Goldin and address any of his concerns as the mission progressed. The administrator was always uneasy during critical phases.

Three and a half years after Hubble had been deployed, astronauts saw it again. It was Hoffman who spied it via binoculars—immediately noting that one of the solar arrays was bent ninety degrees, which might present a problem.

Nevertheless, after the tricky capture and EVA prep, on flight day four, Musgrave and Hoffman ventured outside on the first of what would be five long EVAs. Akers and Thornton did the second and the fourth.

There were some problems with that solar array, with latches and hatches, with Thornton's suit, with Musgrave's suit, with procedures that, no matter how well planned and simulated, took longer on orbit.

Nevertheless, by flight day nine, several of Hubble's instruments had been replaced and the observatory was released.

Endeavour's wheels stopped on the runway at KSC at 12:26 AM on December 13.

It would take weeks to rate the repairs, but they had been successful.

Mission manager Randy Brinkley, too wired to sleep after the fifth and final EVA, had gone to the JSC gym. At 4:30 in the morning, his phone rang—it was Dan Goldin calling from the Watergate apartment asking him to become the new program director for the International Space Station.

And two days later, almost before *Endeavour*'s wheels had stopped on the runway at the Cape, Goldin called Abbey into his office. There was only one person who could make Goldin's vision a reality—not just the new ISS, but support from the vital Johnson Space Center.

Goldin wanted Abbey back in Houston.

51 | A Space Odyssey

GOLDIN WAS DEDICATED TO equality and was always searching for qualified women and minorities that he could consider for leadership roles within NASA. And with Johnson Space Center director Aaron Cohen retiring to become a professor at Texas A&M University, Goldin saw an opportunity to place a woman in the job. Abbey suggested Carolyn Huntoon, head of life sciences at JSC and more recently acting associate director under Cohen and Paul Weitz. At fifty-three, with a master's and a doctorate from Baylor's College of Medicine, Huntoon had joined NASA in 1970 and worked her way up the life sciences ladder, including a tour on the ninth floor as one of Kraft's "bright young people." She had been a mentor to a number of the women selected as astronauts.

However, when Goldin spoke to Huntoon about the job she was reluctant to consider it, noting correctly that the center had always been run by engineers and she was a scientist.

"You don't need to be an engineer," Goldin told her. If she could be a manager, she could get along with people. Of course, Huntoon had no operational or program management experience. She would need a strong new deputy—like George Abbey.

Abbey liked Huntoon and had no objection to working for her, but he wasn't ready to leave NASA headquarters. There were still too many problems and issues there that needed to be resolved. He felt he could be more use to Goldin in Washington.

On Friday afternoon, December 17, 1993, George Abbey picked up the telephone. Tom Stafford was calling from New Zealand, where he and his wife Linda were on a vacation. The time in New Zealand was 3:00 AM. Apparently Mott and Goldin had tracked Stafford down to tell him that Abbey was balking at the return to Houston. And here Stafford made his pitch, citing Abbey's personal hero George Low and the way he accepted a demotion from deputy to Robert Gilruth to become Apollo program manager. He pointed out that ISS was still "in a delicate state. You're the only one who can make ISS work!"

Eventually Abbey relented, agreeing to return to Houston as deputy director—but only if he could find someone to take over his position as Goldin's assistant.

Working on the Synthesis Group, Abbey had found Mini Mott to be a quick learner and and outstanding performer. When that effort disbanded in spring 1991, Mott had gone to work for General Research Corporation, a Beltway firm that was involved in DOD and NASA programs. Abbey wanted him at NASA, and here was the opening.

Dealing with Goldin required special skills, among them multitasking and patience. Mott would say later, "I had about 150 duties assigned by Dan on a daily basis."

Mott took the job, joining the agency on December 28, which freed Abbey to return to Houston. Within a month he packed up his Beltway apartment and drove back to Texas, back to his home on Davon Lane. As for Mott, Abbey said, "I'm not sure he ever forgave me for talking him into taking that job."

On January 6, Goldin unleashed a barrage of new appointments—not only Huntoon as the new director of the Johnson Space Center but also Ken Munechika as new director at NASA Ames Research Center. Goldin also announced that the Dryden Flight Research Center would no longer be under Ames but would once again be a separate center, with Kenneth Szalai as its director. Abbey had recommended the change to Goldin since the cultures of the two facilities were quite different. He also knew that it was essential for NASA to have an aeronautical flight test center like Dryden. Porter Bridwell took over

NASA Marshall from Jack Lee. Donald Campbell, brought in from the secretary of the air force's acquisition team, became director at NASA Lewis.

George Abbey's appointment as deputy director of the NASA Johnson Space Center was announced on January 10, 1994. The same press release noted the retirement of Paul Weitz, the acting director.

Gene Kranz announced his retirement on January 14. He would be succeeded by John O'Neill, a longtime JSC staffer who had worked under Deke Slayton in flight crew operations before moving to Kranz's team when both were consolidated in 1974.

Before the end of January, Huntoon and Abbey had made sweeping JSC leadership changes, officially closing the Space Station Freedom office and moving the joint US-Russian Federation programs office into the new ISS organization. Huntoon added leaders to life and sciences, center operations, education, and public affairs while creating a small New Technology team.

As Abbey moved to JSC, in addition to Mott, he left in place a new space station director at headquarters. Bryan O'Connor had held the job on an acting basis since returning from Russia in September 1993. Will Trafton, an Annapolis graduate and a veteran of eighty-five combat missions in the Vietnam War, was named as the space station director with O'Connor becoming head of the space shuttle program. Both would report to the associate administrator of the office of space flight.

In Houston, Randy Brinkley was now in charge of the new ISS, with astronaut Bill Shepherd as chief engineer and Doug Cooke as deputy.

Under Abbey's guidance, Brinkley and his team crafted a new agreement with Boeing to serve as prime contractor for the "new" International Space Station. Meanwhile, controllers at NASA Goddard and the Space Telescope Institute were pleased with improved imagery downlinked from the repaired Hubble.

Before Abbey could settle in to his corner office on the ninth floor, there was yet another trip to Russia, in March, to nail down steps in ISS phases I and II. Goldin was to have led the delegation, but he had injured himself in a bicycle accident.

The team included Abbey, Mott, Trafton, and Arnauld Nicogossian, head of life sciences at headquarters (and chief medical officer for Apollo-Soyuz).

Their Russian counterparts were Koptev, Alexei Krasnov of the Russian Space Agency, and Yuri Semenov and Valery Ryumin of Energiya. Bryan O'Connor had found Ryumin difficult to deal with on his previous visit.

The existence of formal government-to-government agreements made this negotiation go quickly, though there was one awkward moment when Ryumin complained. Mott and Abbey pushed back, and found that Ryumin's Russian colleagues agreed with them, not the former cosmonaut. The Phase One negotiations were concluded successfully.

Before leaving Washington, Abbey had spoken to Charlie Bolden about the importance of the STS-60 mission, the first flight to include a Russian cosmonaut in its crew. Bolden wanted to fly again, but he had been pilot on STS-30, the Hubble deployment, and was interested in commanding the repair mission, STS-61.

When Dick Covey was named commander of the Hubble repair crew, Bolden, with some reluctance, accepted the STS-60 command. He grew to be pleased with the assignment and developed great respect for his Russian crewmate, cosmonaut Sergei Krikalev. Veteran of two earlier long-duration space station missions, Krikalev was a personable civilian engineer and acrobatic pilot from the Energiya organization, and a capable English speaker. He performed well on Bolden's crew and would eventually become a good friend of Abbey's.

With Krikalev and another cosmonaut, Vladimir Titov, assigned as prime and backup crewmembers for STS-60, NASA officials had to consider which American astronauts would fly to Mir. Norman Thagard had expressed interest early on, in late 1992, and was officially assigned to the Mir mission, with Bonnie Dunbar as his backup.

A NASA team, ten in number, descended on the Gagarin Cosmonaut Training Center (GCTC) in the dead of winter in early 1993—Thagard, Dunbar, a NASA doctor, and several support team members.

Abbey had insisted that the astronaut office provide a flight-experienced astronaut as liaison in charge of the Americans assigned to the GCTC. Shuttle veteran Ken Cameron was the first director, operations—Russia. Many others were to follow, but Cameron and the first group were pioneers.

52 | Return to JSC

ABBEY WAS NOT ONLY back on the ninth floor at JSC, he was back in his Davon Lane home full-time for the first time in six years.

He and Suzanne would start to remodel the house, built in the early 1970s. It became quite a challenge, since Abbey insisted on living in the building as contractors tore it apart.

He had his beloved cars again, and his books, and his handwritten notes—simple entries on the pages of green notebooks. He resumed regular involvement with his church.

And he was still in constant contact with Washington, not only Mott, but also Goldin. According to Suzanne, "Andy and Dad would be out at some football game on Friday night and here's that voice on the phone. 'This is Dan Goldin.' He just seemed to have no idea what other people might be doing.

"Goldin would call in the middle of the night!" But when the conversations did take place, "they would be on the phone for hours, talking about everything, joking and laughing.

"Goldin was just obsessed with Dad—he would always be asking us, 'what does he do in his spare time? Where did he come from?' I think Goldin was a big loner and was intrigued by Dad's big family." When Andy got married that summer, Goldin attended.

At JSC, Abbey and Huntoon accommodated mutual strengths and weaknesses. Huntoon concentrated on center administration, life sciences, and applied science, and left operations—shuttle flights, mission control, ISS development, or aircraft—to Abbey.

Under Brewster Shaw, the shuttle program flew five different science missions during 1994, after STS-60. Abbey attended flight readiness reviews and exercised his own unique way of monitoring activities.

A night owl, Abbey was notorious for showing up in mission control in the wee hours, usually during the planning shift, when the on-orbit shuttle crew was sleeping. He would sit with the flight director and ask—as he had been asking for years by this point—how things were going, what problems the flight director was seeing.

The formal mission management team would arrive later that morning around 6:30 AM, but Abbey had already been there first.

George Low had done the same thing—popping into mission control at odd hours.

The director of flight crew operations in early 1995 was David Leestma, a navy captain and former shuttle mission specialist who had the reputation of being an Abbey favorite. His deputy was Steve Hawley, another Abbey friend—a former bubba. Under them, the official who actually made crew selections was veteran TFNG Hoot Gibson.

Leestma and Hawley took each of Gibson's crew recommendations—a list of names written on a piece of paper—to Abbey. "There were two possible responses," Hawley said. "One was, he marked a name off the list and handed it back to you. Two was, he simply handed it back to you, which meant it was fine.

Hawley remembers that "we might have gotten one past George without a change once, no more than that."

Abbey had noted that, over time, the astronaut office leadership tended to make crew assignments based on seniority, much like airlines, instead of selecting the best qualified astronauts for a particular flight.

Case in point: in the summer of 1994, STS-71, the first shuttle-Mir docking mission, was a year from launch and in need of a crew. It would be the first Russian docking NASA had performed since Apollo-Soyuz in 1975, and the crew would require a skilled, veteran commander. Abbey felt that

Hoot Gibson was the right astronaut for the assignment. He discussed it with Leestma, who in turn talked to Gibson—then chief of the astronaut office. Gibson agreed to accept the assignment if Leestma announced it to the other astronauts.

Accepting the STS-71 assignment meant that Gibson had to give up his post as chief astronaut. Robert Cabana succeeded him.

Cabana recalled his own reviews of crew recommendations with Abbey, and similar rejections. "So I'd just bring the same crew back the second time and explain why. It always worked out. We ended up getting the crews assigned. I really enjoyed interfacing with Mr. Abbey."

When Abbey arrived, Leestma and Hawley were in the process of selecting a new group of ASCANs. The number of people who'd applied was 2,962; the board had winnowed the list to 120 who began arriving in groups of twenty on Sunday, June 25. They had promised the applicants a decision at summer's end and an announcement in the fall.

October and November came and went. Staffers at JSC, and several astronauts, blamed the delay on Abbey. But it was Dan Goldin's fault.

The administrator wanted the new group to include as many women and minorities as possible. Abbey had to take Leestma's preliminary selections to Goldin and justify each one. There were two women selected as shuttle pilots. "Can we get more?"

Confident in his own role in promoting the selection of minorities and women, Abbey thought it was amazing that NASA had found two, Pam Melroy and Susan Still.

When finally announced on December 8, 1994, the group of nineteen—ten pilots, nine mission specialists—included five women. In addition to the two women pilots, there was also navy reservist Kay Hire, the first American woman assigned to a combat aircrew, flying as a navigator/communicator on a navy P-3 Orion crew. She was a mission specialist.

Kalpana Chawla was an aerospace engineer born in India; marine major Carlos Noriega, a pilot and space operator, had been born in Peru. Navy lieutenant commander Robert Curbeam was African American.

This new group, the most diverse selected by NASA to that time, reflected Abbey's beliefs, but also changes in society. For example, just two years earlier Congress had repealed a section of US Code Title 10, which provided the legal framework for the military's "roles, mission and organization," forbidding women from serving in combat roles.

As 1994 became 1995, cosmonauts Titov and Krikalev were still in Houston. Both had experience on long-duration flights.

Abbey went looking for additional expertise.

Enter French astronauts Jean-Loup Chrétien and Michel Tognini. In 1994, Chretien was head of a small French astronaut group which had assigned its members to Soviet and Russian crews for over a decade.

But there hadn't been many flights, and there were few on the horizon. Some of the French astronauts were to transfer to a new unified European Space Agency group, but there they would be competing with other nationalities.

Chretien wanted his team to be active, so he called Abbey to arrange a meeting. Their Houston encounter was supposed to last one hour; it lasted three minutes. According to Chretien, Abbey abruptly said, "I understand, well, there is no more work in France," then added, "there is work here." Did Chretien want to come to Houston?

Chretien agreed, though he wasn't proposing himself. Abbey insisted. "You come first and then we'll take one of your guys in a couple of months."

Chretien returned to France to make arrangements. A couple of days went by; it was now November. Abbey called Chretien and said, "Can you come next week?"

"Yes," Chretien said. "Do you want another meeting?"

"No, no, no. You take your luggage and come down here."

Chretien arrived in mid-December 1994; Michel Tognini shortly thereafter.

Chretien had spent a number of years at the GCTC and had flown two missions with the Russians. His first was a seven-day flight to the Salyut-7 space station in 1982, and the second was a twenty-four-day flight to Mir in 1988. He spoke excellent Russian and was popular and well respected in the Russian space community. Tognini had also trained at GCTC and had spent thirteen days on Mir. Abbey involved both men in the developing relationship between JSC and the Russians, helping to resolve issues and establish trust.

Abbey also added Japanese astronaut Takao Doi and Canadian astronaut Dafyyd "Dave" Williams to the incoming 1995 group.

Beyond matters of personnel and program development, Huntoon and Abbey made strides in improving JSC facilities, which were simply not equipped to support long-duration spaceflight.

Due to tight budgets, mission control teams had been forced to keep using the same mainframe-based systems they'd used for STS-1, a dozen years and several computer generations in the past.

Gene Kranz heard a pitch from a young flight director named John Muratore, a former air force space operator who had come to NASA and worked on Hubble. Muratore had experience with personal computer-style workstations and believed that mission control should have them. "If you'll give me a couple million dollars," he told Kranz, "I will set up a demonstration facility that will show you how to save two hundred M&O [Maintenance and Operations] people. That will pay back that investment in less than a year."

Kranz had found the money and turned Muratore loose. In early 1994 he went to work on the upgrade.

Abbey played a role. According to Muratore, "When I was leading the effort to build the new mission control center in late 1994, I was giving a tour and said that we would have a certain capability available in April of 1995.

"A month later I was giving another tour and said the same thing and Mr. Abbey turned to me and said, 'You've slipped a month, you told me March last time.' I told him no, we were on schedule. He insisted that we had slipped. So I went back to my office and got everybody together and found a way to move it earlier to March.

"The next time I toured Mr. Abbey, I told him that we had managed to pull the milestone a month earlier to March. He said, 'No, you've slipped, you promised me February.' I looked him straight in the eye and told him absolutely no. Then I noticed a little twinkle in his eye. He was just trying to get the best out of me and my team and he was going to keep punching until we pushed back.

"It was a wild ride, and we went from an empty building to flying our first flight in eighteen months."

For years astronauts trained for EVAs in the WET-F, a large pool that had been built as Apollo ended, taking up the circular building 29 that formerly housed the centrifuge.

Twenty years later, WET-F was too small and too old for the number of EVAs required: the Hubble crews had had to travel to Marshall in Alabama in order to complete their hours of training. WET-F was totally inadequate for the number of EVAs required for ISS assembly that would begin in 1998.

While working for Cohen, Huntoon had heard a presentation on building a new tank only to have it turned down for lack of funds.

It wasn't until Abbey arrived that the process revived.

In the early 1990s McDonnell Douglas had built a facility near Ellington Field to process its Space Station Freedom modules. According to Chris Perner, a long-time NASA engineer who had been trying to sell the new facility, "It was going to be where you could bring all the Space Station pieces together, flight hardware, and put it together and then ship it to Florida."

Then, he said, "They decided to do all that assembly work at Florida. So the building wasn't needed anymore. George Abbey said, 'Well, hey, you've got this big building out there. Let's dig a hole in the floor and put your swimming pool in there.' So that's where it is today. It finally happened, and I got to see it before I walked out the door."

Somehow Abbey transformed McDAC's contract to converting the building to a giant pool, a major construction effort. The final result was the Neutral Buoyancy Laboratory (NBL), the world's largest indoor pool, 202 feet long, 102 feet wide, 40 feet deep. It would hold 6.2 million gallons of water—enough to fill eight Olympic swimming pools—generally kept at a temperature of eighty-six degrees Fahrenheit. The weight of that water was 25,800 tons.

From ground-breaking in April 1995, the NBL was completed in October 1996, and after two months of verification and suit-testing, was officially turned over to NASA in December, where it was named for the late astronaut Sonny Carter.

In addition to the large high bay in the McDAC facility occupied by the pool, the building included an excellent machine shop to fabricate and repair ISS training modules that were used in the pool.

STS-63, the first mission of 1995, was the first in the Shuttle-Mir program, with the orbiter *Discovery* flying to a rendezvous with the Russian station—closing to a distance of thirty-three feet in spite of several leaking thrusters on the orbiter, which triggered a serious discussion between Houston and Russian mission control in Kaliningrad.

The crew was led by Jim Wetherbee, with Eileen Collins becoming the first woman to serve as a shuttle pilot. NASA mission specialists were British-born Mike Foale as well as Bernard Harris and Janice Voss. Russian cosmonaut Vladimir Titov rounded out the team, serving as the primary communications link between *Discovery* and Mir.

There were three cosmonauts on Mir at the time, veteran commander Alexandr Viktorenko and flight engineer Elena Kondakova (who happened to be married to Energiya's important Valery Ryumin) and long-duration cosmonaut Valery Polyakov, on the downhill side of a 437-day marathon that had begun in January 1994.

The rendezvous set the table for the first American long-duration mission since 1974.

On March 14, 1995, Norm Thagard became the first American astronaut to ride into space in a foreign vehicle, as flight engineer in the crew of Soyuz TM-21. The Soyuz launch went smoothly, and Thagard and his two crewmates, Vladimir Dezhurov and Gennady Strekalov, spent two quiet days performing the standard chase to the Mir station.

Docking on March 16, the trio joined Viktorenko, Kondakova, and Polyakov, who would spend a week with the new arrivals before departing on March 22.

After spending 112 days in space, Thagard and his crewmates, Vladimir Dezhurov and Gennady Strekalov, welcomed the space shuttle *Atlantis* on the STS-71 mission for the first American-Russian docking in twenty years. *Atlantis*, commanded by Hoot Gibson, carried astronauts Charlie Precourt, Ellen

Baker, Bonnie Dunbar, and the Mir replacement crew of Anatoly Solovyov and Nikolai Budarin.

The orbiter remained docked to Mir for five days, separating on July 4, 1995. Prior to that event, the new Mir crew temporarily left the station in their Soyuz spacecraft in order to photograph the undocking. *Atlantis* landed three days later at the Kennedy Space Center, safely returning the shuttle crew as well as Thagard and his two crewmates and laying the foundation for the construction of the International Space Station later in the decade.

During the summer of 1995, Dan Goldin was growing concerned about JSC leadership, and equally troubled by delays and problems with ISS. He felt he had to make a change, and proposed transferring Carolyn Huntoon to headquarters. Hearing this, Abbey protested but failed to change Goldin's mind. In August 1995 Huntoon moved to Washington and Abbey was named the acting director of JSC. It was obvious, though never stated, that she blamed Abbey for undermining her. Having enjoyed a good relationship with Huntoon that went back over twenty-five years, Abbey tried to send the message that her transfer was not his doing, but doubted that she ever believed him.

Abbey chose to stay in the deputy director's office and never did move into the northeast corner office that traditionally belonged to the center director.

Part VII

“Together to the Stars”

53 | A New Direction

DURING HER TENURE, Huntoon had made dozens of speeches and spent considerable time reaching out to the Houston community. Abbey immediately went further, overseeing the first Johnson Space Center open house on August 26 and creating a Safety Awareness Day for the center's staff that took place the same week. (His first address to the center dealt with safety.)

He also made changes in JSC leadership, his direct reports. First, he appointed astronaut Jim Wetherbee, fresh from command of STS-63, as acting deputy center director. Sue Garman would be retained as assistant to the director (though later promoted to associate director for management).

Then he named John Young as associate director—technical, where he would have responsibility over all center programs. As with Robert Gilruth and George Low, Abbey placed high value on Young's input.

Abbey would come to chair all flight readiness reviews when JSC assumed the lead role for the shuttle program, and Young was always present, playing a key role in identifying critical issues and suggesting corrective actions.

The vital life sciences directorate had had an acting chief for two years—Abbey wanted H. David Short from Baylor College of Medicine to be permanent director. Short had been a member of Dr. Michael DeBakey's heart transplant team and a classmate of astronaut Sonny Carter at medical school. It was a challenge for Abbey to get Short to accept the NASA job. Abbey reshuffled Len Nicholson's engineering directorate, adding James Jaax and John Muratore as deputies. He took steps to create an EVA Projects Office—

ISS would be assembled by astronauts doing EVAs—under astronaut Don McMonagle.

How did he operate? According to Jaax:

> George had a management style where he collected information from anywhere and everywhere. Almost every time you'd go in there he'd have the information you're trying to get to him already from some other source. It caused problems a few times, but basically we'd go in for one-on-ones; last for a half hour, maybe a little longer. John Young, he was always in the meeting, just listening, probably because he was interested in what engineering was doing, because we were doing some neat, creative things at the time, and we were also dealing with shuttle and station issues.
>
> What you tried to do was give him sort of a heads up. You didn't explain. You'd give him the subject and what you're doing—or what the problem was and what you're doing, and that's it. And then go on to the next one; go on to the next one; go on to the next one, because that's really all he needed, because if he wanted more information, he'd ask you.

Jaax found it very effective and emulated it in working with his own division chiefs.

In dealings with the rest of NASA, Abbey still had the full confidence of administrator Dan Goldin, and the grudging respect of deputy administrator Jack Dailey—who frowned at the percentage of the overall NASA budget allotted to JSC ("something like 80 percent").

Jed Pearson had left NASA in November 1984 and had been succeeded as head of manned spaceflight by Wayne Littles, an engineer from NASA Marshall who had worked on redesigning the shuttle's solid rocket boosters. Will Trafton was still leading the ISS team at HQ while Bryan O'Connor led the shuttle.

Abbey's counterparts at other centers included his former bubba, Jay Honeycutt, who had gone to work in the shuttle program office after *Challenger*, and in 1989 moved to KSC to run shuttle operations there. Six years later, in January 1995, Wayne Littles, the associate administrator for spaceflight by then, offered Honeycutt the directorship of the Kennedy Space Center, succeeding Crippen.

Burned out by the pressure and budget cuts, Honeycutt turned him down.

That was a Friday. The next day he received a call from Abbey. "Have you lost your mind? Take the job."

So Honeycutt did, later describing it as the "smartest thing I ever did."

He certainly got along with his fellow JSC director. "We agreed on everything."

And since Abbey had had a hand in selecting or continuing all the other center directors, there were few problems.

One ongoing challenge was a move to commercialize shuttle operations on a scale that far surpassed James Beggs's 1985 effort. The stripped-down Goldin budgets meant that NASA had no new money and, with inflation, less money to perform its core mission. Reducing the cost of shuttle operations would free up needed funds.

The two major contractors performing operational support were Lockheed at the Cape and Rockwell in Houston. As manufacturer of the shuttle orbiter, Rockwell also had a team at the Cape.

Abbey told Goldin that if he wanted to reduce the cost of shuttle operations, he should ask Chris Kraft to study possible options for their commercialization. At Goldin's request, Kraft organized the Space Shuttle Independent Management Team, which included George Jeffs, former vice president of Rockwell and chief engineer for Apollo; Apollo 8 commander Frank Borman, who was also a former president of Eastern Airlines; Ralph Lindstrom, a veteran manager at the NASA Marshall Spaceflight Center who had also been senior vice president of Thiokol Corporation; and several others with experience at NASA and in the commercial world.

Kraft and his team concluded that significant cost reductions could only be found with an innovative management approach that emphasized greater contractor responsibility for standard and daily operations. What emerged was a Space Shuttle Operations Contract that would move civil service jobs and functions to a commercial operator.

United Space Alliance (USA), a joint venture between Lockheed and Rockwell, was formed to bid on the contract. USA would be led by Kent Black, former chief executive of Rockwell, and Jim Adamson, a former astronaut who had retired from government service in 1992 and become president of the Lockheed Engineering and Science Company. USA was awarded the contract, promising to maintain NASA's schedule and service while saving the agency $1 billion over five years.

As the redesigned ISS development proceeded, Abbey oversaw a $190 million fixed price contract between Boeing (as ISS prime contractor) and Russia's Krunichev Research and Production Center (builder of Salyut and Mir modules for the Energiya organization) to produce the Functional Energy Block—the first element of the station ultimately named "Zarya" or "Dawn." In spite of its name and origin, it would technically be American equipment.

The US Node One module, a legacy of the Freedom effort, was nearing completion at this time.

STS-74, the second Shuttle-Mir mission in November 1995, was commanded by Ken Cameron. The crew of five delivered a Russian-built module for future Shuttle-Mir dockings as well as a pair of solar arrays for the station.

At the same time, several astronauts were at the GCTC training to follow Thagard. TFNG Shannon Lucid would launch in March 1996 and would spend a very successful four months aboard.

She would be followed by shuttle commander John Blaha. Then it would be navy captain, and medical doctor, Jerry Linenger, sometime in the fall of 1996.

After Linenger, assignments were still in flux. Astronauts Scott Parazynski and Wendy Lawrence had been enthusiastically supported by Abbey—but after some months of initial training, had been judged by the Russians to be "too tall" (Parazynski) and "too short" (Lawrence).

54 | The Total Equation

GEORGE ABBEY BECAME THE permanent director of the Johnson Space Center on January 23, 1996, ascending to the job held by his idols and mentors Robert Gilruth and Chris Kraft. He was in charge of twenty thousand civil servants and contractors not only in Houston, but in locations such as White Sands.

He had previously made John Young the center's associate director. Since Jim Wetherbee had returned to the astronaut office to make another flight, Abbey appointed astronaut Brian Duffy to replace him as deputy center director.

With Abbey now in place at JSC, Goldin wanted a management change that would make the Houston center responsible for both the space shuttle and International Space Station. He designated JSC as "lead center" for both, putting Abbey directly in the firing line. All too aware of the pressures that came with that responsibility, Abbey wasn't anxious to take on the role. The space shuttle program was undergoing a major restructuring due to the Space Shuttle Operations Contract and the creation of United Space Alliance. That transition had to be implemented safely while conducting an ongoing flight program with a reduced budget.

The International Space Station had been redesigned, yet major components had been carried over from the Freedom program, and the new configuration was not yet final. Boeing had been designated prime contractor but had not yet defined its role. All US contractors were behind schedule and had various problems. At the same time, the Russian configuration and interfaces were also still to be defined. And there were similar problems with the other three major international partners, Canada, Europe, and Japan. Work was being performed in different centers around the world, and yet somehow it all had to come together in Earth orbit.

And, on a personal level, accepting the two leadership jobs would dramatically change Abbey's life, since he would be working all day, every day. Further, Goldin hadn't asked his headquarters staff for recommendations or approval before making this decision. With his own experience in Washington, Abbey knew that the decision wouldn't be universally accepted at headquarters or at other NASA centers. He would have to work with all of these organizations, making each feel essential, and that was no small task. Knowing the challenges, Abbey told Goldin that he didn't want all that responsibility. But Goldin was firm. Abbey and JSC took on the lead center role for space shuttle and ISS.

Once JSC became the lead center, Tommy Holloway became shuttle program manager and Randy Brinkley took on ISS. Abbey had great confidence in Holloway and would leave day-to-day shuttle management to him. Abbey would take part, as needed, in the safe transition of shuttle operations to USA; ISS required more of his time and attention.

In January 1996, Abbey and JSC announced that astronaut Bill Shepherd and cosmonaut Sergei Krikalev, with a Russian Soyuz commander to be named later, would be the first ISS "expedition" crew. The early assignment allowed Shepherd and Krikalev, both very experienced and respected, to provide oversight and crew perspective to the redesign and serve as pathfinders for the crews that would follow them—to find flaws in training materials and schedules, for example, or even living arrangements in Russia and Texas.

On the program development front, however, Abbey knew that he faced real challenges.

By early 1996, a team from Boeing had been serving as prime contractor for over a year and was experiencing great difficulty in implementing its responsibilities. In August 1996, Boeing announced the acquisition of Rockwell International's space and defense units, including its ISS activity (Space Systems Division and Rocketdyne). And in December the company announced that it had also acquired McDonnell Douglas. These moves would eventually allow Boeing to better fulfill its role as prime contractor, but until then Abbey was forced to work with a dysfunctional organization and group of contractors.

Somehow he had to get Boeing, McDAC, and Rocketdyne, as well as NASA and the ISS international partners, to work as a team.

The big problem was that the NASA engineers and their contractor counterparts had little experience in designing and developing spacecraft for human flight. The designers and builders of Mercury, Gemini, and Apollo were long retired, and so were many who had done similar work on the space shuttle. In reviewing ongoing ISS work, Abbey found NASA and the contractors repeating mistakes made in the 1960s and 1970s. There had been no carryover of corporate knowledge; the lessons of the past would have to be relearned.

Abbey knew where he could find experienced engineers who worked with piloted spacecraft and understood the challenges. He called his friend Jay Honeycutt, director of the Kennedy Space Center, and asked for his help. Abbey wanted experienced engineers from KSC to be assigned to every contractor facility where ISS modules were being assembled, to ensure that the work matched the standards employed at the launch center. Honeycutt and one of his deputies, John J. "Tip" Talone, deployed experienced engineers to oversee operations at Rocketdyne, Boeing, and McDAC facilities around the nation. Thanks to Talone's team, modules and systems were delivered to KSC in improved condition.

Abbey also called for a series of weekly meetings that would be officially known as the George Abbey Saturday Review or GASR—soon to be called GASSERS. Each meeting would cover the work planned for the previous week and its status—had it been completed, and if not, why not? What corrective action would be taken and how would this action affect delivery dates? In addition to contractor and NASA personnel, representatives from Canada, the European Space Agency, Japan, and Russia were also welcome to attend. In addition to the weekly GASSERS, Abbey instituted a more comprehensive monthly review known as the Station Development and Operations Meeting (SDOM). In order to identify problems that would affect NASA's ability to fly and operate the station, Abbey had the SDOM agenda prepared by flight directors assigned to the ISS program. These meetings would be held at a different location every month, either at one of the contractor facilities or at JSC, KSC, or NASA Marshall. Senior contractor executives and the respective center directors were required to attend.

The SDOMS would replicate the means and methods of George Low's Apollo configuration control board in which only those empowered to make decisions were allowed to be present. (In addition to NASA managers and contractors, these meetings also included representatives from the ISS international partners.) Participation by senior executives from the different organizations raised awareness of problems and issues, and led to corrections that might otherwise have been delayed or deferred.

Following Boeing's acquisitions of Rockwell and McDAC, the prime contractor was responsible for all ISS work through its Information, Space & Defense Systems unit. In February 1997 Alan Mulally was named its president. He had previously worked in Boeing's Commercial Airplane group, serving as vice president and general manager of the Boeing-777 program. Abbey welcomed Mulally's involvement, and his participation in the monthly SDOMS. Under the new Boeing leader there was a noticeable improvement in the company's performance on ISS. (When Mulally later moved back to the Commercial Airplane Group as president, he was succeeded by Jim Albaugh, who continued to provide effective leadership for Boeing's ISS work.)

GASSERS were unpopular within NASA and especially the contractor community. "Too many highly paid executives had to give up their Saturday golf," one participant said. Even Abbey acknowledged "Everybody hated the idea." In his view, however, if the GASSERS ran long, it was because planned work had not been done, and participants then had a difficult time explaining the reason and what corrective action would be taken. Ideally, these should have been short meetings. One astronaut remembered them as "brutal," going from 8:00 in the morning to early afternoon on Saturdays. Each product group, whether it was ISS modules, communications, environmental systems, or, more and more, computers and software, had to send a representative to discuss the week's work.

Steve Hawley, Leestma's deputy in flight crew operations, remembered one department head briefing Abbey on a presentation that was still incomplete on Friday. Abbey stared at him. "There are seven days in the week, you know." Meaning, spend Saturday and Sunday getting it done.

Jay Greene found the GASSERS to be so stressful that he began running rehearsals on Thursdays, hoping to identify and solve major issues—or at least prepare team members for the likely complaints.

In these meetings, the image or myth of the sleepy-eyed, quiet or mumbling Abbey died. He could be, and frequently was, outspoken and belligerent, snapping at presenters to make things better, faster. Calling back to his first NASA experiences, he would frequently have to referee a shouting match between individuals—in many cases, John Young and Jay Greene.

Hawley admitted, "The GASSERS and SDOMS gave us focus and purpose."

Brian Dailey said, "Everything George worked on in the space council and later worked on directly in NASA, working for Dan Goldin after working for me, was to find some way to get humans back into space. Even space station, for him, was a mere stepping stone, so to speak, and what we really had to do was go back to the Moon, then on to Mars."

In the 1990s, according to Congress—and to its own hypersensitive internal guidelines—NASA was not pursuing a flight to Mars. *No, sir, not us.* But Abbey was enabling technologies that would make that journey easier.

One was the Assured Crew Return Vehicle, aka the X-38. In appearance and structure it was a wingless lifting body largely based on the X-24 from the 1960s. Launched in the payload bay of a shuttle orbiter, it would be docked to ISS, where it would remain for an indefinite period, serving as a lifeboat for an International Space Station crew of seven.

Who would lead this program? The selection was typical George Abbey.

As soon as his work on the mission control upgrade was complete, John Muratore was summoned to Abbey's office. "I thought he was going to yell at me about something else."

But Abbey had this question: "How would you build a spacecraft?"

Muratore eagerly accepted the challenge, deciding that he would use the same principles he'd used in the control center project. "I'd come up with a spacecraft design that we could test in the atmosphere, test in pieces, and then slowly move up into spaceflight. I'd start building very primitive vehicles and then build up to much more sophisticated vehicles. And I'd try to use as much commercial, off the shelf, and reused components as I could.

"He said, 'When will you have it ready?' I said, 'Well, I'm going to need money,' and I went off and put a budget together."

And soon Muratore had some money, and the new X-38 program.

As for the deep space habitat, in the spring of 1997 Abbey's head of engineering, Leonard Nicholson, asked his team to develop a "lighter, cheaper" habitat for a Mars mission. Such a mission required six hundred cubic meters of habitation, and creating a single or even multiple aluminum modules that were large enough yet strong enough to withstand launch resulted in fantastic penalties.

William Schneider came up with TransHab, an inflatable module with a foot-thick skin made largely of Nextel and foam and other materials. TransHab could be launched in a relatively compact state (fourteen inches in diameter, thirty-six inches in length), then inflated once on orbit to a full diameter of twenty-seven feet. It would be easier and cheaper to launch and, when inflated, as strong as any metal module.

The TransHab team, working swiftly, was able to test a prototype in JSC's Simulation Chamber A on December 21, 1998. Abbey also found funding for VASIMR (Variable Specific Impulse Magnetoplasma Rocket), a plasma drive conceived by astronaut Franklin Chang-Díaz and of potential use on interplanetary missions.

All programs showcased Abbey's style—narrow goals and focus, small in-house teams, disregard for larger agency concerns.

Looking back years later, NASA's James Jaax said, "Nobody could believe that you could produce what we did with the minimal amount of resources that we said it took."

As soon as the delayed 1994 astronaut candidates were announced in January 1995, Abbey put Duane Ross and his team to work on the next biannual selection.

To outsiders, it appeared as though Abbey were just fattening the astronaut office for mysterious reasons of his own. Even some astronauts thought so. Janice Voss said, "George could hire all this engineering and scientific talent without increasing the JSC headcount."

Abbey was indeed happy to have this "free" talent, but there were forcing factors, the biggest one being the need for ISS expedition crews. In 1996 NASA and its partners hoped to launch the first crews in 1999, and then to rotate them every three months.

So ISS ops would require twelve crew members each year, at least half of those from NASA.

ISS training was projected to require four years. Hence around twenty-five NASA astronauts at any given time.

The shuttle was still flying six missions a year, with four to five NASA crew members each time. Call it another twenty-five, and since astronauts rarely flew every year, the number in the shuttle flow was closer to forty at any given time.

That added up to sixty-five astronauts in active training.

There would be twenty astronaut candidates. So minimal staffing ran to eighty-five.

Then there were branch chiefs, astronauts on details to other departments, and those in support jobs—Capcom, Cape Crusaders, SAIL. Easily another twenty-five.

In 1996, an internal study showed that with shuttle and ISS, NASA needed 110 astronauts.

The new group numbered thirty-five, and among them were eight women—three of them African American women. There were space twins, navy test pilots Scott and Mark Kelly, and an unrelated third Kelly, air force test pilot James Kelly. And also some astronauts Abbey knew from past associations, like Don Pettit from the Synthesis Group,

Abbey would add nine international astronauts to the group.

And then move on to a selection for 1998.

Abbey had frequently been frustrated by NASA's failure to record, store, and catalogue important data. (Years later he would still be wondering where some of the Apollo medical data had gone.)

The agency and center had done a good job of archiving material and writing in-house histories, the same way military units did. But, as Duane Ross said, "when they started cutting budgets" in the 1970s, "history was the first to go."

One day in the fall of 1995, Ross was in Abbey's office discussing plans for the next selection. As he left, Abbey said, "Mr. Ross, there's something I want you to do." (Even though they had worked together for two decades, Abbey occasionally called Ross that.)

Abbey wanted Ross to go to the US Naval Institute at Annapolis "and look at their oral history program." He wanted JSC to have a similar program.

Ross and his associate, Rich Dinkel, soon teamed up with William Larsen, former supervisor of the Apollo history, and they began collecting documents from dispersed sources—basements, Rice University—and creating a list of individuals to be approached.

"We also decided early on that this program was not just going to be for all the upper level management, because you don't get all the stories. We talked to the upper level management folks. We talked to suit technicians. We talked to engineers. We talked to secretaries, and all kinds of folks to get their take on the space business."

The program would continue for two decades. By 2017, there were a thousand oral histories available to researchers or anyone with internet access.

Around this time certain NASA and contractor individuals began referring to themselves as "FOGs" (Friends of George). In one sense, the term implied favoritism, something Abbey had been criticized for. Abbey happily conceded that he made friends with astronauts and other people in the space business, and naturally many of those friends had upwardly mobile careers.

But he also employed and worked with dozens with whom he had no extracurricular connection. He was perfectly capable of promoting those with whom he had no social interactions at all. Astronaut Jerry Ross, for example, didn't participate in beer call and yet made seven spaceflights, tying him with Franklin Chang-Díaz for the most by any astronaut.

The term might better be rendered as "the Family of George."

From his experiences in the air force, in Apollo and especially Apollo-Soyuz, everywhere in NASA, and in his family life, Abbey realized that human spaceflight didn't belong to the military or civilians, or commercial entities—to one nation or one gender. "It had to be everybody," he said.

It was obvious to astronauts that working in orbit or walking on the Moon had value, but what did it mean to those on Earth? Some would be excited by the idea or the history, but most wouldn't, especially after pioneering "firsts." Abbey wanted citizens not just of Houston or Texas or the United States of America, but the whole world to be aware of and invested in what happened in space.

After all, their television programs and phone calls had gone through satellites for years. Weather forecasts became a science thanks to spaceflight. So did navigation.

For years, of course, NASA had tried to make Americans aware of the ways their lives had been improved by spinoffs from space research. Abbey wouldn't deny that, but felt that this connection required more.

In addition to the open house, Abbey began observing Black History Month by hosting a program on the Tuskegee Airmen, and gave more emphasis to Women's History Month and to American Heritage Week.

He also reached out to John Wilson, superintendent of the Clear Creek Independent School District, and offered to make open land on the JSC site available to the district's high school agricultural students for raising longhorn cattle. Abbey brought the famed Houston Livestock Show and Rodeo into the project. And with the additional help of the Longhorn Breeder Association, longhorn cattle were grazing on open JSC land, to the annoyance if not mystification of Dan Goldin.

Superintendent Wilson was also seeking to purchase a parcel of land for a new intermediate school in his district. Abbey was able to arrange a long-term lease from the government to place the school—later known as Space Center Intermediate—on unused JSC land.

In 1998, Abbey was asked by businessman and former Secretary of Commerce Robert Mosbacher to serve on the board of a charter school Mosbacher was starting in Galveston, Texas. Named Odyssey Academy and originally a middle school aimed at students who were economically disadvantaged, it grew into a pre-K through eighth grade school.

As JSC director, he also created a program to bring students from other countries to Houston each summer to learn about spaceflight from engineers, scientists, and astronauts. Working with astronaut Bonnie Dunbar, he established a similar exchange with schools in Scotland, too.

Abbey encouraged team-building events like the chili cook-off, which kept expanding in scope. JSC staffer Tom Diegelman remembered, "When I first started working on it, we did the scoring with pen and paper. George came by and saw this, and complained that these were hardly the sophisticated tools you expected at a NASA event.

"I said, we could computerize this. 'I like the concept,' he said. So we did, and things went much better.

"George also thought we needed to have more teams. 'How about fifty?' At this time we had around thirty-two. But based on his urging, we all went out recruiting, and in the next year we had fifty-eight or fifty-nine teams. I asked George the next year, should I give him credit? 'No need. Let's aim for one hundred next year.' And by god, next year we got over a hundred teams."

JSC was part of the Houston family, Houston part of Texas, Texas (in spite of all appearances) part of the United States, and the United States part of the whole global family, as all-embracing and engaged as the Abbeys of Seattle.

That was Abbey's vision.

How did it work in practice? In 1995, Robert Yowell was a young flight controller with the opportunity to take a year's leave from JSC and get a master's degree. He was applying to Purdue and other aero engineering programs when Abbey's office called him.

Yowell had never been to Abbey's office. "He was just this mysterious deputy director in the big building."

Abbey asked about Yowell's subject for his master's. Yowell was hoping to explore the challenges of communications between an expedition to Mars, and Earth. "Have you considered attending the Air Force Institute of Technology?" Abbey asked. This was his alma mater.

Yowell had never served in the military. "AFIT doesn't take civilians."

"Let me worry about that," Abbey said. Sure enough, Yowell learned that he was one of the few civilians admitted to AFIT, where he spent a year happily researching the challenges of relaying messages and data from the surface of Mars to Earth.

He was barely back in mission control when his phone rang again: Mr. Abbey, asking how the year had gone, and was Yowell happy with his research?

Yowell assured Abbey that he was happy on both fronts. "Great," Abbey said. "Send your thesis over to Wendell Mendell." Mendell was the head of planetary science at JSC. And Yowell realized that he had been part of some grand scheme all along.

After every shuttle mission, Abbey and Young held a private two-on-one with the commander before the general debrief with crew.

Rick Searfoss commanded STS-90, a sixteen-day flight of the Neurolab, a medical mission examining the human nervous system, in April 1998.

Already planning to leave NASA, Searfoss walked into the meeting with one goal firmly in mind: don't hurt my crewmates—pilot Scott Altman and mission specialists Kay Hire, Dave Williams, and Rick Linnehan. (There were two payload specialists, Jay Buckey and Jim Pawelczyk, but their futures weren't going to be affected by what Searfoss said.)

Mission specialist Linnehan was a veterinarian by training who had resigned himself to being a "pin cushion," an astronaut assigned only to medical missions due to his background. He had complained to Searfoss, "I'm never going to get to make an EVA."

In the debrief, Searfoss praised all his crew members but singled out Linnehan. "We had some problems with the animals aboard, and he had saved their lives," Searfoss said. "I told George and John that all he needed was the chance to develop his operational and technical skills, as in being a Cape Crusader."

Linnehan's next assignment? Cape Crusader, which put him in line for assignment to a more technically demanding mission. Which turned out to be making EVAs on Hubble repair mission years later, a mission commanded by Scott Altman, the STS-90 pilot.

That wasn't the only repercussion. Searfoss said, "I noted some real issues between our mission and the JSC life sciences directorate. I thought there was a lack of leadership, that they were negative, too bureaucratic." Life sciences director David Short had previously told Abbey that he wanted to return to medical practice; for several months the directorate had been under acting leadership. "We had not gotten the support we should have gotten. I just made this comment as, 'oh, by the way', with the ISS coming up, this is something you might look into."

Within weeks, Dave Williams, the Canadian STS-90 mission specialist and physician, was named JSC director of life sciences, filling the vacancy created by Short's departure several months earlier. "He did the job for four years and was terrific." Williams was the first non-American to hold a senior management job at JSC, a fact that pleased the Canadian Space Agency and other ISS international partners.

These are just two of hundreds of examples. The mechanics of Abbey's leadership are summed up by astronaut Jim Wetherbee, who worked for Abbey in a variety of roles beginning in 1984.

"He made decisions independent of possible personal repercussions, damage to his career. His work ethic was superb; he was tireless. For example, we had a long series of tough negotiations on the final contract with Boeing for ISS—multiple billions of dollars at stake. Friday night at 10:30 we finally get signatures, and I'm thinking, time for champagne. Instead, George went back to his desk to deal with his inbox, all the material that had piled up during the negotiations and needed to be dealt with. He treated every decision—large and small—with the same amount of attention and integrity."

Jay Honeycutt: "In all the time I worked with George I never ever knew him to make a decision for his own benefit, either financially or politically. It was always, will this be good for NASA, for the center, for the program? His whole philosophy was, do what's right. . . . Don't worry about whether it knocks the train off the tracks."

Or Joseph Rothenberg, who succeeded Trafton as associate administrator for spaceflight at NASA HQ. "George was usually described as secretive—'no fingerprints'—but really he was just shy. He didn't seek attention. He was the opposite of people who run around taking credit for things they didn't do.

"He went a long way toward establishing good relationships with the Russians—not at every level, but certainly with the top people in their program.

"Whatever George did—and sometimes he was wrong—he never did anything for personal gain. As for the astronauts, he tried to watch out for them."

Abbey's friend Constable Bill Bailey would also come to him privately if any one of the astronauts got into trouble.

Not that Abbey was above claiming a few perks.

Abbey's mother, Brenta, had grown up in Laugharne, Wales, a small village that became home to famed poet Dylan Thomas in the 1940s. Thomas moved into a cottage, the "Boathouse," in the same area where Brenta and her sisters played as children. As Abbey remembered, "Thomas's wife, Caitlin, would lock her husband in the 'Boathouse' to work every day until early evening, when she would release him and they would walk to Browns Hotel. There they would meet up with Dick Lewis, who was my mother's cousin. These three would drink until closing time. Lewis's family ran the Laugharne dairy, but he had gone away to college, earning a law degree. Before he could practice, World War II broke out, and Lewis wound up with the British code-breakers at Bletchley Park."

Returning to Laugharne after the war, this lawyer code-breaker took over the family dairy and milk delivery, becoming good friends with the Thomases. Abbey would meet Lewis on trips to Wales in the late 1990s and early 2000s, and arranged to fly a portrait of Dylan Thomas on the first ISS assembly mission in 1998.

55 | Crises on Mir

ON SUNDAY, FEBRUARY 23, there were six space travelers aboard Mir: the out-going resident crew of Valery Korzun and Alexandr Kaleri, the incoming Mir 24 team of Vasily Tsibliyev and Alexandr Lazutkin, ESA astronaut Reinhold Ewald—a short-term visitor who had arrived with Mir 24 and would be departing with Korzun and Kaleri. And NASA astronaut Jerry Linenger, the fourth Shuttle-Mir crew member, who had been delivered to the Russian station by STS-81 on January 14.

Chief astronaut Hoot Gibson and flight crew operations director Dave Leestma had asked for volunteers to fly on Mir missions, and were having little luck. Experienced astronauts like Charlie Bolden and Tom Jones turned them down. Some astronauts who were willing, like Bernard Harris, were too tall. (NASA astronauts who flew to Mir had to meet strict anthropometric standards for the Soyuz spacecraft, and these were more stringent than those for shuttle.)

Linenger, forty-one, was a US Navy flight surgeon by training and had joined NASA in 1992. He had quickly volunteered for the Shuttle-Mir program, and in order to qualify (the Russians were adamant about only flying experienced astronauts) had been rushed into a shuttle crew, STS-64.

With that eleven-day mission in 1994 as his qualifying experience, in January 1997 Linenger launched aboard *Atlantis* for a much longer—a 132-day—stay aboard Mir. One month into his Mir stay he was already at odds with Tsibliyev and Lazutkin.

Around 10:00 PM Mir time on February 23, the entire crew was gathered in the core module for dinner that included red caviar, a special treat brought up by the new residents. Linenger excused himself to head to the Spektr module while Lazutkin drifted into Kvant-1 to ignite an oxygen candle—a lithium perchlorate cylinder that gave off oxygen, used for decades on Russian submarines and on Mir when additional crew members were aboard.

This time, however, the candle exploded and set fire to its casing. Flames shot out of the wall, and Kvant began to fill with green smoke. The crew began to take emergency procedures, searching for masks and possible escape routes . . . but then the candle burned itself out. Nevertheless, it took hours for Mir's environmental system to clear the smoke.

Making matters worse, word of the fire didn't reach NASA for six hours.

The reaction in the United States, especially in Congress, and even within JSC, was swift: Mir was inherently unsafe, and its emergency procedures flawed.

Abbey granted that the fire had been a serious problem and encouraged an investigation by an independent American and Russian commission of experts, cochaired by Tom Stafford and Russian academician (and missile designer) Vladimir Utkin. The commission had originally been established by Dan Goldin and Yuri Koptev to advise both space agency leaders on Shuttle-Mir activities. (The commission was still in existence in 2018, with Stafford, who outlived two of his Russian counterparts, still serving as cochair). Linenger returned safely to earth with STS-84, the crew commanded by Charles Precourt that delivered Mike Foale to Mir.

ISS faced a major hurdle of its own in the spring of 1997. The Russians were proceeding nicely with the construction of the Zarya module for Boeing. It would be one of the first two elements of the station launched in 1998. But the vital Zvezda (Star) module, scheduled to provide early station control and habitation, was little more than an aluminum shell on the floor of the Krunichev Space Center factory in northwest Moscow. None of its subsystems had been installed.

There were two reasons for this. One was a financial crisis in Russia, with the government starved for revenue and cutting back everywhere, especially in civilian space activities.

The real reason was . . . NASA wasn't ready, either. Because of their presence in the GASSERS, Russian representatives knew exactly how far along Boeing's modules were, and the state of vital ISS software. "The Russians aren't big on doing things until they have to," Abbey said years later. "They knew there was no need for the service module until 2000."

Abbey's awareness wasn't enough to stop a visit by Tom Stafford to the Russian Space Agency and to Krunichev in April 1997. He had discussions with Koptev, and with Russian prime minister Viktor Chernomyrdin, restating Russia's commitment to the ISS.

Four months later there was another crisis aboard Mir, and this one threatened the whole program.

Mike Foale had embraced Abbey's holistic approach to space training, especially with the Russians, moving his family to Star City and immersing them all in the local customs and language. He was working happily with Tsibliyev and Lazutkin as they approached the end of their tour on June 25, 1997. Tsibliyev's last major task was to guide the Progress M-34 to a redocking with the station as a test of a remote piloting system.

It went badly, with Progress approaching Mir too fast and failing to respond to command, smashing into one of the solar arrays and damaging the Spektr module, which began to leak.

The crew managed to cut or remove cables running through the open hatch to Spektr and seal it off, stabilizing the situation aboard Mir for the moment.

But the incident raised more questions about Mir's safety. Congressman James Sensenbrenner, chairman of the House Science Committee, held heated public hearings at which Frank Culbertson, head of Shuttle-Mir, testified. Sensenbrenner challenged NASA and especially Dan Goldin to stop sending Americans to the trouble-prone Mir or explain why.

Abbey recalled, "We held the flight readiness review two weeks prior to the launch of STS-86. Dave Wolf, the next Shuttle-Mir astronaut, was at the

Cape. Everyone agreed we were ready to go, then Goldin called and said he had not yet approved our proceeding." Normally NASA administrators, including Goldin, treated the FRR as the final authority on whether or not to proceed with a launch and mission.

At the Cape, Abbey, the astronauts, and other officials could only head to KSC to await further word from Goldin. It finally came on Thursday, September 26, 1997, just hours before STS-86's liftoff at 10:34 AM. According to the account by Kathy Sawyer in the *Washington Post*, Goldin "said he had agonized over his decision until the last moment—Wednesday night alone in his Washington headquarters office—because he 'wanted to do it right.'"

As for the collision that triggered this confrontation, Abbey would say, "the accident took place because the Russians put their Mir crew—Tsibliyev in particular—in an impossible situation trying to dock the Progress in those conditions. The investigation of the collision by the Stafford-Utkin commission exonerated Tsibliyev and the crew. They had done all they could.

"We believed that Mir was still safe. People seemed to find it easy to think that the Russians cared less about safety than we did, which is just wrong. In some areas they were more sensitive to safety than we were."

Wolf launched and served his time aboard Mir into January 1998, when he was replaced by Andy Thomas, the seventh and last NASA Shuttle-Mir astronaut.

56 | First Element

ON NOVEMBER 20, 1998, a Russian Proton booster lifted off from the Baikonur Cosmodrome, placing Zarya, the first element of the International Space Station, in orbit.

Two weeks later, on December 4, the space shuttle *Endeavour* rocketed off its pad at the Kennedy Space Center on STS-88, with Node One, the large docking element now named "Unity."

The crew of six, including Russian cosmonaut Sergei Krikalev, was led by former chief astronaut Bob Cabana. Following rendezvous with Zarya, remote manipulator operator Nancy Currie successfully linked it with Unity.

Then EVA astronauts Jim Newman and Jerry Ross performed a pair of space walks hooking up forty different cables and connectors and installing exterior equipment like antennae.

Krikalev and Cabana were the first to enter the modules on orbit, and were soon joined by the other astronauts.

Endeavour departed the linked Zarya-Unity vehicle, a rudimentary International Space Station, on December 13. Over the next two years there would be three visiting shuttle missions (shades of Max Faget's "man-tended" concept) that helped to train teams in NASA's mission control and Russia's flight control center to work together.

On the political front, however, the Zvezda delay was causing problems. According to Jeff Bingham, who was then NASA's legislative liaison, "NASA began considering alternatives with a Tiger Team that included John Aaron and Chet Vaughan. One option was the Interim Control Module (ICM) from the Naval Research Lab. The other was a Marshall program, the Propulsion Module. This idea had a lot of political support, from Alabama congressmen, naturally, and others on the Hill—it was all-American. But it was also more expensive than the ICM and would take longer to develop and would, if used, take the Russians out of the critical path."

Congressman Sensenbrenner was involved again, and Abbey fought back. "George needed the Russians in the critical path to keep the Clinton administration supporting the program," Bingham said. He believed that they would drop it otherwise.

Abbey failed to stop Marshall's module, but he would get it delayed—and in 2001 it would be shut down altogether.

Abbey knew that Goldin's unilateral decision to put JSC in charge of the shuttle and ISS was not universally popular at headquarters or other NASA centers. Now his critics were making the case that he was too independent and that no single center should have program responsibility that rightly belonged in Washington.

His methods were not only costing him support within NASA, at headquarters, and especially Marshall, but in Congress. He was safe, however, as long as the ISS was still incomplete. There was simply no one else who would or could take on that job and the responsibility.

The Proton launch of the ISS service module was scheduled for July 12, 2000. The launch had been threatened by the October 1999 failure of a Proton. Roskosmos, the Russian Federal Space Agency, decided to use modified engines on the second and third stages of the launcher and scheduled two flights to validate the improvements prior to the Zvezda attempt. Only July 5, 2000, the second such launch—the one carrying a Geyzer military relay satellite—succeeded, and after analyzing Proton's performance, Russian officials decided to proceed with Zvezda.

Abbey flew to Russia a week before the July 12 target date. Tommy Holloway, the ISS program manager, was already in Moscow, where he had been briefed by the Russians on their readiness to proceed with the Zvezda launch. Dan Goldin had also arrived in Moscow by then, and after meeting with the Russians to discuss the matter, raised several concerns. Holloway felt that the Proton problems had been resolved, but Goldin was not yet convinced.

In spite of Goldin's reservations, on July 8, at Baikonur, Roskosmos rolled Proton and its Zvezda payload to the launch pad. On July 10, Abbey boarded a GCTC airplane and flew to Baikonur; Goldin arrived two days later, the morning of the scheduled launch. As Abbey stood with Russian officials at the viewing site, Mike Hawes, one of Goldin's headquarters staffers, handed him a piece of paper. He told Abbey that Goldin would not approve the Proton launch unless Abbey signed the paper, accepting full responsibility for success or failure. Abbey said he would sign it, but only if he received a copy. There was some delay, and a copy was produced. Abbey signed.

He knew that the decision to launch was a Russian one. Roskosmos had determined that it was safe to launch Proton, and was not asking for NASA approval. The countdown proceeded, and minutes later Proton lifted off, safely delivering Zvezda to orbit, where it would soon be linked with the Zarya and Unity modules to form the core of the International Space Station.

With Zvezda safely on-orbit, Goldin wrote a letter to Abbey apologizing for the anxiety he had shown.

The Zvezda docking with the Zarya-Unity combination allowed the ISS partners to go ahead with the rest of phase I assembly.

First up, the October 11 launch of STS-92 with a crew of seven. They delivered the Z1 truss element and a pressurized mating adapter, installing them (with other key equipment) during four EVAs.

This set the stage for the October 31 launch aboard Soyuz of the first ISS crew, Bill Shepherd, Sergei Krikalev, and Yuri Gidzenko. By this time, Goldin was trying to meet with every crew prior to launch, or at least speak with them by phone. With Shepherd, Krikalev, and Gidzenko at Baikonur, it was necessary for the administrator to contact them via phone from NASA Ames in California, where he was visiting. Abbey would not be able to participate. After speaking with the crew, Goldin called Abbey with new concerns, apparently because of something said or implied. Abbey was forced to follow up with Goldin's chief of staff, Ed Heffernan, who had been part of the call between the administrator and crew. Heffernen thought the conversation had gone well, so Abbey asked him to reassure Goldin, which he did.

Shepherd, Krikalev, and Gidzenko launched aboard Soyuz TM-31 as planned and reached the International Space Station two days later, commencing the first ISS expedition.

57 | End Game

AS HEAD OF JSC, Abbey demanded performance from his contractors. Most of them, including Boeing, Lockheed, and Rockwell, received award fees for their performance. Abbey didn't hesitate to reward contractors that performed, but he also invoked zero fees if performance was lacking. At one time or another, zero fees were given to the same three companies, and others, where such penalties weren't well received by executives.

"You screwed up," Abbey told one executive from an underperforming contractor. "Stop screwing up and you'll get your award."

Dick Cheney, the former Secretary of Defense, now chairman of Halliburton, responded to a zero fee for Halliburton subsidiary Brown and Root by visiting JSC to personally assure Abbey that he understood the company's failure and would make sure performance improved.

In 1994, Dan Goldin created a Reusable Launch Vehicle Office, beginning the X-33 and X-34 programs. X-34 was to be a low-cost, unmanned test bed. X-33, however, was a substantially more ambitious program aimed at building an actual flying prototype for a single-stage-to-orbit vehicle. Goldin announced the goal: "to build a vehicle that takes days, not months, to turn around; dozens, not thousands, of people to operate; with launch costs that are a tenth of what they are now. Our goal is a reusable launch vehicle that will cut the cost of getting a pound of payload to orbit from $10,000 to $1,000."

NASA would contribute 80 percent of the costs; the balance would come from industry, a new type of commercial space partnership.

To Abbey, Goldin confessed that he started the X-33 because "I want my own launcher." Since Congress would not let him build an expendable vehicle, he chose the reusable.

After a bidding contest, Lockheed Martin was selected as the prime contractor in July 1996. Its X-33, it was hoped, would evolve into a manned reusable single-stage vehicle called VentureStar in five years.

Goldin wanted Abbey and JSC to manage the X-33, but Abbey was reluctant to take on the task. The development, testing, and integration of the International Space Station, and the safe operation of the space shuttle, were so demanding that Abbey didn't feel that JSC could also take on this additional challenge.

The X-33 concept of a single-stage-to-orbit (SSTO) vehicle was a long-standing goal of spacecraft developers. But building a fully reusable SSTO presented great technical challenges, such as developing structural material that would support the SSTO's weight; any material would have to be stronger than that used in traditional expendable "staged" vehicles like Titan and Saturn.

Then there were the propulsion options. One would be a combined cycle system that would "breathe" air as it climbed through the atmosphere. At high altitude, the air-breather would close its intakes and become a rocket, relying on an internal oxidizer. This approach required less physical volume for propellants, allowed for simpler tankage, and resulted in a lower takeoff weight. Its disadvantage was that the combined cycle propulsion system would be ambitious new technology.

The other approach envisioned an SSTO that used traditional rocket power, was incredibly lightweight structurally, and carried all its propellant. The propulsion system would be simpler, but the vehicle would require the use of advanced materials, since the basic structure of the vehicle could account for no more than ten percent of its total mass.

Abbey suggested that Goldin give the program to the Dryden Flight Research Center and the Marshall Spaceflight Center. Between Dryden's experience with the X-15 and Marshall's long history of propulsion and materials research, those centers had the expertise needed to support the X-33.

Abbey also felt that the X-33's technology should be proven before the program attempted to actually build a vehicle. Goldin appreciated Abbey's feelings and didn't disagree, but asked the JSC director to keep his feelings about the X-33 to himself. The strain in their relationship was starting to show.

With the crew of Shepherd, Krikalev, and Gidzenko living and working aboard ISS, the STS-97 team arrived on December 3, 2000, delivering solar arrays.

STS-98 followed in February. During the twelve-day mission, a crew of five astronauts had delivered, then installed the vital Destiny laboratory.

Phase I was now complete.

At the same time, the new George W. Bush administration was taking over in Washington after a bruising and controversial election.

As always, eager to survive in the new environment, Goldin sought to find financial victories. He was aided by a new chief of staff, Courtney Stadd, who had worked with Abbey on the Space Council.

Stadd recalled, "George and I had to attend the annual Goddard Space Dinner. This was a formal event, tuxedos and such, a couple of thousand government and space industry people in attendance. I was complaining about it one day: 'We have to dress up and smile at the same people we see all the time. We should have an anti-Goddard Dinner!'

"The next day, in my inbox, I found a biography of a woman named Mary Goddard, who had been the first woman postmaster in the 1800s, quite an accomplished woman. It came from George, and led to our Mary Goddard Dinner, where jeans were the required dress and the dining was casual.

"Abbey had heard me, said nothing, but taken action. It was vintage George."

Stadd valued Abbey's experience and knowledge, but by 2000 also felt that the JSC director had also made too many enemies and was too powerful. And he knew that Goldin was no longer a "Friend of George."

Mark Albrecht had worked with both Abbey and Goldin for nine years after his departure from the Space Council. "George changed. From being the guy who would never ever openly disagree with his boss, for the first time in his career, he went head to head with Dan.

"This is a relationship that ultimately turned out exactly like Henry II and Thomas À Becket. Henry II wants to appoint Becket archbishop of Canterbury.

'Please, sire, don't!' Becket knows what this will mean: to do that job properly he will be forced to be in conflict with the King. Sure enough, he takes the job, does what he must, and Henry II is saying, 'Who will rid me of this archbishop?'

"That was what happened with Dan and George."

On Abbey's next visit to headquarters in early 2001, Goldin asked him to leave JSC and work again in Washington. Abbey replied that ISS was still at a critical stage and suggested that they return to the matter after the completion of the next assembly mission, STS-98, which was to deliver the US laboratory module to the station in early February 2001.

Abbey realized that Goldin's desire to transfer him out of JSC was also motivated by a desire to show that he was taking action to avert projected cost increases to the ISS program.

The International Space Station was limited to a fixed amount each year, but projections showed that station development would be extended with an increase in the program's total cost. JSC under Abbey had never exceeded its annual budget, but to mollify Congress, Goldin felt that he needed to address the projected overrun, and he had a list of proposed cuts.

Ever since becoming center director, Abbey had held open houses at JSC, and the 2000 event had brought in almost two hundred thousand people to view the work being done there. He had also instituted an annual "NASA Inspection," similar to events held at the old NACA centers, where industry personnel would be invited to learn about new research being performed at JSC. All of this had been done without spending additional money; JSC employees welcomed the chance to show the public and industry reps what they were doing.

Work on the X-38 and TransHab programs had also been conducted in-house by JSC civil servants, with no increase in the center's budget. And Abbey had initiated a program allowing student "scientists" to fly experiments aboard JSC's zero-G KC-135 aircraft. He had also supported astronaut Franklin Chang-Diaz's VASIMR research program for advanced plasma propulsion.

Goldin and Joe Rothenberg, the head of spaceflight at headquarters, wanted Abbey to cancel all these activities. Goldin even wanted to do away with the Longhorn Project being done in partnership with the local school district. Abbey tried to explain that such cancellations would only save a small amount of money, but without success.

Abbey had Bill Parsons, a former Cape engineer who was now his deputy center director, prepare a letter outlining the cancellations.

Abbey signed the letter but didn't release it. He would try one more time to talk to Joe Rothenberg.

After STS-98, having successfully delivered the Destiny lab module to ISS, landed on February 20, 2001, Goldin telephoned Abbey again, repeating his hope that Abbey would return to headquarters to work directly for him, claiming, "It will be like old times."

Given his current relationship with Goldin, Abbey knew he could no longer be effective at JSC or provide leadership to the ISS and shuttle programs. It was time to move on. STS-102, the next ISS assembly mission, was scheduled for early March 2001. Commanded by Jim Wetherbee, the mission would deliver the second ISS expedition crew as well as the first multipurpose logistics module. Abbey told Goldin that the best time to reassign him would be after STS-102. The mission's flight readiness review was scheduled to be held soon, and Abbey wanted continuity for it to ensure success.

Goldin did not agree, so Abbey could only wait for his reassignment. It came on Friday, February 23.

Late that afternoon, NASA headquarters issued an announcement that, effective immediately, George W. S. Abbey would be "special assistant for international programs," reporting directly to the administrator.

That evening, free of his responsibilities at the Johnson Space Center, no longer having to worry about the press of a full calendar of work in the upcoming week, Abbey joined his friend Jay Honeycutt and his wife, Peggy, for dinner at a local restaurant. Abbey's "space brother" Yuri Glazkov had once told him it was harder on those who sent astronauts and cosmonauts into space than it was on the crews that went there. Abbey would have to agree, but it was no longer his challenge.

On Monday morning Abbey went in early to clean out his office. His secretary, Mary Lopez, and El Onizuka's widow, Lorna, were there to help him.

58 | The Journey Continues

ABBEY HAD MADE IT clear to Goldin that he did not want to return to Washington but would be willing to work on international programs if his duty station was Houston. Further, he did not want to be based on the JSC site, so through former astronaut Brewster Shaw, now with Boeing, he found office space in the Boeing building near the center.

Goldin gave Abbey the job of recommending possible international partnerships for ISS as well as other NASA programs. Abbey had always felt that Australia had great potential as a space partner, especially since its location near the equator made it an excellent launch site as well as a recovery site for the shuttle and other spacecraft, like the X-38. Australia had long been home to NASA command, telemetry, and communications sites at Carnavon, Woomera, Parkes, and other locations. Abbey made arrangements to visit the country, especially the communications site at Tidbinbilla, and met with Australian officials in Canberra, where he found them enthusiastic about new partnerships with NASA.

Returning to the United States with this good news, he was dismayed by the lack of response from NASA headquarters. He realized that he needed to find another way to make a contribution.

One promising area was working with the academic community. The World Space Congress was scheduled to be held in Houston in October 2002, and Abbey felt that Rice University's James B. Baker Institute of Public Policy could play a role in the event. Abbey also believed he could help strengthen relationships between NASA and minority universities in Texas, notably the

University of Texas, El Paso (UTEP). He convinced NASA to assign him to support both Rice and UTEP as his final agency task.

Soon after the conclusion of the World Space Congress, with fifty-two-and-a-half years of government service, George W. S. Abbey officially retired from NASA on January 3, 2003. The ceremony was held at Space Center Houston (adjacent to JSC), attended by hundreds of friends and colleagues and capped off by tributes and presentations. As those ended, and accompanied by a Scottish piper playing one of his favorite tunes, Abbey left the stage.

A month later, on February 1, *Columbia* was lost on reentry at the end of STS-107, its crew of seven killed as the orbiter broke up at an altitude of two hundred thousand feet. The leading edge of the shuttle's left wing had been struck on launch by a piece of insulating foam shed by the external tank. NASA managers had discussed the event but judged it to be minor—certainly no threat to the orbiter. They were wrong.

Soon after the accident, Jim Wetherbee called Abbey to ask if anyone at JSC had contacted him for advice about recovery of wreckage or the investigation. They had not. Wetherbee would spend the next three months in east Texas as the senior JSC flight crew operations representative on the *Columbia* search and rescue team. He invited Abbey there to observe.

Sean O'Keefe replaced Goldin as administrator in late 2001, and two years later he was succeeded by Mike Griffin. Abbey had his own preference for NASA administrator—Boeing's Jim Albaugh. He had also wanted Charles Precourt to run JSC, and astronaut Don Pettit to become the center's chief engineer. His wishes were ignored.

Abbey accepted a position as a senior fellow at the James B. Baker Institute. In collaboration with Neal Lane, another senior fellow there, Abbey would publish a number of papers on space policy, science, and technology.

In 2007, Abbey and his colleague Dr. Bobby Alford of the Baylor College of Medicine, and chairman of the National Space Biomedical Institute, put together the International Space Medicine Summit to be held on the Rice campus. The summit brought and continues to bring together leading physicians, space biomedical scientists, engineers, astronauts, cosmonauts, and educators from the world's spacefaring nations to not only identify goals in space medical research but also further international cooperation and collaboration.

Abbey worked in support of the congressional campaign of Gabrielle Giffords, married to astronaut Mark Kelly, in Arizona—and was shocked by the January 8, 2011, attempted assassination on Giffords that killed six people and left Giffords herself with brain injuries that forced her retirement from Congress.

Abbey could also note that individuals he had mentored during his NASA career had assumed positions of great responsibility in the world's human spaceflight programs. Charles Bolden became NASA administrator in 2009, and Sergei Krikalev became director of the Gagarin Cosmonaut Training Center in 2011. (Krikalev would later become head of human spaceflight for Roskosmos.) Former astronauts Michael Coats and Ellen Ochoa would become directors of the Johnson Space Center, and Robert Cabana would lead KSC. There were many others in key positions.

By 2017 ISS was in its seventeenth year of continuous human occupation.

In January 2004, President George W. Bush proposed that NASA refocus its programs and resources with the goal of returning to the Moon. This plan called for the completion of ISS assembly by 2010, and the retirement of the space shuttle from service.

Abbey and Lane took exception to this idea in a 2005 policy paper, writing: "The space station's full potential will be realized when it is completely assembled and when all of the modules, including those of our international partners, are in orbit. To accomplish meaningful science, the station requires both up-mass (delivering payloads from Earth to orbit) and down-mass (returning payloads from orbit to Earth) capability. If the shuttle is retired in 2010, that down-mass capability will clearly be unavailable." They called for the continuation of the shuttle program in support of ISS operations, and repeated that call in subsequent papers.

As plans for shuttle retirement proceeded, the most significant scientific experiment to be flown to ISS, Alpha Magnetic Spectrometer (AMS), was removed from the manifest. The AMS was designed to use the unique environment of space to advance knowledge of the universe and its origins by searching for antimatter and dark matter, and to measure cosmic rays.

Abbey and Lane supported the AMS principal investigator, Nobel Laureate Samuel Ting, a physicist at MIT, in discussions with Rep. Nick Lampson of Texas to restore AMS to the shuttle schedule. Lampson was subsequently able to garner overwhelming congressional support in requiring NASA to fly an additional shuttle mission to deliver AMS to the station. Launched aboard STS-134 in May 2011, AMS proved its value, collecting fifty-seven billion cosmic ray events in its first forty months of operation.

Abbey was less successful in his quest for continued shuttle support to ISS. STS-135, the flight that followed the AMS delivery, would be a final resupply mission, delivering an Italian-built multipurpose logistics module loaded with supplies and spare parts. Because of new flight safety rules, NASA would only assign four astronauts to the final flight—veteran commander Chris Ferguson, pilot Doug Hurley, and mission specialists Sandy Magnus and Rex Walheim. This flight was launched from the Cape on July 8, 2011. Abbey followed its launch, docking, and on-orbit activities via the NASA Select television channel. As the mission drew to a close after almost thirteen days, Abbey chose to witness its final hours from mission control on July 21, 2011—forty-two years and a day after he had watched Armstrong and Aldrin land on the Moon.

Once the reentry burn was complete, he left mission control and drove home. The successful landing would be celebrated at JSC, but Abbey didn't feel that the last flight was an occasion for celebration.

His daughter Suzanne was visiting from her home in Dallas that night, and when Abbey entered his living room he found her still awake, watching *Atlantis*'s reentry on television.

Father and daughter sat together as Ferguson and Hurley guided the orbiter into its final flare and touchdown in the dark at KSC. *Atlantis*'s wheels stopped at 4:57:54 Houston time as Chris Ferguson radioed, "Mission complete."

The most capable spacecraft the world had ever launched had made its last flight.

Houston days and nights offer extremes of weather, from tropical heat and humidity to bitter ice storms. No matter: on many of those days and nights a late model Chevy SUV can be seen on busy I-45, or on NASA Road One. Its destinations will be as varied as the weather: north to Rice University, west to buildings housing tech start-ups, south to Odyssey Academy in Galveston, east to restaurants or to the NASA Johnson Space Center, and frequently to House of Prayer Lutheran Church.

The driver is a silver-haired man in his mideighties still hard at work promoting his vision of international cooperation in spaceflight, and in education in science, math, and engineering.

Both are on display at Rice University's Baker Institute, where George Abbey has been a fellow since 2001. In addition to policy papers and lectures, he shepherds programs such as a student exchange between Rice and the Russian Bauman Moscow State Technical University in Moscow. That effort began in 2011 and found Abbey working with cosmonaut Vladimir Solovyev, the senior flight director in the Russian mission control center, and Victoria Mayrova, director of Bauman's Youth Space Center. Each July ten students from Rice and other American universities are sent to Moscow to participate in a two-week space course at Bauman, and each spring ten students visit Rice in return, to participate in a similar program on space and science and engineering.

In 2008, after ten years on its board, Abbey became chairman and president of the Mosbacher Odyssey Academy, a charter school in Galveston. One of his first challenges was dealing with the destruction of the school by Hurricane Ike on September 13 of that year. After finding Odyssey a temporary home at Moody Methodist Church, Abbey and the board were able to issue bonds and raise enough money to purchase and modify a different building, opening the school in a new location for the 2009–2010 academic year.

In 2013, the board expanded Odyssey to include high school grades, and in 2015 added three new schools in Galveston Bay serving pre-K through twelfth, with an enrollment of over eleven hundred students. Odyssey's curriculum emphasizes science and math, and encourages college education for all its graduates, many of them the first in their families to attend college.

An international space school program, started by Abbey in 1998 with two students from a single nation, flourishes too. Conducted at the University of Houston, Clear Lake campus, it now hosts fifty students from twenty-five countries, including those from Tagore High School in Karnal, India, attended by *Columbia* astronaut Kalpana Chawla.

The NASA-Scotland exchange Abbey developed with former astronaut Bonnie Dunbar thrives under the auspices of the University of Strathclyde in Glasgow. One hundred fifty students selected from high schools all over Scotland attend engineering and science seminars by NASA JSC specialists and astronauts at Strathclyde each summer. Ten of the most outstanding students then attend a two-week program in Houston each October.

In a departure from his days as a High-Ranking NASA Official, when he remained out of the public eye, these days Abbey is a frequent speaker at space conferences in locations like Shanghai and Zhuzhou in China. He has lectured at universities in Scotland and Wales, and has been known to combine his European trips with an annual visit to Brittany. Other destinations include Moscow and Baikonur, most recently at the invitation of cosmonaut Yury Malenchenko, to view the launch of Soyuz TMA-19M and its international crew in December 2015.

Abbey is still engaged with the space community, and was invited to tour Jeff Bezos's Blue Origin facility in Seattle, for example. And since 2005 he has served as a board member for Franklin Chang-Diaz's Ad Astra Rocket Company, which is developing an advanced plasma rocket propulsion system.

Travels, presentations, and meetings have convinced Abbey that the world's spacefaring nations, including China, would support a new program returning humans to the Moon, and that such an international effort would provide an excellent opportunity for the United States to continue the cooperation demonstrated by the International Space Station program.

Even get-togethers with old friends and colleagues such as Abbey's mentor, Chris Kraft, and Kraft's wife, Betty Anne, turn on the same subject: the next logical step for human endeavors in space is a return to the Moon, not yet a mission to Mars.

There is time for family, of course. Abbey and his children, most of them living in or near Houston, get together each Sunday evening for dinner. There are grandchildren to spoil, and occasionally guide.

There are moments of sadness. Abbey remembers his lost family—Sam and Brenta, his brothers and sister—and the growing list of departed friends.

In late 2017 and early 2018 alone he lost Constable Bill Bailey, Mary Lopez, and John Young.

He can often be found as a solitary visitor to the grove he had planted at the Johnson Space Center beginning in 1996, with a tree and plaque for each deceased astronaut. The plaque for *Challenger* pilot Mike Smith contains a line from the movie *Bridges at Toko-Ri*, which inspired Smith to become a naval aviator. It reads: "Where do we get such men?" Those words could be inscribed on each plaque in the grove.

His son Andy says, "Dad's priorities were God, family, space, in that order." Without making an issue of his religious beliefs, Abbey honors them, with keen awareness of the opportunities he has had in his life, growing up with a loving father and mother, and with a caring sister and brothers. He values his attendance at the naval academy with a group of outstanding classmates, and his service in the air force, where he made new friends and embraced new challenges.

Then there was NASA, where he was privileged to take part in the space program, working with genuine geniuses and pioneers, making lifelong friends from Russia and other nations. When asked, he credits his longevity and success to them.

In September 1963, two months before he died, President John F. Kennedy addressed the United Nations. There he spoke of America's efforts in space, noting that "members of the United Nations have foresworn any claim to territorial rights in outer space or on celestial bodies" and wondering why the United States and Soviet Union should treat a flight to the Moon as "a matter of national competition." He said, "Surely, we should explore whether the scientists and astronauts of our two countries—indeed, of all the world—cannot work together in the conquest of space, sending some day in this decade to the Moon not the representatives of a single nation, but the representatives of all our countries."

Hearing those words as a young air force officer in 1963, Abbey saw little chance for international cooperation. George Low's initial efforts to reach out to the Soviet Union in 1970 changed that; the Apollo-Soyuz mission five years later fulfilled Kennedy's vision.

For a generation, George Abbey worked to make human spaceflight available to all, regardless of citizenship, gender, color, or ethnic background, to embody the motto: *Simul Per Ardua ad Astra* . . . Together to the Stars.

Acknowledgments

I AM GRATEFUL TO George William Samuel Abbey for generosity and patience as he endured dozens of phone calls and many more emails, in addition to a considerable number of in-person interviews. It still astonishes me that, given his insistence that his role within NASA was merely "member of a team" and his initial suggestion that I would be better off writing about George M. Low, he didn't have me ejected from the Villa Capri that first dinner in January 2011.

With one minor exception, he answered every question I asked—and that sole exception sent me to a direct source.

The Abbey family—George Jr., Joyce BK, Suzanne Abbey Fair, James, and Andrew shared memories and photos. A special thanks is due to Suzanne, the family archivist.

This book would not have been possible without the assistance of Thomas P. Stafford, who somehow convinced the legendary NASA official to speak at all.

The team at the Museum of Flight in Seattle—John Little, Jessica Jones—guided me through the George W. S. Abbey papers in summer 2016.

Thank you to those who took time to be interviewed.

Dwayne Day's invitation to the 2011 National Research Council panel on human spaceflight allowed me to not only make my pitch to Mr. Abbey, but to meet several of his associates, some of whom became sources for the book.

Tony Reichhardt of *Air & Space* magazine accepted my proposal for an article about Mr. Abbey, a vital first step toward this book.

Andrew Chaikin provided encouragement, guidance, and perspective on the history of NASA and the Apollo program. Lynn Sherr was generous with support and information about the Thirty-Five New Guys.

Francis French, coauthor of *Into That Silent Sea* and other books on human spaceflight, was the perfect beta reader. David Fowler also helped.

Ed Hengeveld provided valuable aid in obtaining high-resolution imagery.

Special thanks are due my agent, Stephen Barr, at Writers House, who devoted hours to multiple versions of the original proposal and found *The Astronaut Maker* a happy home, and to Lisa Reardon and Jerome Pohlen, my editors at said happy home, Chicago Review Press.

Finally, thanks to Cynthia Cassutt, Ryan Cassutt, and Alex Cassutt, for love and patience.

Notes

THIS BOOK IS BASED on a series of in-person, telephone, and email interviews and exchanges with George W. S. Abbey conducted between January 2011 and November 2017.

Richard Truly once said, "The real book about the space program would be a book about George Abbey." In order to write that book, he continued, "You would have to interview hundreds of people." That goal was beyond me, and probably any individual writer. Nevertheless, this half a hundred took part in interviews or Abbey-related conversations between 2011 and 2017: Joyce BK Abbey, Suzanne Abbey Fair, George W. S. Abbey Jr., James Abbey, Andrew Abbey, James Adamson, Mark Albrecht, Constable Bill Bailey, James Banke, Patrick Baudry, Douglas Beason, Jeff Bingham, Cheryl Boullion, Michael Collins, Robert Crippen, Thomas Diegelman, Robert Gibson, George Jeffs, Steven Hawley, Jay Honeycutt, Millie Hughes-Fulford, Christopher Kraft, Alan Ladwig, Frank Martin, Thomas K. Mattingly II, John Muratore, Story Musgrave, George Nelson, William Parsons, Joseph Rothenberg, Jerry Ross, Rick Searfoss, Paul Sefchek, Brewster Shaw, Lee Silver, Courtney Stadd, Thomas P. Stafford, Richard Truly, Brett Watterson, James Wetherbee, and Robert Yowell.

Some information derives from interviews conducted for earlier projects, including the autobiographies of Deke Slayton and Tom Stafford.

I tried four different times to interview Daniel Goldin, but he never responded to queries. I refrained from a fifth attempt. Several others, many of them fairly described as not "Friends of George," also declined to speak.

I made free use of interviews archived in the NASA Johnson Space Center Oral History Project—referred to as JSC Oral History throughout the notes—which was started by Abbey himself in 1998. In some cases I supplemented the oral histories by conducting follow-up interviews.

Unless otherwise noted, all quotes are George Abbey's.

Accounts of spaceflights or spacecraft development are drawn from a wide variety of published accounts—see bibliography—all supported or in some cases supplanted by Abbey's memories.

MC
Los Angeles 2017

Epigraph

"The real book about the space program": Richard Truly, quoted in Bryan Burrough, *Dragonfly: NASA and The Crisis Aboard Mir* (New York: HarperCollins, 1998), 18.

Prologue

"fiefdom": Joseph Rothenberg, interview, September 23, 2015.
"like children": Robert Gibson, interview, February 20, 2009.
"I've never known": Thomas K. Mattingly, interview, February 9, 2011.
"the entire equation of motion pictures": F. Scott Fitzgerald, *The Last Tycoon* (New York: Scribner, 1941), 4.
"a godfather type": Norman Thagard, quoted in Burrough, *Dragonfly*, 23.
"dictator": Andrew Gaffney, quoted in Burrough, *Dragonfly*, 23.
"rapacious power monger": Mike Mullane, *Riding Rockets: The Outrageous Tales of a Space Shuttle Astronaut* (New York: Scribner, 2006), 93.
"George Abbey saved": Christopher Kraft, interview, February 11, 2011.
"George Abbey was the single human being": James Wetherbee, interview, February 15, 2011.
"the best program manager": Thomas Stafford, interview, July 28, 2006.
"real advocate": Michael Mott, interview, April 12, 2000.

Chapter 2: The Abbeys

"If there was a war": Suzanne Abbey Fair, interview, April 26, 2016.

Chapter 3: Annapolis

"Take a young boy": Robert A. Heinlein, *Grumbles from the Grave* (New York: Del Rey, 1989), 33.
"Lists among his likes": Text from *Lucky Bag*, 1954 Naval Academy yearbook, courtesy of George Abbey Jr.

Chapter 4: Air Force

"West Point of the Air": Jody Cook and Dr. John H. Sprinkle Jr., "Significance of Property, Overview," National Historic Landmark Nomination, Randolph Field Historic District, US Department of the Interior, National Park Service, February 26, 2001, 43, www.nps.gov/nhl/find/statelists/tx/RandolphHD.pdf.

"the best pilot": Bob Albert, quoted in Suzanne Abbey Fair, interview, July 10, 2016.

Chapter 7: Destination: Moon

"the first Canadian": Owen Maynard, JSC Oral History, April 21, 1999.

"before this decade is out": John F. Kennedy, "Special Message to Congress on Urgent National Needs," May 25, 1961, www.nasa.gov/vision/space/features/jfk_speech_text.html.

Chapter 8: North American Aviation

"who pretty much left": Alan B. Kehlet, JSC Oral History, September 30, 2015.

"There was a meeting": Wetherbee, interview.

Chapter 9: Houston and the Manned Spacecraft Center

"He was a genius": Maynard, JSC Oral History.

"a team of undomesticated": J. R. Dempsey, W. A. Davis, A. S. Crossfield, and Walter C. Williams, "Program Management in Design and Development," in *Aerospace Reliability and Maintainability Conference, Society of Automotive Engineers, Third Annual* (Washington, DC: Society of Automotive Engineers, 1964).

Chapter 10: Astronaut Abbey

"There's a hell of a thunderstorm": Richard F. Gordon, interview, May 6, 2011.

Chapter 11: The Phillips Report

"We believe that S&ID": "The Phillips Report," NASA Historical Reference Collection, NASA History Office, https://history.nasa.gov/Apollo204/phillip1.html.

"I am definitely not satisfied": "Phillips Report."

"I submit that the record of these": "Phillips Report."

"Too bad, you didn't make it": William Anders, quoted in Andrew Chaikin, *A Man on the Moon: The Voyages of the Apollo Astronauts* (New York: Viking, 1994), 40–41.

"We found a number": George Jeffs, JSC Oral History, May 21, 1998.

"He was an unmanned": Maynard, JSC Oral History.

"a piece of cake": John Disher, quoted in Charles Murray and Catherine Bly Cox, *Apollo: The Race to the Moon* (New York: Simon & Schuster, 1999), 159.

Chapter 12: Flight

made the program sound too "frivolous": NASA HQ to Center Public Affairs Officers, Memo, May 12, 1966, *The Apollo Spacecraft: A Chronology*, vol. 4, April–June 1966, by Ivan D. Ertel and Roland W. Newkirk with Courtney G. Brooks (Washington, DC: NASA, 1978).

"simplicity itself": Michael Collins, *Carrying the Fire: An Astronaut's Journey* (New York: Bantam, 1983), 257.

"It isn't that we don't": Apollo 1 crew photo caption, quoted in Murray and Bly Cox, *Apollo*, 184.

Chapter 13: Countdown

"If all three of us get back": Gus Grissom, quoted in George Leopold, *Calculated Risk: The Supersonic Life and Times of Gus Grissom* (West Lafayette, IN: Purdue University Press, 2016), 238.

"How are we going to get to the Moon": Apollo 1 recording, January 27, 1967, https://commons.wikimedia.org/wiki/File%3AApollo_One_Recording.ogg.

"Hey! . . . We've got a fire": "Apollo 204 Review Board: Final Report," NASA Historical Reference Collection, NASA History Division, last updated February 3, 2014, www.hq.nasa.gov/office/pao/History/Apollo204/content.html.

Chapter 15: Commitment

"The response we have given": Frank Borman, quoted in Courtney G. Brooks, James M. Grimwood, Lloyd S. Swenson, *Chariots for Apollo: A History of Manned Lunar Spacecraft* (Washington, DC: NASA, 1979), 224.

"My father adored": Joyce BK Abbey, interview, March 17, 2016.

Chapter 16: From the Earth to the Moon

"We built mockups": George M. Low, "The Spaceships," in *Apollo Expeditions to the Moon*, edited by Edgar Cortright (Washington, DC: NASA, 1975), 59–79.

"To err is human": Glynn Lunney quoting Christopher Kraft, quoted in Craig Nelson, *Rocket Men: The Epic Story of the First Men on the Moon* (New York: Penguin, 2010), 69.

"sat in all the meetings": Michael Collins, interview, May 6, 2011.

"He had a memory like three elephants": Thomas Stafford, interview, March 20, 2012.

"They had org charts": Thomas K. Mattingly, JSC Oral History, April 22, 2002.
"He was this nondescript guy": Mattingly, JSC Oral History.
"two Georges": Mattingly, JSC Oral History.

Chapter 17: Challenge

"As a result of the command problems": Gene Kranz, *Failure Is Not an Option* (New York: Simon & Schuster, 2000), 220.
"It went into orbit": Murray and Bly Cox, *Apollo*, 312

Chapter 18: One Week in August

"This makes C-prime": Low, quoted in Murray and Bly Cox, *Apollo*, 323.
"It's your neck": Apollo 7 Air-to-Ground Voice Transcriptions, www.jsc.nasa.gov/history/mission_trans/AS07_TEC.PDF.
"a detailed analysis and review": NASA News Release MSC 68-78, October 28, 1968.
"the most strenuous": Joseph Loftus, JSC Oral History, October 27, 2000.
"essentially gray": Cortright, ed., *Apollo Expeditions*, 200.
"In the beginning": Apollo 8 reading by astronauts, quoted in Chaikin, *Man on the Moon*, 121.

Chapter 19: Mission Accomplished

"He's the best engineer": James McDivitt, JSC Oral History, June 26, 1999.

Chapter 20: Apollo 13

"We never got that beer": Thomas K. Mattingly, JSC Oral History, November 6, 2001.

Chapter 21: Stability

"Everyone was either": Abbey Fair, interview, April 26, 2016.
"We rescued this poor duck": Abbey Fair, interview.
"Your own children don't recognize you!": James Abbey, interview, September 6, 2016.
"the quintessential staffer": George Jeffs, interview, April 15, 2011.

Chapter 22: Gravity

"Chris was brilliant": Richard Truly, interview, November 28, 2015.
"feel free": Donald K. Slayton, quoted in Al Worden with Francis French, *Falling to Earth: An Apollo 15 Astronaut's Journey to the Moon* (Washington, DC: Smithsonian Institute Press, 2011), 247.

Chapter 23: The Space Shuttle

"Solid-fuel rockets are not suitable": Louis Goodman and Rufino Ignacio, *Engineering Project Management: The IPQMS Method and Case Histories* (Boca Raton FL: CRC Press, 1999), 159.

"The Grumman proposal": Albert H. Crews Jr., JSC Oral History, August 6, 2007.

Chapter 24: The Soviets

"I was kind of a tomboy": Abbey Fair, interview, April 26, 2016.

"at first denied having signed the stamps": Walter Cunningham, *The All-American Boys* (New York: Macmillan, 1977), 238.

Chapter 25: Skylab

"better at matching humans with technical challenges": Lee Silver, interview, April 1, 2015.

"33-day fire drill": Edward Gibson, quoted in Ben Evans, "All the King's Horses: The Final Mission to Skylab (Part Three)," *AmericaSpace* (blog), November 30, 2013, www.americaspace.com/2013/11/30/all-the-kings-horses-the-final-mission-to-skylab-part-3.

"We need more time to rest": Gerald P. Carr, quoted in Evans, "All the King's Horses."

Chapter 26: The Ace Moving Company

"We generally start our days": Richard Truly, interview, December 28, 2015.

"Cheryl, what do I do?": Cheryl Bouillion, interview, December 29, 2015.

"We ran, we played": Robert Crippen, interview, May 4, 2016.

"I want to be": Kraft, interview.

Chapter 27: Apollo-Soyuz

"We had spent hours": Richard Truly, interview, February 24, 2016.

"Soyuz and Apollo are shaking hands": Edward Clinton Ezell and Linda Neuman Ezell, *The Partnership: A History of the Apollo-Soyuz Test Project* (Washington, DC: NASA, 1978), 328.

"the worst personnel decision": Kraft, interview.

Chapter 28: The Right Stuff

"I never thought of George": Mattingly, interview.

"I thought he preferred": Truly, interview, December 28, 2015.

"It was just a new assignment": George Abbey Jr., interview, April 26, 2017.
"one stop short of open warfare": Mattingly, interview.
"Right after we got the project": Charles Harlan, JSC Oral History, November 14, 2001.
"Why are you so mad at us?": Harlan, JSC Oral History.
"So you've got this cylinder": Robert Thompson, JSC Oral History, October 3, 2000.
"ten consecutive miracles": John Young quoted by Abbey.
"Y'all are going": John Young, quoted in Musgrave, interview.
"I *was the specialist"*: Henry W. Hartsfield Jr., JSC Oral History, June 12, 2001.

Chapter 29: The Next Generation

"willingness to accept hazards": Lloyd S. Swenson Jr., James M. Grimwood, and Charles C. Alexander, *This New Ocean: A History of Project Mercury* (Washington, DC: NASA, 1989), ch. 5, "Project Astronaut."
"old aviation person": Joseph D. Atkinson Jr. and Jay M. Shafritz, *The Real Stuff: A History of NASA's Astronaut Recruitment Program* (New York: Prager, 1985), 35.
"pretty much handpicked": Deke Slayton, interview, August 17, 1991.
"Why isn't there a Negro astronaut?": Thomas Stafford, interview, March 15, 2016.
"a good pilot but not a great": Charles Yeager, quoted in Colin Burgess, Kate Doolan, and Bert Vis, *Fallen Astronauts: Heroes Who Died Reaching for the Moon* (Lincoln: University of Nebraska Press, 2003), 15.
"We want to assure": Charles Berry, JSC Oral History, April 24, 1999.
"We're in a down time": Henry Clements, JSC Oral History, August 31, 1998.
"pretty much knew": Slayton, interview.
"along with about sixty percent of Edwards": Loren Shriver, JSC Oral History, December 16, 2012.
"Who are you?": Frederick D. Gregory, JSC Oral History, April 29, 2014.
"When I answered the phone": Duane Ross, quoted in Atkinson and Shafritz, *Real Stuff*, 157.
"Some people actually put résumés": Abbey Fair, interview, July 10, 2016.
"Aren't you supposed to ask": Lynn Sherr, *Sally Ride: America's First Woman in Space* (New York: Simon & Schuster, 2014), 88–89.
"Here comes the criminal!": Jeffrey Hoffman, JSC Oral History, September 24, 2009.
"Jeez, that didn't": Brewster Shaw, JSC Oral History, April 19, 2002.
"It's one appointment": Kathryn Sullivan, JSC Oral History, May 10, 2007.
"Halfway through the interview": George Nelson, JSC Oral History, May 6, 2004.
"His future was assured": Steven Hawley, interview, November 16, 2015.
"We were all tucked": Robert Piland, JSC Oral History, August 21, 1998.

Chapter 30: Approach and Landing

"What the hell was that?": Stafford, interview, March 15, 2016.

Chapter 31: Thirty-Five New Guys

"How's the weather": Jon McBride, in Henry S. F. Cooper Jr., *Before Lift-Off: The Making of a Space Shuttle Crew* (Baltimore, Johns Hopkins University Press, 1987), 25.
"Remember that job": Sherr, *Sally Ride*, 91.
"started asking about the weather": Daniel Brandenstein, JSC Oral History, January 19, 1999.
"Why don't you come down here": Jerry Ross, interview, September 30, 2015.
"How'd you like to fly": Crippen, interview.
"didn't want to be considered junior": Joseph Allen, JSC Oral History, March 16, 2004.
"I told you there were thirty-five": Allen, JSC Oral History, March 16, 2004.
"welcomed us with a few forgettable words": Mullane, *Riding Rockets*, 48.
"There were no responsibilities": John Fabian, JSC Oral History, February 10, 2006.
"questioning several of us men": Mullane, *Riding Rockets*, 51
the name evolved to "Thirty-Five New Guys": Sherr, *Sally Ride*, 98–99.
"I see that you made a C": Norman Thagard, JSC Oral History, April 23, 1998.
"I had gone to Catholic schools": Mullane, *Riding Rockets*, 36
"very much disillusioned": Shaw, JSC Oral History.
"Abbey said, 'You're going to be'": Alan Bean, JSC Oral History, February 23, 2010.
"George Abbey would just get that camel's nose": Robert Gibson, JSC Oral History, November 1, 2013.
"Joe Algranti was extremely talented": Robert Gibson, JSC Oral History, January 22, 2016.

Chapter 32: Problems

"I didn't know whether": Christopher Kraft, *Flight: My Life in Mission Control* (New York: Dutton, 2001), 43.
"He couldn't just go off to work": Abbey Fair, interviews, April 26 and July 10, 2016.

Chapter 33: Those Other Astronauts

"Our job was to integrate DOD": Brett Watterson, interview, April 15, 2016.
"We were scouts": Paul Sefchek, interview, November 4, 1987.
"came aboard much quicker": Mattingly, JSC Oral History, April 22, 2002.
"to settle on a realistic launch date": Richard Nygren, JSC Oral History, March 9, 2006.

"space station was right around the corner": Edward Gibson, JSC Oral History, December 1, 2000.

Chapter 34: STS-1

"patting each other": Crippen, interview.

Chapter 35: The Used Spaceship

"minimize using NASA assets": Bean, JSC Oral History.
"I look at the latest spacecraft": Cunningham, *All-American Boys*, 309.
"You'd sit around your office": Jay Greene, JSC Oral History, December 8, 2004.
"I thought it was shit-hot": John Young, quoted in James Bagian, email to the author, March 16, 2018.
"was absolutely correct": Glynn Lunney, JSC Oral History, January 13, 2000.
"boyfriend in tow": Joyce BK Abbey, interview.
"Which only made me nervous": Abbey Fair, interview, April 26, 2016.
"Let's just say it was not the best set": Andrew Abbey, interview, April 28, 2016.

Chapter 36: Operational Flight

"kind of a wheelie": Gordon Fullerton, JSC Oral History, May 6, 2002.
"Chris and a lot of other people": Neil Hutchinson, JSC Oral History, January 21, 2004.
"outstanding public presence": Henry Clements to Christopher Kraft, memo, April 14, 1982.
"Can I fly with you guys?": Guion Bluford, JSC Oral History, August 2, 2004.
"With Abbey's words, TFNG comraderie vaporized": Mullane, *Riding Rockets*, 100.
"he was that sharp": Robert Gibson, JSC Oral History, January 22, 2016.
"rinky-dink collection of junk": Mattingly, interview.
"What do we use?": Jeff DeTroye, interview, September 5, 1987.
"do Tab November": Hartsfield, JSC Oral History.
"it was slower than normal": Mattingly, JSC Oral History, April 22, 2002.
"It's not selfless": Josheph Allen, JSC Oral History, March 18, 2004.
"the impedance matching device": Allen, JSC Oral History, March 18, 2004.
"I'd like to be capcom": Robert Gibson, JSC Oral History, January 22, 2016.
"You can fly the MMU": Bruce McCandless, interview, February 20, 2009.
"Every time I'd go in": Gerry Griffin, JSC Oral History, March 12, 1999.
"Mission designations consist of": NASA Johnson Space Center press release 83-036, September 21, 1983.
"Gerry did not want": Clements, JSC Oral History.

Chapter 37: Frequent Flying

"an E-ticket": Lynn Sher, *ABC Evening News*, June 18, 1983.
"I probably ought to get": Anna Fisher, JSC Oral History, February 17, 2009.
"Bears for the bairn": Fisher, JSC Oral History.
"Who did you have": Charles D. Walker, JSC Oral History, November 19, 2004.
"We had what we called": John Tribe, JSC Oral History, July 13, 2011.

Chapter 38: 1984

"directing NASA to develop a permanently manned space station": President Ronald Reagan, State of the Union address, February 4, 1986.
"You walk in and there's John Young": Wetherbee, interview.
"We're going to select more astronauts": Robert Cabana, JSC Oral History, November 17, 2015.
"I thought we'd be higher": Steven Hawley, quoted in Wayne Hale, ed., *Wings in Orbit: Scientific and Engineering Legacies of the Space Shuttle, 1971–2010* (Washington DC, NASA, 2011), 160.
"more serious question": Jeff Bingham, JSC Oral History, November 9, 2006.
"Young is generally considered": Cooper, *Before Lift-Off*, 21.
"There were many times": Hawley, interview.
"Charlesworth never could": Crippen, interview.
"Do you want this crew": Watterson, interview.
"But that's why you test": Crippen, interview.
"An unnamed NASA spokesperson": Joseph Allen, JSC Oral History, November 18, 2004.
"If we can get both": Allen, JSC Oral History, November 18, 2004.

Chapter 39: The Breaking Point

"Mr. Abbey, in his infinite wisdom": Charles Bolden, JSC Oral History, January 6, 2004.
"hammer-type pressure": Gary Johnson, JSC Oral History, May 4, 2010.

Chapter 40: The Day the Earth Stood Still

"was a model payload specialist": Nelson, JSC Oral History.
"You know, you're sitting out there": Bolden, JSC Oral History.
"We were a little worried": Nelson, JSC Oral History.
"Bill had a really hard time": Nelson, JSC Oral History.
"Pete Conrad, Stu Roosa, and Charlie Duke were big country music fans": Bill Bailey, interview, April 15, 2016.

Chapter 41: Recovery

"He never came home": Joyce BK Abbey, interview.

"in utter turmoil": Truly, interview, February 24, 2016.

"34 safety items and seven studies": Steve Bale, quoted in William Harwood, "Young Memo Charts Shuttle Concern," UPI, March 9, 1986, www.upi.com/Archives/1986/03/09/Young-memo-charts-shuttle-concern/6099510728400.

"You guys will get re-nerded": Mark Lee, quoted in Mullane, *Riding Rockets*, 253.

"an accident rooted in history": Report to the president by the Presidential Commission on the Space Shuttle *Challenger* Accident (Rogers Report), June 6, 1986, p. 121.

"Please don't take": Frederick Hauck, JSC Oral History, November 20, 2003.

Chapter 42: Marooned

"I want you to be the new chief": Daniel Brandenstein, interview, August 21, 2008.

"Hooter, what do you think": Robert Gibson, interview, August 14, 2008.

"drink beer, play guitars": George Nelson, interview, August 14, 2008.

"Brewster played rhythm": Gibson, interview, August 14, 2008.

"We weren't good": Gibson, interview, August 14, 2008.

"We were doing this on our own time": Brewster Shaw, interview, September 23, 2008.

"But it's not my turn": Robert Gibson, JSC Oral History, January 22, 2016.

"I believed that my center directors": Truly, interview, February 24, 2016.

"It's time for you to move on": Crippen, interview.

"a new Orient Express": President Ronald Reagan, State of the Union address, February 4, 1986.

"It was like having your own": Nygren, JSC Oral History.

Chapter 43: PEPCON

"I talked with George": Hawley, interview.

"After George left": Brandenstein, interview.

"didn't have the sort of negatives": Dan Quayle, *Standing Firm: A Vice Presidential Memoir* (New York: HarperCollins: 1994), 181.

"I'm just getting to the point": William Lenoir, JSC Oral History, November 18, 2004.

Chapter 44: Space Exploration Initiative

"The Soviet Union was falling": Mark Albrecht, interview, October 9, 2015.

"Where is the money": Thomas P. Stafford with Michael Cassutt, *We Have Capture: Tom Stafford and the Space Race* (Washington, DC: Smithsonian Institution Press), 220.

"Give us a Cadillac": Albrecht, interview.

"like selecting five different ways": Dwayne Day, "Aiming for Mars, Grounded on Earth, Part Two," *Space Review*, February 23, 2004, www.thespacereview.com/article/106/1.

"NASA blew it": Albrecht, interview.

Chapter 45: Synthesis

"You're being assigned": Douglas Beason, interview, September 28, 2015.

"You're supposed to ask the dumb questions": Bingham, JSC Oral History.

"But you quickly realized": Beason, interview.

"Stafford paid for the beer": Beason, interview.

"Sometime during the Synthesis Group's work": Albrecht, interview.

"I believe that George Abbey knew": Beason, interview.

Chapter 46: Space Council

"While he had his problems": Albrecht, interview.

"here was a guy": Frank Martin, interview, September 25, 2015.

"I would be ranting": Albrecht, interview.

"Meet Dan Goldin": Albrecht, interview.

Chapter 47: Dan Goldin

"I know there's a mythology": Albrecht, interview.

"I think they wanted somebody": John R. Dailey, JSC Oral History, July 24, 2012.

"carried off on gurneys": Charles Bolden, JSC Oral History, January 15, 2004.

"thought he was wonderful": Lennard Fisk, JSC Oral History, September 9, 2010.

"contrived by test pilots": Kenneth Chang, "$79 for an Out-of-Date Book About a Modern NASA Logo," *New York Times*, September 1, 2015.

"I got a call": Thomas Stafford, interview, December 28, 2012.

"Within a few years": Albrecht, interview.

"it wasn't so much the idea": Brian Dailey, JSC Oral History, April 19, 1999.

"Let Hoot be chief": Richard Covey, interview, March 9, 2011.

Chapter 48: Space Cowboys

"We're facing a national quasi-emergency": Stafford, interview, December 28, 2012.

"We're going to have a meeting": Michael Mott, quote from George W. S. Abbey.

"That weekend saved human spaceflight": Stafford, interview, March 20, 2012.

Chapter 49: Redesign

"stand by": Bryan O'Connor, JSC Oral History, March 24, 2010.

"tell new stories": Douglas A. Levy, "NASA to Cut Space Station Cost in Half," UPI, March 10, 1993, www.upi.com/Archives/1993/03/10/NASA-to-cut-space-station-cost-in-half/5060731739600.

Chapter 51: A Space Odyssey

"You don't need to be an engineer": Carolyn Huntoon, JSC Oral History, June 5, 2002.

"in a delicate state": Stafford, interview, March 20, 2012.

"I had about 150": Michael Mott, JSC Oral History, April 23, 1999.

Chapter 52: Return to JSC

"Andy and Dad would be out": Abbey Fair, interview, July 10, 2016.

"There were two possible responses": Hawley, interview.

"we might have gotten": Hawley, interview.

"So I'd just bring the same crew": Cabana, JSC Oral History.

"I understand, well, there is no more": Jean-Loup Chrétien, JSC Oral History, May 8, 2002.

"If you'll give me a couple million": John Muratore, JSC Oral History, May 14, 2008; John Muratore, interview, February 4, 2011.

"It was going to be where you could bring": Chris D. Perner, JSC Oral History, July 26, 2001.

Chapter 53: A New Direction

"George had a management style": James Jaax, JSC Oral History, November 16, 2006.

"Have you lost your mind?": Jay Honeycutt, interview, October 15, 2015.

Chapter 54: The Total Equation

"There are seven days": Hawley, interview.

"The GASSERS and SDOMS gave us focus": Hawley, interview.

"Everything George worked on": B. Dailey, JSC Oral History.

"I thought he was going to yell": Muratore, JSC Oral History.

"Nobody could believe": James Jaax, JSC Oral History, October 17, 2006.

"George could hire all this engineering": Janice Voss, interview, June 2, 2001.

"when they started cutting": Duane Ross, JSC Oral History, April 19, 2013.

"When I first started working": Thomas Diegelman, interview, October 3, 2016.

"He was just this mysterious deputy director": Robert Yowell, interview, May 10, 2014.
"pin cushion": Rick Searfoss, interview, March 11, 2011.
"He made decisions independent of possible": Wetherbee, interview.
"In all the time I worked with George": Honeycutt, interview.
"George was usually described as secretive": Rothenberg, interview.

Chapter 57: Crises on Mir

"said he had agonized": Kathy Sawyer, "NASA Decides to Send Another Astronaut to Mir," *Washington Post*, September 26, 1997.

Chapter 56: First Element

"NASA began considering": Jeff Bingham, interview, December 7, 2015.

Chapter 57: End Game

"to build a vehicle": Daniel Goldin, public announcement of the X-33 award, July 2, 1996, https://spinoff.nasa.gov/spinoff1996/14.html.
"George and I had to attend": Courtney Stadd, interview, February 8, 2016.
"George changed": Albrecht, interview.

Chapter 58: The Journey Continues

"The space station's full potential": George W. S. Abbey and Neal Lane, *United States Policy: Challenges and Opportunities* (Cambridge, MA: American Academy of Arts and Sciences, 2005), https://scholarship.rice.edu/handle/1911/91895.
"Dad's priorities were God, family": Andrew Abbey, interview.
"members of the United Nations": John F. Kennedy, Address Before the 18th General Assembly of the United Nations, September 20, 1963, www.presidency.ucsb.edu/ws/?pid=9416.

Bibliography

Abbey, George W. S., and Neal Lane. *United States Policy: Challenges and Opportunities*. Cambridge, MA: American Academy of Arts and Sciences, 2005. https://scholarship.rice.edu/handle/1911/91895.

Atkinson, Joseph D., Jr., and Jay M. Shafritz. *The Real Stuff: A History of NASA's Astronaut Recruitment Program*. New York: Prager, 1985.

Borman, Frank, with Robert J. Serling. *Countdown: An Autobiography*. New York: Morrow, 1988.

Brooks, Courtney G., and Ivan D. Ertel. *The Apollo Spacecraft: A Chronology*. Vol. 3. Washington, DC: National Aeronautics and Space Administration, 1976.

Brooks, Courtney G., James M. Grimwood, and Loyd S. Swenson Jr. *Chariots for Apollo: A History of Manned Lunar Spacecraft*. Washington, DC: National Aeronautics and Space Administration, 1979.

Burgess, Colin. *Moon Bound: Choosing and Preparing NASA's Lunar Astronauts*. New York: Springer Praxis, 2013.

Burgess, Colin, with Kate Doolan and Bert Vis. *Fallen Astronauts: Heroes Who Died Reaching for the Moon*. Lincoln: University of Nebraska Press, 2003.

Burrough, Bryan. *Dragonfly: NASA and the Crisis Aboard Mir*. New York: HarperCollins, 1998.

Cassutt, Michael. "Mr. Inside," *Air & Space*, August 2011.

———. "I'm with the (Astronaut) Band," *Air & Space*, March 2009.

Chaikin, Andrew. *A Man on the Moon: The Voyages of the Apollo Astronauts*. New York: Viking, 1994.

Collins, Michael. *Carrying the Fire: An Astronaut's Journey*. New York: Bantam, 1983.

Compton, W. David, and Charles D. Benson. *Living and Working in Space: A History of Skylab*. Washington, DC: National Aeronautics and Space Administration, 1983.

Cooper, Henry S. F., Jr. *Before Lift-Off: The Making of a Space Shuttle Crew*. Baltimore: Johns Hopkins University Press, 1987.

Cortright, Edgar, ed. *Apollo Expeditions to the Moon*. Washington, DC: NASA, 1975.

Cunningham, Walter. *The All-American Boys*. New York: Macmillan, 1977.

Dethloff, Henry C. *Suddenly Tomorrow Came . . . : A History of the Johnson Space Center.* Washington, DC: National Aeronautics and Space Administration, 1993.

Ertel, Ivan D., and Roland W. Newkirk with Courtney G. Brooks. *The Apollo Spacecraft: A Chronology.* Vol. 4. Washington, DC: National Aeronautics and Space Administration, 1978.

Ezell, Edward Clinton, and Linda Neuman Ezell. *The Partnership: A History of the Apollo-Soyuz Test Project.* Washington, DC: National Aeronautics and Space Administration, 1978.

Geiger, Clarence J. *History of the X-20A Dyna-Soar.* Vol. 1. Wright-Patterson Air Force Base, Dayton, OH: Historical Division, Aeronautical Systems Division, 1963.

Hale, Wayne, ed. *Wings in Orbit: Scientific and Engineering Legacies of the Space Shuttle, 1971–2010.* Washington DC: National Aeronautics and Space Administration, 2011.

Heppenheimer, T. A. *The Space Shuttle Decision: NASA's Search for a Reusable Space Vehicle.* Washington, DC: National Aeronautics and Space Administration, 1999.

Hogan, Thor. *Mars Wars: The Rise and Fall of the Space Exploration Initiative.* Washington, DC: National Aeronautics and Space Administration, 2007.

Jenkins, Dennis. *Space Shuttle: The History of Developing the National Space Transportation System.* 2nd ed. Cape Canaveral: Dennis R. Jenkins, 1996.

Johnson, Stephen B. "Samuel Phillips and the Taming of Apollo." *Technology and Culture* 42, no. 4 (October 2001): 685–709.

Kelly, Thomas J. *Moon Lander: How We Developed the Apollo Lunar Module.* Washington, DC: Smithsonian Institution Press, 2001.

Kraft, Christopher. *Flight: My Life in Mission Control.* New York: Dutton, 2001.

———. "Report of the Space Shuttle Management Independent Review Team." National Aeronautics and Space Administration, February 1995.

Mark, Hans. *The Space Station: A Personal Journey.* Durham, NC: Duke University Press, 1987.

Morgan, Clay. *Shuttle-Mir: The United States and Russia Share History's Highest Stage.* Houston: National Aeronautics and Space Administration, 2001.

Mullane, Mike. *Riding Rockets: The Outrageous Tales of a Space Shuttle Astronaut.* New York: Scribner, 2006.

Murray, Charles, and Catherine Bly Cox. *Apollo: The Race to the Moon.* New York: Simon & Schuster, 1989.

NASA Safety Center. "From Rockets to Ruins: The PEPCON Ammonium Perchlorate Plant Explosion." *System Failure Case Study* 6, no. 9 (November 2012).

Nelson, Craig. *Rocket Men: The Epic Story of the First Men on the Moon.* New York: Penguin, 2009.

Oberg, James. *Star-Crossed Orbits: Inside the U.S.-Russian Space Alliance*. New York: McGraw-Hill, 2002.

Presidential Commission on the Space Shuttle *Challenger* Accident. "Report to the President." Washington, DC: National Aeronautics and Space Administration, 1986.

Quayle, Dan. *Standing Firm: A Vice Presidential Memoir*. New York: HarperCollins, 1994.

Ross-Nazzal, Jennifer. "Legacy of the Thirty-Five New Guys." *Houston History* 6, no. 1 (Fall 2008).

Saturn V Flight Evaluation Group. "Saturn V Launch Vehicle, Flight Evaluation Report–AS-502, Apollo 6 Mission." NASA Marshall Space Flight Center, 1968.

Sherr, Lynn. *Sally Ride: America's First Woman in Space*. New York: Simon & Schuster, 2014.

Slayton, Donald K., with Michael Cassutt. *Deke!—U.S. Manned Space: From Mercury to the Shuttle*. New York: Forge, 1994.

Stafford, Thomas P., with Michael Cassutt. *We Have Capture: Tom Stafford and the Space Race*. Washington, DC: Smithsonian Institution Press, 2002.

Swenson, Loyd S., Jr., James M. Grimwood, and Charles C. Alexander. *This New Ocean: A History of Project Mercury*. Washington, DC: National Aeronautics and Space Administration, 1966.

Trento, Joseph, with Susan B. Trento. *Prescription for Disaster: From the Glory of Apollo to the Betrayal of the Shuttle*. New York: Crown, 1987.

Wetherbee, Jim. *Controlling Risk in a Dangerous World*. New York: Morgan James, 2017.

White, Rowland. *Into the Black: The Extraordinary Untold Story of the First Flight of the Space Shuttle Columbia and the Astronauts Who Flew Her*. New York: Touchstone, 2016.

Worden, Al, with Francis French. *Falling to Earth: An Apollo 15 Astronaut's Journey to the Moon*. Washington, DC: Smithsonian Institution Press, 2011.

Wright, Lawrence. "From Texas to the Stars." *Texas Monthly*, July 1981.

Wright, Rebecca, Sandra Johnson, and Steven J. Dick, eds. *NASA at 50: Interviews with NASA's Senior Leadership*. Washington, DC: National Aeronautics and Space Administration, 2008.

Index